V. W. Cairns, P. V. Hodson, J. O. Nriagu
Contaminant Effects on Fisheries   Volume 16 1984

J. O. Nriagu and C. I. Davidson
Toxic Metals in the Atmosphere   Volume 17 1986

Allan H. Legge and Sagar V. Krupa
Air Pollutants and Their Effects on the
Terrestrial Ecosystem   Volume 18 1986

J. O. Nriagu and J. B. Sprague
Cadmium in the Aquatic Environment   Volume 19 1987

Volume

19

in the Wiley Series in

Advances in Environmental Science and Technology

JEROME O. NRIAGU, Series Editor

**CADMIUM IN THE
AQUATIC ENVIRONMENT**

CADMIUM IN THE AQUATIC ENVIRONMENT

Edited by
Jerome O. Nriagu
National Water Research Institute
Burlington, Ontario, Canada

John B. Sprague
Department of Zoology
University of Guelph
Guelph, Ontario, Canada

A WILEY-INTERSCIENCE PUBLICATION
JOHN WILEY & SONS
New York • Chichester • Brisbane • Toronto • Singapore

Copyright © 1987 by John Wiley & Sons, Inc.

All rights reserved. Published simultaneously in Canada.

Reproduction or translation of any part of this work beyond that permitted by Section 107 or 108 of the 1976 United States Copyright Act without the permission of the copyright owner is unlawful. Requests for permission or further information should be addressed to the Permissions Department, John Wiley & Sons, Inc.

Library of Congress Cataloging-in-Publication Data:
Cadmium in the aquatic environment.

  (Advances in environmental science and technology, ISSN 0065-2563; v. 19)
  "A Wiley-Interscience publication."
  Includes index.
  1. Cadmium—Environmental aspects. 2. Water—Pollution—Environmental aspects. 3. Aquatic organisms—Effect of water pollution on.
I. Nriagu, Jerome O.   II. Sprague, John B.
III. Series.

TD180.A38   vol. 19   628 s   87-2198
[QH545.C37]   [574.5'263]
ISBN 0-471-85884-6

Printed in the United States of America

10 9 8 7 6 5 4 3 2 1

CONTRIBUTORS

BARRY, P. J., Chalk River Nuclear Laboratories, Atomic Energy of Canada Ltd., Chalk River, Ontario, Canada

BEWERS, J. M., Department of Fisheries and Oceans, Bedford Institute of Oceanography, Dartmouth, Nova Scotia, Canada

DABEKA, R. W., Food Research Division, Health Protection Branch, Health and Welfare Canada, Ottawa, Ontario, Canada

FÖRSTNER, U., Arbeitsbereich Umweltschutztechnik, Technische Universität Hamburg-Harburg, Hamburg, West Germany

IHNAT, M., Chemistry and Biology Research Institute, Research Branch, Agriculture Canada, Ottawa, Ontario, Canada

KERSTEN, M., Arbeitsbereich Umweltschutztechnik, Technische Universität Hamburg-Harburg, Hamburg, West Germany

LUM, K. R., Environment Canada, National Water Research Institute, Burlington, Ontario, Canada

MACGREGOR, D. J., Environmental Canada, Commercial Chemicals Branch, Hull, Quebec, Canada

MCCRACKEN, I. R., Atlantic Research Laboratory, National Research Council of Canada, Halifax, Nova Scotia, Canada

MCLEESE, D. W., Fisheries and Environmental Sciences, Fisheries Research Branch, Department of Fisheries and Oceans, Biological Station, St. Andrews, New Brunswick, Canada

RAY, S., Fisheries and Environmental Sciences, Department of Fisheries and Oceans, Biological Station, St. Andrews, New Brunswick, Canada

SPRAGUE, J. B., Department of Zoology, University of Guelph, Guelph, Ontario, Canada

WONG, P. T. S., Great Lakes Fisheries Research Branch, Department of Fisheries and Oceans, Canada Center for Inland Water, Burlington, Ontario, Canada

YEATS, P. A., Atlantic Oceanographic Laboratory, Bedford Institute of Oceanography, Department of Fisheries and Oceans, Dartmouth, Nova Scotia, Canada

INTRODUCTION
TO THE SERIES

The deterioration of environmental quality, which began when mankind first congregated into villages, has existed as a serious problem since the industrial revolution. In the second half of the twentieth century, under the ever-increasing impacts of exponentially growing population and of industrializing society, environmental contamination of the air, water, soil, and food has become a threat to the continued existence of many plant and animal communities of various ecosystems and may ultimately threaten the very survival of the human race. Understandably, many scientific, industrial, and governmental communities have recently committed large resources of money and human power to the problems of environmental pollution and pollution abatement by effective control measures.

Advances in Environmental Sciences and Technology deals with creative reviews and critical assessments of all studies pertaining to the quality of the environment and to the technology of its conservation. The volumes published in the series are expected to service several objectives: (1) stimulate interdisciplinary cooperation and understanding among the environmental scientists; (2) provide the scientists with a periodic overview of environmental developments that are of general concern or that are of relevance to their own work or interests; (3) provide the graduate student with a critical assessment of past accomplishment, which may help stimulate him or her toward the career opportunities in this vital area; and (4) provide the research manager and the legislative or administrative official with an assured awareness of newly developing research work on the critical pollutants and with the background information important to their responsibility.

As the skills and techniques of many scientific disciplines are brought to bear on the fundamental and applied aspects of the environmental issues, there is a heightened need to draw together the numerous threads and to present a coherent picture of the various research endeavors. This need and the recent tremendous growth in the field of environmental studies have clearly made some editorial adjustments necessary. Apart from the changes in style and format, each future volume in the series will focus on one particular theme or timely topic, starting with Volume 12. The author(s) of each pertinent section will be expected to critically review the literature and the most important recent developments in the particular field; to critically evaluate new concepts, methods, and data; and to focus attention on important unresolved or controversial questions and on probable future trends. Monographs embodying the results of unusually extensive and well-rounded investigations will also be published in the series. The net result of the new editorial policy should be more integrative and comprehensive volumes on key environmental issues and pollutants. Indeed, the development of realistic standards of environmental quality for many pollutants often entails such a holistic treatment.

JEROME O. NRIAGU, Series Editor

PREFACE

Since the turn of this century, the introduction of cadmium into natural waters has increased at a drastic rate. Indeed, recent studies point to the fact that the concentrations and fluxes of cadmium in many bodies of water have been overwhelmed by inputs from anthropogenic sources. Aquatic ecosystems are particularly sensitive to cadmium pollution mainly because of (1) the very low levels of this toxicant in waters and biomass of prehistoric times and (2) the strong tendency of the food web to bioaccumulate this element. Cadmium thus is widely regarded as a priority contaminant, and, historically, guidelines have been established for its levels in public water supplies and for other designated uses.

This volume presents a comprehensive account of current research on the chemistry and toxicology of cadmium in natural waters. The first four chapters focus on the sources (natural vs. anthropogenic), behavior, and fate of cadmium in natural waters. These are followed by three chapters on the biocycling and effects of cadmium on freshwater biota. The ecotoxicity of cadmium to marine biota are covered by the next two chapters. The final chapter presents a succinct review of the analytical chemistry of cadmium in natural waters. A common goal of the biological chapters has been to relate the uptake and toxicity of cadmium to the key biogeochemical processes. Biological monitoring of the sensitivity of ecosystems and the biochemical interactions of cadmium with other metals are also extensively covered in various chapters. The authors have made a concerted effort to emphasize the general principles rather than present a comprehensive documentation of the literature.

This volume provides the scientific rationale for the continuing effort to establish water quality criteria and standards for cadmium. The technical information provided should be invaluable to environmental managers and regulators and should be of much interest to the chemists and biologists concerned with the fate and effects of toxic metals in aquatic ecosystems.

This volume grew out of a report that was being prepared for the now defunct Associate Committee on Scientific Criteria for Environmental

Quality (National Research Council of Canada). Any credit for the volume belongs to our distinguished group of contributors. We thank Wiley-Interscience for invaluable editorial assistance.

JEROME O. NRIAGU
JOHN B. SPRAGUE

June, 1987

CONTENTS

1. Distribution and Cycling of Cadmium in the Environment 1
   J. M. Bewers, P. J. Barry, and D. J. MacGregor

2. Evidence for Anthropogenic Modification of Global Transport of Cadmium 19
   P. A. Yeats and J. M. Bewers

3. Cadmium in Fresh Waters: The Great Lakes and St. Lawrence River 35
   K. R. Lum

4. Cadmium Associations in Freshwater and Marine Sediment 51
   M. Kersten and U. Förstner

5. Biological Cycling of Cadmium in Fresh Water 89
   I. R. McCracken

6. Toxicity of Cadmium to Freshwater Microorganisms, Phytoplankton and Invertebrates 117
   P. T. S. Wong

7. Effects of Cadmium on Freshwater Fish 139
   J. B. Sprague

8. Effects of Cadmium on Marine Biota 171
   D. W. McLeese, J. B. Sprague, and S. Ray

9. Biological Cycling of Cadmium in Marine Environment 199
   S. Ray and D. W. McLeese

10. Methods of Cadmium Detection 231
    R. W. Dabeka and M. Ihnat

Index 265

**CADMIUM IN THE
AQUATIC ENVIRONMENT**

DISTRIBUTION AND CYCLING OF CADMIUM IN THE ENVIRONMENT

J. M. Bewers

*Department of Fisheries and Oceans
Bedford Institute of Oceanography
Dartmouth, Nova Scotia, Canada*

P. J. Barry

*Chalk River Nuclear Laboratories
Atomic Energy of Canada
Chalk River, Ontario, Canada*

D. J. MacGregor

*Environment Canada
Hull, Quebec, Canada*

1.1. Source Rocks
1.2. Atmosphere
1.3. Terrestrial and Freshwater Environments
1.4. Ocean Waters
1.5. Marine Sediments
References

This chapter reviews information on the distribution and geochemical behavior of cadmium in the environment with particular emphasis on the

2   Distribution and Cycling of Cadmium in the Environment

marine environment. It is divided into sections each dealing with a particular compartment of the geosphere.

1.1. SOURCE ROCKS

The concentrations of cadmium in some common rock types, derived from Simpson (1981), Fairbridge (1972), and Page and Bingham (1973), are depicted in Table 1.

Only a few minerals contain cadmium at exploitable concentrations. The primary source of cadmium is sphalerite (zinc sulfide). Other ores include greenockite (a hexagonal crystalline form of cadmium sulfide), hawleyite (a cubic crystalline form of cadmium sulfide), otavite (a mineralized form of calcium carbonate that was the main source of cadmium in the nineteenth century), cadmoselite (a hexagonal crystalline form of cadmium selenide), monteponite (a cadmium oxide), and saukovite (a cubic cadmium metacinnabar containing cadmium substituted for mercury).

Table 1. Cadmium Concentrations in Some Common Rock Types

Rock Type	Cadmium Concentration (mg/kg)	
	Range	Mean
Igneous		
Granite	0.001–0.60	0.12
Granodiorite	0.016–0.10	0.07
Pitchstone	0.05–0.34	0.17
Obsidian	0.22–0.29	0.25
Basalt	0.006–0.6	0.22
Gabbro	0.08–0.20	0.11
Syenite	0.04–0.32	0.16
Metamorphic		
Ecologite	0.04–0.26	0.11
Garnet schist	—	1.0
Gneiss	0.12–0.16	0.14
Sedimentary		
Bituminous shale	<0.3–11	0.8
Bentonite	<0.3–11	1.4
Marlstone	0.4–10	2.6
Limestone	—	0.035
Sandstone	—	0.01
Shale	—	0.3

1.2. ATMOSPHERE

Cadmium has been described as "the dissipated element" (Fulkerson and Goeller, 1973), and indeed, the element has been widely mobilized by both natural and anthropogenic processes. Natural processes, including aeolian transport of weathered rock particles, forest fires, volcanic emissions, transpiration, and also possibly volatilization of cadmium from soils, all give rise to the injection of cadmium into the atmosphere. Gases, vapors, and particles are deposited from the atmosphere by wet and dry processes. Wet processes include scavenging of airborne substances by falling precipitation or during the growth of precipitation droplets within clouds. The efficiency of these processes depends on the relative sizes of the particles of dust and precipitation or, in the cases of gases and vapors, on the molecular diffusivity and solubility in water and on the intensity of precipitation. Dry deposition occurs by impaction of particles or by absorption of gases and vapors onto water and vegetation surfaces. For large particles, gravitational settling is the dominant process, while for small particles, gases, and vapors, the principal agency is the turbulent velocities in the wind. The rates depend on particle size or molecular diffusivity, the nature of the surface, and the intensity of turbulence. Although these physical processes are reasonably well understood, the information needed to solve the equations is rarely available.

In general, however, most particles and reactive gases and vapors are deposited within one month of injection and probably in less than half that time. This means that most of the material released in one year will be deposited in the year. It also means that interlatitudinal dispersion is relatively small so that most of the material injected is deposited within the latitudinal bands of the sources. In particular, transequatorial transport will be limited, except in the cases of sources injecting material into the high atmosphere.

The aggregate natural flux of cadmium through the atmosphere has been estimated to be 8.3×10^5 kg/yr (Nriagu, 1979). The concentrations of cadmium in aerosols and wet and dry atmospheric precipitation vary widely both in time and space, but it appears that cadmium is considerably enriched relative to the major conservative constituents of source rocks, such as aluminum. Whether this enrichment is the consequence of natural processes or due to the increased dissemination of cadmium through anthropogenic activities remains unclear. Thus, cadmium is found in arctic snows and ice fields in concentrations that are far greater than would be expected on the basis of source rock compositions (Cd/Al ratios) (Herron et al., 1977). The extent to which the large estimated excess of cadmium introduced into the atmosphere through human activities has resulted in further enrichment of cadmium in arctic and antarctic snows and ice fields is a matter of considerable debate (Nriagu, 1979; Landy et al., 1980; Nriagu, 1980b; Boutron,

1980), but it seems that evidence for a recently increased rate of environmental dissemination of heavy metals is restricted to the case of lead and does not exist in the cases of either cadmium or zinc. In some senses, this is surprising since the flux of anthropogenically mobilized cadmium is believed to be comparable with (Simpson, 1981) or to exceed by nearly an order of magnitude (Nriagu, 1980a) the natural flux.

The atmospheric distributions and fluxes of cadmium have been reviewed by Nriagu (1980a), and therefore, the topic does not merit detailed discussion here. In summary, remote areas of the atmosphere contain cadmium concentrations in air in the 10–1000-pg/m^3 range with cadmium enrichments, based on a comparison of Cd/Al ratios in air and source rocks of between 500 and 5000. Average figures for these two variables might be 50 pg/m^3 concentration and 1000-fold enrichment. It is conceivable that the major contributor to the enrichment of cadmium in the atmosphere is the production of aerosols from ocean surface water. However, since cadmium is depleted in surface waters by biological processes, substantial enrichment factors for the sea-to-air transfer process would need to be invoked. Urban air can contain considerably higher concentrations than those in remote areas and is generally in the nanogram-per-cubic-meter range. As expected, there exists a wide degree of inhomogeneity in the concentration of cadmium in air due to the effects of industrially released cadmium. There is reason to believe that large proportions of the cadmium emitted to the atmosphere from industrial sources are removed fairly quickly by either dry precipitation or washout. Distance scales for the removal of the vast majority of industrially derived cadmium are in the range 100–1000 km depending on the height of injection and regional atmospheric conditions. This in turn would explain the lack of evidence of anthropogenically released cadmium in remote areas. It is worth noting in this context that a recent paper on the cadmium content of snows in the eastern Arctic (Mart, 1983) shows that fresh snow is very low in cadmium content (about 0.4 ng/L) but that the deposited snow rapidly becomes augmented in cadmium to levels of about 5 ng/L presumably due to particulate contamination. Whether this contamination is largely due to particles derived from industrial activities, as concluded by Mart (1983), or whether there is an important cadmium component of marine-derived aerosol remains to be determined. However, in view of the absence of any clear evidence for any industrially related influence on cadmium in polar snow and ice cores (Herron et al., 1977), it seems likely that the source of the additional cadmium is predominantly natural rather than anthropogenic.

The concentration of cadmium in rain is a very difficult topic to discuss largely because of the uncertainties in the reliability (accuracy) of the sampling and analytical methods employed in field studies. In general, all measurements of cadmium, and other metals, in seawater made prior to 1975 are regarded as unreliable. In most cases, the most reliable and geochemically consistent data have been published since 1979, and there is strong reason to believe that the same problems of inadequate analytical sensitivity and susceptibility to contamination that invalidate the early marine data might also invalidate contemporary freshwater (rain, snow, lake, and river) measurements. As noted above, recent measurements of cadmium in fresh snow in the eastern arctic ocean yield a concentration of 0.4 ng/L (Mart, 1983). The reported concentrations of cadmium in rainwater are generally in the range 100–50,000 ng/L (Nriagu, 1980a), but it is virtually impossible to place a great deal of faith in most of these data because of uncertainty regarding the adequacy of the sampling and analytical procedures. Often, an equally important difficulty is the collection of representative samples of atmospheric precipitation. This can be assessed by the degree of correlation between volumes of water collected in chemical sampling devices and precipitation measured with adjacent conventional rain gauges.

A fairly careful study in temperate rural environments (Struempler, 1973) yielded values for the cadmium concentration in atmospheric precipitation in the 0.1–1.0 ng/L range. Struempler's work also shows the effects of particulate washout in the higher mean values associated with intermittent showers (720 pg/L) as compared to continuous rain (260 pg/L). It seems that remote areas should yield cadmium concentrations in precipitation generally less than 1 ng/L except where significant contamination by industrially derived particles occurs. Clearly, precipitation in industrialized urban areas must contain a dominant proportion of washout-derived cadmium originally in atmospheric particulate form. Based on Struempler's (1973) experience and that of Mart (1983), it can be predicted that the rainout concentration of cadmium in most areas should be relatively small and on the order of 0.1–1 ng/L.

1.3. TERRESTRIAL AND FRESHWATER ENVIRONMENTS

Freshwater and unconsolidated geologic sediments together form a complex interacting subsystem that controls the fluxes of cadmium through, on the one hand, agricultural produce and, on the other, rivers and lakes. Soils form from the weathering of basement geologic materials and, in the process, are continually leached and eroded by drainage waters derived from precipitation. Rivers and lakes are fed by water, most of which has spent more or less of its time on land in intimate contact with the soil and subsoil. In the natural system, we may therefore reasonably expect the cadmium content of soil to be related to the parent rock from which the soil is derived, though somewhat subsequently modified by leaching, erosion, and biological activity. Since soil is a ternary mixture, virgin weathering products such as soils should contain less cadmium than the underlying bedrock. Similarly, the cadmium content of rivers and lakes should be related to the types of soil through which the drainage water has passed, though again subject to later modifications due to physicochemical and biological processes.

A summary of cadmium concentrations in common rocks is presented in

Table 2. Cadmium Content of Soils (mg/kg)

Soils	Range	Mean
36 soils, United States: 12 profiles	0.12–1.82	0.56
33 topsoils, British Columbia	<0.1–4.67	0.88
254 topsoils, Michigan, United States	—	0.56
7 topsoils, Australia	0.032–0.212	0.092
8 soils, Ontario	0.55–1.72	1.14
39 topsoils, Canada	0.40–1.7	0.97
37 topsoils, England	<1.0–4.0	0.3
51 topsoils, Wales	0.40–2.3	1.78
7 soils, Wales: 1 profile	0.4–0.9	0.60
296 topsoils, Ontario	0.1–8.1	0.56
12 topsoils, Australia	0.013–0.56	0.28
361 topsoils, Sweden	0.03–2.3	0.22
15 topsoils, Norway	0.02–3.7	0.57
51 topsoils, Denmark	—	0.26
10 topsoils, Scotland	<0.3–1.5	0.77
689 topsoils, England and Wales	0.08–10	(Median 1.0)
23 soils, Scotland: 4 profiles	<0.005–1.4	0.83
121 topsoils, Wales	—	1.1
10 topsoils, England and Wales	0.27–1.04	0.63
62 topsoils, Scotland	<0.25–1.4	0.66
209 soils, Scotland: 31 profiles	<0.25–2.4	0.7

Table 1. The widest range and highest concentrations (0.3–11 mg/kg) are generally to be found in sedimentary rocks, with igneous rocks showing more uniformity in concentration at levels of 0.1–0.3 mg/kg. The ranges within a given rock type are, however, wide (e.g., 0.001–0.6 mg/kg in granite), and there is considerable overlap between rock types. The earth's crust has been estimated to contain an average cadmium concentration of 0.2 mg/kg (Taylor, 1964), while continental crust has been estimated to contain about half that (0.098 mg/kg; Heinrichs et al., 1980). Analyses of the cadmium content of 1642 soils from different parts of the world have been reviewed by Ure and Berrow (1982), and a summary is presented in Table 2. The mean cadmium content of the 1642 soils is 0.62 mg/kg, or 3–6 times greater than that estimated for the total or continental crusts, respectively. The soils analyzed are derived from many different basement rocks, but the data have not been stratified to reveal any systematic differences in soil concentration arising from this source. Vinogradov (1959) has analyzed soils from the USSR and concluded that the main factor determining the cadmium concentrations in soils is the nature of the parent rock. However, Garcia-Maragaya and Page (1977) found that cadmium is strongly adsorbed by soils having a pH above 6. Thus, artificially induced pH changes (e.g., through

liming) may have an effect on cadmium retention in soils. Most measurements of cadmium content have been made on agricultural soils over sedimentary basement rock, which may account for the values reported being somewhat higher than the concentrations reported for the global or continental crusts. The possibility of contamination of agricultural soils through the application of phosphate and sewage sludge fertilizers, although perhaps unlikely on the scale implied by the soils represented in Table 2, cannot be entirely ruled out. The information included in Table 2 is pertinent to estimating the fluxes of cadmium to humans through terrestrial food webs. However, to estimate the fluxes of this metal to surface waters, both fresh and marine, it is the weighted mean abundances in all areas that are most appropriate.

The cadmium content of soils for our present purposes carries a threefold interest: first, to obtain information about the behavior of cadmium during weathering of the parent rock; second, to determine the transfer parameters relating cadmium soil content and its content in agricultural produce; and third, to determine the flux of cadmium to surface fresh water. The current literature contains little that is useful for these purposes. Only a few systematic studies to find the area-weighted mean concentrations in bedrock, soil, farm produce, and stream water seem to have been carried on a drainage basin scale. The literature on this topic is, however, confusing, with several different methods having been used for treating the soil complex prior to analysis. Some workers have reported "total" cadmium concentrations by digesting the sample with hydrofluoric/perchloric acid. Others have used cold dilute hydrochloric, or hot dilute nitric acid, or concentrated hydrochloric/nitric acid extracts to estimate "extractable" cadmium concentrations. Still others have used a variety of "weak" extractants (such as DPTA, EDTA, and HOAc) in an attempt to gain information about the "biologically available" constituents. Finally, the wide range of measured concentrations reported in Table 2 (0.02–10 mg/kg overall and, in the case of the 689 topsoils from England and Wales, 0.08–10 mg/kg) suggests that a wide range of soil and parent rock types, a broad spectrum of localized agricultural practices, and different sampling and handling protocols and analytical methods have confused the results. Whatever the cause of the wide range observed among the individual samples, its very presence raises important questions about the possible meaning and usefulness of the simple mean concentration.

The chemistry of cadmium in freshwater is highly complex depending on its speciation, which in turn is dominated by the oxidation status and the pH of the containing medium, as well as on the concentrations of numerous organic and inorganic anions and other metal cations. The mobility of cadmium in soil pore water and groundwater is strongly controlled by precipitation (as $CdCO_3$ and CdS) and by adsorption by soil particles, clay minerals, organic matter, and the hydroxides of iron and manganese. Adsorbability is generally greatest for iron hydroxides and least for clay minerals. The presence of complexes of cadmium, however, can considerably reduce adsorption by all of the above materials, and the presence of other cations can further reduce cadmium adsorption. One frequently used measure of adsorption of ions from solution is the equilibrium distribution coefficient K_d (v/w) defined as

$$K_d = \frac{\text{concentration of ion on solid (w/w)}}{\text{concentration of ion in solution (w/v)}}.$$

Using batch laboratory measurements, Gardiner (1974) found values of K_d for cadmium to vary from a few hundred for the clay mineral kaolinite to nearly 20,000 cm^3/g for organic matter. The equilibrium coefficient for iron hydroxides is not so easily measured the same way, and there is evidence that adsorption by this substrate is irreversible (Lee, 1975). In general, complexation and high ionic strength tend to reduce K_d while a high pH (i.e., >6.0) tends to increase it, though useful numbers to quantify these processes are not available. There appear to be no widespread systematic measurements of cadmium concentrations in uncontaminated groundwater or soil pore water. A survey of cadmium levels in Canadian drinking water derived from raw well water by Health and Welfare Canada (Meranger, Subramanian, and Chalifoux, 1979) gives a range from <10 to 100 ng Cd/L water.

References to the complex chemistry of cadmium in groundwater apply equally and for much the same reasons to its chemistry in fresh surface waters. The importance of cadmium speciation in surface freshwaters has been recognized as a necessary prerequisite for understanding and predicting cadmium availability and toxicity to biota. Several detailed schemes for separating cadmium in solution into different fractions have been devised. None have succeeded in identifying individual chemical species with any certainty, and it is likely that each separated fraction contains more than one species. Even the primary separation of a "water-soluble" fraction is ambiguous. What is determined by the separation process is the mixture of chemical forms that can pass through a filter of given pore size, usually 0.45 μm. Besides Cd^{2+} ions and a variety of soluble complexes, various colloidally dispersed forms with humic acids or clay particles, to which cadmium may be adsorbed, can pass through the filter.

Soluble cadmium forms are removed from the water column by interaction or adsorption onto sediments and by biota. The extent of such removal depends on factors similar to those controlling adsorption by soil particles. The presence and nature of complexing ligands, other metals, oxidation potential, and pH are important, but little understood, influences. Remobilization of cadmium adsorbed by sediment particles is also an unknown variable that affects the ability of lakes to act as permanent traps and so reduces the flux of cadmium to the oceans and its availability to biota. Concentrations of cadmium in surface waters have been measured by many workers but are difficult to interpret. Wide variation in the reported concentrations reflect biases introduced in the collection and sample handling and preparation protocols followed by different workers as well as real differences due to the variety of rock types underlying the drainage basins. Most large rivers draining the industrialized countries of Europe, North America, and Asia are contaminated, while measurements on smaller uncontaminated tributaries are highly variable from natural causes and hardly representative of a regional or global average.

Waters of the rural western Australian Helena River have been found to contain cadmium concentrations in the range 6–40 ng/L, while those in other interior (upstream of major population centers such as Perth) western Australian rivers ranged from 10 to 170 ng/L (Rosman and De Laeter, 1976, 1977). Rivers in eastern Australia contained cadmium in the range 20–50 ng/L (Doolan and Smythe, 1973). A survey of remote streams in northeast Minnesota showed concentrations in the range 6–34 ng/L with a mean of 19 ng/L in 18 samples (Poldoski and Glass, 1978). Recent data (E. A. Boyle, private communication; Trefry et al., 1983) suggest that the pristine river concentrations of cadmium are in the range of 5–20 ng/L. The latter paper even suggests that relatively radical changes have already occurred in the cadmium content of the Mississippi as a result of controls on cadmium dissemination. However, the extent to which such reductions are real and not related to improvements in sampling and analytical procedure remains unclear. Data for other, less pristine rivers (The St. Lawrence: Yeats and Bewers, 1982; The Rhine: Duinker and Nolting 1978; and some other European rivers: J. C. Duinker, private communication) suggest that much higher levels of cadmium in river discharge occur that may be related to the intensity of industrial activity within their drainage basins. Further work on this topic is urgently required, just as it is for elucidating the source of the excess cadmium found in remote snow and ice fields. Knowing that many of these data are likely to reflect inaccuracies due to sample contamination or other analytical artifacts, we might assume that the average dissolved cadmium concentration in world rivers is 10 ng/L and that river concentrations considerably in excess of this might arise as a consequence of anthropogenic emissions in the associated drainage basin.

1.4. OCEAN WATERS

Current understanding of the incidence and distribution of cadmium in pelagic areas of the ocean is extremely good. It is possible to give accurate information regarding the spatial distribution of cadmium in seawater in open-ocean areas and to describe in some detail the processes that govern this distribution. The corresponding picture for near-shore areas is somewhat less clear. Some aspects of the near-shore behavior of cadmium appear to be universal, but others appear to vary between estuaries, and results of additional studies are needed to create a universally consistent understanding of the near-shore marine geochemistry of cadmium.

In freshwater, cadmium can exist as Cd^{2+} ions and $Cd(OH)_2$ and $CdCO_3$ complexes and in various organic complexes depending on pH and the prevalence of soluble organic material. As salinity increases, cadmium complexation with chloride ion increases until, in normal seawater, cadmium exists almost entirely as chloride ($CdCl^+$, $CdCl_2$, $CdCl_3^-$) with minor proportions as Cd^{2+} (Raspor, 1980). The extent of cadmium association with organic ligands in seawater is still a matter of question. Bruegmann (1974) concluded from speciation studies of a coastal water sample rich in dissolved and suspended organic material that only 40% of the total cadmium was in labile form. This would imply that substantial amounts of cadmium in filtered samples might be associated with organic ligands or with inorganic ligands that are nondissociable by ion exchange resins. As will be seen from the discussion of the behavior of cadmium in pelagic areas, there exists a close relationship between cadmium and marine organic matter, and therefore, such a conclusion is consistent with the current understanding of cadmium geochemistry. The speciation of cadmium is discussed more fully by Raspor (1980) and Nuernberg and Valenta (1983). The inorganic speciation pattern shown in Table 3 calculated by Nuernberg and Valenta (1983) ignores contributions due to chelate formation with strong organic ligands that are associated with dissolved organic matter (DOM). Nevertheless, the elucidated speciation is clearly valid for the major part of the ocean in which biological productivity is low.

More complicated speciation will occur when the levels of DOM and organic colloids are relatively high, such as in upwelling areas and near-shore areas that are either very productive or receive considerable organic material discharges. Based on the stability constant K at seawater ionic strength given by Sillen and Martell (1971) for a number of potential ligands among the DOM components and the prevailing low concentrations of DOM in most ocean areas (less than 2 mg C/L), it may be concluded that no signifi-

TABLE 3. Distribution of Predominant Species of Cd (II) in Seawater with and without Correction of Ion Pairing of Major Salinity Components[a]

Species	Without Ion-Pairing Correction Distribution (%)	With Ion-Pairing Correction Distribution (%)
Cd^{2+}	2.7	1.9
$CdCl^+$	34.6	29.1
$CdCl_2$	36.8	37.2
$CdCl_3^-$	25.6	31.0
Other	0.3	0.8

[a] After Nuernberg and Valenta (1983).

cant contribution to cadmium speciation is to be expected for DOM components belonging to the classes of carbohydrates and of amino acids, excepting perhaps sulfur-containing amino acids and protein degradation products. The situation may be somewhat different in certain inshore waters and estuaries provided there exist substantially elevated DOM levels.

The vertical distribution of cadmium in the ocean is largely influenced by marine biological activity. Cadmium is incorporated into and released from marine biogenic detritus in direct proportion to the regeneration of the labile nutrients phosphate and nitrate (Knauer and Martin, 1981). Thus, the surface layer of the ocean, in which the photosynthesizing primary producers predominate, contains reduced levels of cadmium compared with deeper layers in which the biogenic detritus undergoes decomposition. There generally exists a characteristic gradient in concentration with depth with low concentrations at the surface, where primary production is greatest, increasing to maximum values at the predominant depths of nitrate and phosphate regeneration, and then decreasing to relatively constant values in the intermediate and deep waters of the open-ocean water column. Vertical profiles of cadmium, phosphate, and nitrate (Fig. 1) and the linear relationship between cadmium and phosphate (Fig. 2) obtained in the eastern Pacific by Bruland et al. (1978) exemplify this distribution pattern. The magnitude of such vertical concentration gradients in the mixed layer depends on the physical oceanographic conditions in the area, particularly the intensity of vertical mixing and the presence or absence of vertical convection. Nevertheless, a remarkably strong relationship is generally found between dissolved cadmium, dissolved nitrate, and dissolved phosphate. Recent studies in the Pacific Ocean (Bruland et al., 1978) gave the following relationship of these three constituents:

$$Cd : P : N \text{ (atom)} = 3.5 \times 10^{-4} : 1 : 15.2.$$

On a mass basis, this relationship becomes

$$Cd : P : N = 1.3 \times 10^{-3} : 1 : 6.9.$$

This relationship reflects the removal of cadmium in the surface layer from the dissolved phase into particulate organic material within phytoplankton and its subsequent regeneration at greater depths both within and below the mixed layer. For every cadmium atom passing through this cycle, there exist a corresponding 300,000 atoms of carbon acting as a carrier. Thus, the internal biogeochemical cycle of carbon, associated with the growth and decay of photosynthesizing organisms, dominates the vertical distribution of cadmium within the mixed layer and for some distance beneath it. The supply of cadmium and labile nutrients to primary producers in the surface waters is maintained by atmospheric precipitation and vertical mixing and advection, with the latter two processes resupplying the regenerated constituents back to the euphotic zone. It appears that the residence time for

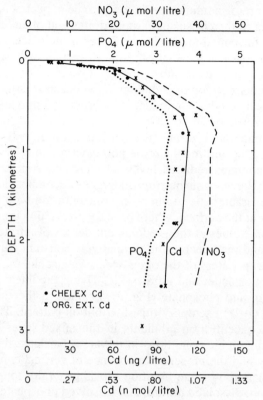

Figure 1. Vertical profiles of cadmium, phosphate, and nitrate at a northeast Pacific station (after Bruland et al., 1978). Cadmium profile line represents the average of the phosphate-predicted and nitrate-predicted cadmium distribution.

cadmium in the mixed layer of the northeast Pacific is about 0.1 yr. This is very short in comparison with the surface layer residence times of water and most other metals.

Much of the cadmium settling within biogenic detritus is quickly regenerated and returned to the euphotic zone by vertical mixing. Cadmium-containing particles that settle below the thermocline without complete regeneration constitute a source of cadmium for the deep waters and oceanic sediments. Aside from fractions of phytoplanktonic detritus, such materials include zooplankton detritus and fecal matter. Dissolution and regeneration of such organic detritus continues below the thermocline, but the faster settling particles, such as fecal pellets, may well succeed in carrying a significant fraction of the cadmium removed from the mixed layer to the pelagic sediments.

The removal of cadmium from the deeper layers of the ocean is less well understood. While removal from the dissolved state in deep waters by settling particles, particularly clays, is a potential mechanism for the sedimentation of cadmium that enters the deep water through residual regeneration,

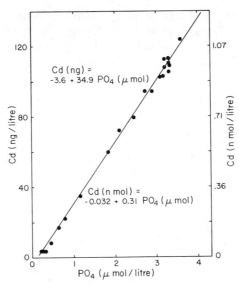

Figure 2. Relationship of cadmium and phosphate at stations in the northeast Pacific (after Bruland et al., 1978) ($r = 0.998$; $n = 21$).

it also seems possible that manganese chemistry plays a more important role in the removal of cadmium from the deep water than it does elsewhere (Simpson, 1981; Li, 1982). There is some evidence that the distribution of deep-water manganese may be controlled by the rate of oxidation of Mn^{2+} to MnO_2. Certainly, any exposed and accreting manganese oxide surfaces on suspended particles in the deep water or at the sediment–water interface will act as sites for cadmium removal. However, the degree to which such a mechanism competes with removal of cadmium by ambient clay particles needs to be determined.

The natural sources of oceanic cadmium are predominantly continental runoff and atmospheric deposition. The majority of the suspended particulate material discharged in runoff is sedimented out on the continental shelves, often relatively close to shore. Only about 5% of the particulate material discharged by rivers and streams is able to survive near-shore precipitation and enter the pelagic areas. Thus, the majority of cadmium that is tightly bound in mineral lattices of rock-weathering products is retained in near-shore and continental shelf sediments. Furthermore, in the case of some elements, such as iron, the dissolved forms discharged from rivers are precipitated from solution in estuarine environments because of the change in physicochemical conditions between the freshwater regime and the marine regime. While the extent of such removal from solution, which occurs through precipitation, flocculation of colloids, and adsorption processes, is generally known for elements whose estuarine geochemistry has been widely studied, such as iron and manganese, in the case of cadmium there has been insufficient evidence of consistently universal behavior to provide authoritative assessments of the estuarine behavior of the element

(Bewers and Yeats, 1980). However, in two independent field studies on differing estuaries (Windom et al., 1976; Bewers and Yeats, 1978), approximately 15% of river-borne dissolved cadmium was found to be removed during estuarine transport. Laboratory studies of estuarine processes involved in particulate-dissolved metal exchanges by Sholkovitz (1978) and Sholkovitz and Copland (1981) estimate the removal of dissolved cadmium in estuaries to be approximately 3–5% of the total. However, it appears likely that a considerable proportion of the dissolved cadmium (in such cases, the term *dissolved,* which is operationally defined by filtration procedures, includes colloidal forms) remains in solution and can enter the pelagic environment after undergoing biological cycling processes on the continental shelves. It is, of course, still possible that the flux of dissolved cadmium into the ocean from estuaries is augmented by cadmium desorbed from riverine particles. This reverse process of desorption of cadmium from particles in estuaries is still a matter of some speculation because the nature of the particulate cadmium–salinity relationship is complicated by the concomitant mixing between freshwater-derived and marine particles, having very different compositions in most cases. The extent to which the particle composition changes are due to simple mixing between particles of different origins or due to actual changes in the composition of river-derived particles as they pass into the estuary is still a matter for scientific discussion (De Groot, 1973; Salomons and De Groot, 1977; Mueller and Foerstner, 1975; Duinker and Nolting, 1977).

1.5. MARINE SEDIMENTS

Cadmium associated with sedimentary solids may be present in several chemical forms, depending on the sediment composition and the physicochemical properties of the sedimentary environment. Cadmium bound within crystalline mineral lattices is essentially unavailable, and in the literature, this fraction is often referred to as the lithogenic, detrital, or residual fraction. Cadmium can be adsorbed by electrostatic attraction to negatively charged ion exchange sites on mineral colloids, clay particles, organic particles, and hydrous oxides. These phases are in equilibrium with the sediment interstitial water and are exchanged with the aqueous phase as the physicochemical conditions change, although the rates of reaction may be very slow. Cadmium may also be associated with oxides, hydroxides, and hydrous oxides of iron and manganese through coprecipitation. When the iron and manganese are solubilized under reducing conditions in the sediments, they are able to migrate into oxidized layers, where, to differing extents, oxidation and reprecipitation occur. Through coprecipitation, other metals can be affected by these processes, but since it appears that cadmium is an intrinsically weak competitor for adsorption onto hydrous metal oxides, this may not be as important a mechanism for the fixation of cadmium in seawater as it is in fresh water (Jenne, 1968; Gadde and Laitinan, 1974; Lee, 1975) except where other processes of fixation are of reduced importance, such as in the deep waters of the ocean. There exist, however, highly insoluble and stable complexes of cadmium sulfides in reduced sediments (Krauskopf, 1956). It is likely that some proportion of the cadmium released during the reduction of oxides of manganese and iron to mobile forms would be precipitated as insoluble sulfides in the reduced layer itself. Oxidation of reduced sediments results in the transformation of sulfide-bound cadmium into more mobile forms that are in equilibrium with the pore water (Khalid, 1980).

A range of values for the concentration of cadmium in pelagic clays and calcareous ooze has been reported (Heinrichs et al., 1980). Respective mean values for cadmium in these two sedimentary phases have been given as 0.21 and 0.23 mg/kg by Chester and Aston (1976). However, Simpson (1981), in his review of cadmium, gives a mean cadmium concentration in marine sediments of 0.5 mg/kg, which is inexplicably larger than previous estimates. As with other metals, cadmium concentrations in sediments show a relationship to grain size. Recently available data from Baffin Bay (D. H. Loring, private communication) show a range in concentration for cadmium from 0.06 mg/kg in sands to 0.15 mg/kg in sandy muds and muds in the area. The only sedimentary phases to show considerable enrichment of cadmium, relative to aluminum, are phosphorites and manganese concretions and nodules. It must be stressed, however, that the increased concentrations of cadmium in manganese nodules (about 10 mg/kg) relative to pelagic clays do not represent enrichments when the concentrations of cadmium are compared with those of manganese. In sediments having a significant proportion of fine-grained material, a large proportion of the cadmium is relatively loosely bound, and greater than 75% can be extracted from the sediments with a weak acid (25% v/v acetic or 1 N hydrochloric) leach. This has been used to suggest that large proportions of the cadmium associated with marine sediments might be potentially available to marine biota.

The primary source of cadmium that is remobilized in the sediments will be organic material that is undergoing oxidation. The organic phase of sediments will comprise components of varying stability from the easily decomposable and mobile forms to forms that are very resistant to decomposition. Cadmium complexed with the organic fraction may be roughly divided into the chelated and organically bound fractions. Chelated cadmium is the fraction that is loosely attached to immediately mobile and easily decomposable organic material and is a good indicator of the easily bioavailable cadmium form (Gambrell et al., 1977a,b). Organically bound cadmium is the fraction incorporated into the less soluble organic material and can only be solubilized after intense oxidation (Gambrell et al., 1977a). Clearly, the majority of cadmium regeneration from easily decomposable forms of organic material occurs in the water column. Deep sediments will contain lower proportions of easily regeneratable cadmium. Nevertheless, the more slowly decomposing components of biogenic material will reach the sediments, and some release of residual cadmium at the sediment–water interface will occur.

Since it appears that there are not strong mechanisms available for the fixation of cadmium in oxidized sediments, one might reasonably expect, where there occurs organic matter oxidation, a significant flux of cadmium from marine sediments to the overlying waters.

The association of cadmium and organic matter in the marine environment raises the possibility that, in shallow environments, cadmium may be released in substantial quantities from coastal sediments through the oxidation of organic matter that is incompletely regenerated in the water column. Chen et al. (1976) studied the release of sediment-bound cadmium to the overlying seawater in laboratory experiments. The results of this work show that significant releases of cadmium only occur under oxidizing conditions. In the laboratory lysimeter experiments, cadmium concentrations in the overlying water were raised from 30 to 500 ng/L during a 4-month period. During this experiment, there occurred an increase in the water-soluble, exchangeable, and carbonate fractions at the expense of the organic and sulfide fractions. Little release was evident under slightly oxidizing and reducing conditions. This is consistent with the view that the primary forms of cadmium that are susceptible to dissolution from sediments are the oxidizable organic forms and that releases from iron–manganese oxide phases are very limited.

REFERENCES

Bewers, J. M., and Yeats, P. A. (1978). Trace metals in the waters of a partially mixed estuary. *Est. Coast. Mar. Sci.,* **7,** 147–162.

Bewers, J. M., and Yeats P. A. (1980). Behaviour of trace metals during estuarine mixing. In *River Inputs to Ocean Systems,* Martin, J. M., Burton, J. D. and Eisma, D. (eds.) UNESCO, Paris, pp. 103–115.

Boutron, C. (1980). Comments on the letter from Landy et al. *Nature,* **284,** 575–576.

Bruegmann, L. (1974). Die Bestimming von Zink, Kadmium und Blei in der Ostsee durch inverse Voltammetrie. *Beitr. Meeresk.,* **34,** 9–21.

Bruland, K. W., Knauer G. A., and Martin, J. H. (1978). Cadmium in northeast Pacific waters. *Limnol. Oceanogr.,* **23,** 618–625.

Chen, K. Y., Gupta, K. A., Sycip, A. Z., Lu, J. C. S., Knezevic, M., and Choi, W. W. (1976). Research study on the effect of dispersion, settling, and resedimentation on migration of chemical constituents during open-water disposal of dredged materials. U.S. Army Engineer Waterways Experiment Station, Contract Report, D-76-1, CE, Vicksburg, MS.

Chester, R., and Aston, S. R. (1976). The geochemistry of deep-sea sediments. In *Chemical Oceanography,* Riley J. P., and Chester, K. (eds.). Academic, London, pp. 281–390.

De Groot, A. J. (1973). Occurrence and behaviour of heavy metals in river deltas with special reference to the rivers Rhine and Ems. In *North Sea Science,* Goldberg, E. D. (ed.). MIT Press, Cambridge, MA, pp. 308–325.

Doolan, K. J., and Smythe, L. E. (1973). Cadmium content of some New-South-Wales waters. *Search,* **4,** 162–163.

Duinker, J. C., and Nolting, R. F. (1977). Dissolved and particulate trace metals in the Rhine Estuary and the Southern Bight. *Mar. Pollut. Bull.,* **8,** 65–69.

References

Duinker, J. C., and Nolting, R. F. (1978). Mixing, removal and mobilization of trace metals in the Rhine estuary. *Neth. J. Sea Res.*, **12**, 205–223.

Fairbridge, R. W. (ed.). (1972). *Encyclopedia of Earth Sciences Series*, Volume IVA, *The Encyclopedia of Geochemistry and Environmental Sciences*. Van Nostrand Reinhold, New York.

Fulkerson, W., and Goeller, H. E. (1973). Cadmium, the dissipated element. Oak Ridge National Laboratory, Report No. ORNL-NSF-EP-21, Oak Ridge, TN.

Gadde, R. R., and Laitinan, H. A. (1974). Studies of heavy metal adsorption by hydrous iron and manganese oxides. *Anal. Chem.*, **46**, 2022–2026.

Gambrell, R. P., Khalid, R. A., Verloo, M. G., and Patrick, W. H. Jr. (1977a). Transformations of heavy metals and plant nutrients in dredged sediments as affected by oxidation-reduction potential and pH. Vol II: Materials and Methods, Results and Discussion. U.S. Army Engineer Waterways Experiment Station, Contract Report, D-77-4,CE, Vicksburg, MS.

Gambrell, R. P., Collard, V. R., Reddy, C. N., and Patrick, W. H. Jr. (1977b). Trace and toxic metal uptake by marsh plants as affected by Eh, pH, and salinity. U.S. Army Engineer Waterways Experiment Station, Technical Report D-77-40, CE, Vicksburg, MS.

Garcia-Maragaya, J., and Page, A. L. (1977). Influence of exchangeable cation on the sorption of trace amounts of cadmium by montmorillonite. *J. Soil Sci. Soc. Am*, **41**, 718–721.

Gardiner, J. (1974). The chemistry of cadmium in natural water: II. The adsorption of cadmium on river muds and naturally occurring solids. *Water. Res.*, **8**, 157–164.

Heinrichs, H., Schulz-Dobrick, B., and Wedepohl, K. H. (1980). Terrestrial geochemistry of Cd, Bi, Tl, Pb, Zn and Rb. *Geochim. Cosmochim. Acta*, **44**, 1519–1533.

Herron, M. H., Langway, C. C. Jr., Weiss, H. V., and Cragin, J. H. (1977). Atmospheric trace metals and sulfate in the Greenland Ice Sheet. *Geochim. Cosmochim. Acta*, **41**, 915–920.

Jenne, E. A. (1968). Controls of Mn, Fe, Co, Ni, Cu, and Zn concentrations in soils and water: The significant role of hydrous Mn and Fe oxides. In *Trace Inorganics in Water*, Baker, R. A. (ed.), Advances in Chemistry Series 73. American Chemical Society, Washington, DC, pp. 337–388.

Khalid, R. A. (1980). Chemical mobility of cadmium in sediment–water systems. In *Cadmium in the Environment*, Part 1, *Ecological Cycling*, Nriagu, J. O. (ed.). Wiley, New York, pp. 257–304.

Knauer, G. A., and Martin, J. H. (1981). Phosphorus-cadmium cycling in northeast Pacific waters. *J Mar. Res.* **39**, 65–76.

Krauskopf, K. B. (1956). Factors controlling the concentration of thirteen rare metals in sea water. *Goechim. Cosmochim. Acta*, **9**, 1–32.

Landy, M. P., Peel, D. A., and Wolff, E. W. (1980). Trace metals in remote Arctic snows: Natural or anthropogenic. *Nature*, **284**, 574–575.

Lee, G. F. (1975). Role of hydrous metal oxides in the transport of heavy metals in the environment. In *Heavy Metals in the Aquatic Environment*, Krenkel, P. A. (ed.). Pergamon, Oxford, England, pp. 137–154.

Li, Y-H. (1982). Interelement relationship in abyssal Pacific ferromanganese nodules and associated pelagic sediments. *Geochim. Cosmochim. Acta*, **46**, 1053–1060.

Mart, L. (1983). Seasonal variations of Cd, Pb, Cu and Ni levels in snow from the eastern Arctic Ocean. *Tellus*, **35B**, 131–141.

Meranger, J. C., Subramanian, K. S., and Chalifoux, C. (1979). A national survey for cadmium, chromium, copper, lead, zinc, calcium and magnesium in Canadian drinking water supplies. *Environ. Sci. Technol.*, **13**, 707–711.

Mueller, G., and Foerstner, V., (1975). Heavy metals in sediments of the Rhine and Elbe estuaries: Mobilization or mixing effect. *Environ. Geol.*, **1**, 33–39.

Nriagu, J. O. (1979). Global inventory of natural and anthropogenic emissions of trace metals to the atmosphere. *Nature,* **279,** 409–411.

Nriagu, J. O. (1980a). Cadmium in the atmosphere and in precipitation. In *Cadmium in the Environment* Part 1, *Ecological Cycling,* Nriagu, J. O. (ed.). Wiley, New York, pp. 71–114.

Nriagu, J. O. (1980b). Trace-metals in remote Arctic snows: Natural or anthropogenic: A reply. *Nature,* **284,** 575.

Nuernberg, H. W., and Valenta, P. (1983). Potentialities and applications of voltammetry in chemical speciation of trace metals in the sea. In *Trace Metals in Seawater,* Wong, C. S., Boyle, E., Bruland, K. N., Burton, J. D., and Goldberg, E. D. (eds.). Plenum, New York, pp. 671–697.

Page, A. L., and Bingham, F. T. (1973). Cadmium residues in the environment. *Resid. Rev.,* **48,** 1–44.

Poldoski, J. E., and Glass, G. E. (1978). Anodic stripping voltammetry at a mercury film electrode: Baseline concentrations of cadmium, lead and copper in selected natural waters. *Anal. Chim. Acta,* **101,** 79–88.

Raspor, B. (1980). Distribution and speciation of cadmium in natural waters. In *Cadmium in the Environment,* Part I, *Ecological Cycling,* Nriagu, J. O. (ed.). Wiley, New York, pp. 147–236.

Rosman, K. J. R., and De Laeter, J. R. (1976). Low level determinations of environmental cadmium. *Nature,* **261,** 685–686.

Rosman, K. J. R., and De Laeter, J. R. (1977). The cadmium content of some river systems in Western Australia. *J. Roy. Soc. West. Austr.,* **59,** 91–96.

Salomons, W., and De Groot, A. J. (1977). Pollution history of trace metals in sediments, as affected by the Rhine River. Delft Hydraulics Laboratory, Publication No. 84, Delft, The Netherlands.

Sholkovitz, E. R. (1978). The flocculation of dissolved Fe, Mn, Al, Cu, Ni, Co and Cd during estuarine mixing. *Earth Planet. Sci. Lett.,* **41,** 77–86.

Sholkovitz, E. R., and Copland, D. (1981). The coagulation, solubility and adsorption properties of Fe, Mn, Cu, Ni, Cd, Co and humic acids in a river water. *Geochim. Cosmochim. Acta,* **45,** 181–189.

Sillen, L. G., and Martell, A. E. (1971). Stability constants of metal complexes. Chemical Society Special Publication 17, Chemical Society, London.

Simpson, W. R. (1981). A critical review of cadmium in the marine environment. *Progr. Oceanogr.,* **10,** 1–70.

Struempler, A. W. (1973). Adsorption characteristics of silver, lead, cadmium, zinc and nickel borosilicate glass, polyethylene and polypropylene container surface. *Anal. Chem.,* **45,** 2251–2254.

Taylor, S. R. (1964). Abundance of chemical elements in the continental crust: A new table. *Geochim. Cosmochim. Acta,* **28,** 1273–1285.

Trefry, J. H., Metz, S., and Trocine, R. P. (1983). Decreased inputs of cadmium and lead to the Gulf of Mexico from the Mississippi River. *EOS,* **64,** 244 (Abstract).

Ure, A. M., and Berrow, M. L. (1982). *The Elemental Constituents of Soils in Environmental Chemistry,* Vol. 2, Bowen, H. J. M. (ed.). The Royal Society of Chemistry, London, 94–204.

Vinogradov, A. P. (1959). *The Geochemistry of Rare and Dispersed Chemical Elements in Soils.* Consultants Bureau, New York.

Windom, H. L., Gardner, W. S., Dunstan, W. M., and Paffenhofer, G. A. (1976). Cadmium and mercury transfer in a coastal marine environment. In *Marine Pollutant Transfer,* Windom, H. L., and Duce, R. A. (eds.). Lexington Books, Lexington, MA, pp. 135–157.

Yeats, P. A., and Bewers, J. M. (1982). Discharge of metals from the St. Lawrence River. *Can. J. Earth Sci.,* **19,** 982–992.

2

EVIDENCE FOR ANTHROPOGENIC MODIFICATION OF GLOBAL TRANSPORT OF CADMIUM

P. A. Yeats
J. M. Bewers

Department of Fisheries and Oceans
Atlantic Oceanographic Laboratory
Bedford Institute of Oceanography
Dartmouth, Nova Scotia, Canada

2.1. Introduction
2.2. Mobilization of Cadmium in World Rivers
2.3. Natural Mass Balance for Cadmium in Ocean
2.4. Potential Magnitude of Anthropogenically Mobilized Cadmium
2.5. Consequences of Anthropogenically Augmented Influx of Cadmium to Ocean
References

2.1. INTRODUCTION

Within the last decade, there have been substantial improvements in the precision and accuracy of measurements of trace metals in the environment. Consequently, substantially better understanding of the distribution and transport of these elements has been obtained, especially within the marine environment. The natural environmental transport of some trace metals may have been substantially altered by anthropogenic activities. Such a situation clearly applies to lead (Schaule and Patterson, 1983), and it has been hypothesized that both cadmium (Simpson, 1981; Yeats and Bewers, 1983) and zinc (Yeats and Bewers, 1983) may be mobilized by human and industrial activities in amounts that rival, or exceed, the natural circulation of these elements. This chapter focuses on cadmium and attempts to assess the evidence for increased dissemination of cadmium in the environment.

Cadmium is, with mercury, one of the metals most commonly given special status in environmental management and regulation, largely for historical reasons associated with the incidence of Itai Itai disease in Japan. Thus, the element appears in the Annex I list (the so-called black list) of substances that are proscribed from dumping in the ocean under the Convention on the Prevention of Marine Pollution by Dumping of Wastes and Other Matter, London, 1972 (referred to as the London Dumping Convention) and has been given similar status under certain regional agreements. This status for cadmium has been questioned (Taylor, 1984), but it is, nevertheless, of some regulatory interest to assess the degree to which man and other organisms are likely to be exposed to the element under natural conditions and the extent to which anthropogenic activities might have increased such exposures.

The distribution of cadmium in the ocean reflects biological uptake and regeneration processes (Boyle et al., 1976; Bruland, 1980) such that cadmium is depleted in ocean surface waters, has a maximum concentration at the depth of the phosphate and nitrate maxima, and has a deep-water concentration that varies by about a factor of 4 throughout the oceans. The ratio of the concentration of cadmium to that of phosphate is fairly constant at a value of 3×10^{-5} mol Cd/mol P. Until recently, assessments of cadmium concentrations in rivers generally gave values considerably greater than the concentration in the ocean, usually in the hundred-nanogram-per-liter range. Such high concentrations cannot be easily rationalized with the rate of cadmium sedimentation from the ocean (Bewers and Yeats, 1977). While some of these high riverine concentrations may be due to industrial activities within the corresponding drainage basins, the recognition that sampling and analytical artifacts were largely responsible for the high concentrations of metals reported in the ocean up until the latter years of the 1970s has raised suspicions about the quality of many of the measurements of river water composition. It can be assumed that similar problems to those experienced in trace analysis of seawater have also applied to river determinations and have had a deleterious effect on the quality of such measurements (Edmond, 1980). A rather striking example of this is the difference between the dissolved cadmium concentration of 100 ng/L reported for the Mississippi River in 1976 (Trefry and Presley, 1976) and that reported by the same senior author for the same river in 1983 of 15 ng/L (Trefry et al., 1983). In the latter paper, this difference was attributed to a decline in cadmium concentrations due to the imposition of regulatory controls on waste discharges to the river. It now appears equally likely that the difference between the two measurements is largely due to analytical improvements achieved in the intervening years. Within the last few years, a wider body of more reliable data on cadmium in rivers has become available. These data lend themselves to a closer examination of the heterogeneity in river composition and the transport of cadmium into the ocean.

2.2. MOBILIZATION OF CADMIUM IN WORLD RIVERS

Table 1 summarizes existing data on cadmium in dissolved and particulate phases in river water. These data have been chosen on the basis of the ability of the authors to conduct reliable measurements of cadmium in seawater. Consequently, this represents a small subset of the reported concentrations of cadmium in rivers. It can be seen immediately that there exists considerable variability in the riverine concentration of cadmium in both dissolved and particulate form. Nevertheless, with one exception (the Rhine), where calculable, the distribution coefficient between particulate and dissolved cadmium is fairly constant in the $1 \times 10^4 - 4 \times 10^4$ range and is comparable with laboratory measurements of the distribution coefficient (K_d) for cadmium between suspended particulate and dissolved phases conducted by Li et al. (1984) of between 1×10^4 and 2×10^4. This tends to lend credence to the quality of the data. These results indicate that the cadmium concentration in rivers is extremely variable. Furthermore, those rivers that might be expected to show the effects of intense anthropogenic activity within their drainage basins, namely, the Elbe, Weser, Rhine, Scheldt, St. Lawrence, and Hudson, contain higher concentrations relative to the (assumed) more pristine rivers such as the Amazon and Orinoco. It is to be admitted that the above assessment is relatively superficial since it is hard to quantify the degree to which individual rivers are likely to be affected by human and industrial activities. However, if the hypothesis of anthropogenic influence is considered, it is reasonable to assume that grossly affected rivers might show increases in their cadmium content from source to mouth. We can now turn to examining the case of two rivers that are likely to reflect anthropogenic activity in the central and lower parts of their drainage basins and for which there exist suitable data.

Table 2 presents measurements of the concentration of cadmium in the St. Lawrence drainage basin from Lake Superior to the mouth of the St. Lawrence River at Quebec City. In assessing these data, it is necessary to remember that there exist proportionally minor amounts of industrial activity on the shores of Lake Superior, increased activity in the southern basin of Lake Michigan, and intense industrialization on the shores of lakes Erie and Ontario. There seems to be a corresponding gradient in the concentration of cadmium from low values in Lake Superior to high concentrations at Quebec City at the mouth of the St. Lawrence River. The St. Lawrence drainage basin is dominated by the igneous and metamorphic geology of the Canadian

Table 1. Concentration of Cadmium in Rivers

River	Suspended Particulate Matter (mg/L)	Particulate Cadmium (ng/L)	Particulate Cadmium (mg/kg)	Dissolved Cadmium (ng/L)	Particulate/Dissolved Distribution Coefficient	References
Elbe	75	150	2	100–200	$1 \times 10^4 - 2 \times 10^4$	Duinker et al., 1982a
Weser	50	85	$(1)^a$	35	3×10^4	Mart et al., 1985
		100	2	100	2×10^4	Duinker et al., 1982b
Varde Å	40	80	2	70		Mart et al., 1985
Rhine	30	1800	60	70	3×10^4	Duinker et al., 1980
				400	1×10^5	Duinker and Nolting, 1978
Amazon				8		Boyle et al., 1982
Mississippi			0.6	15	4×10^4	Trefry et al., 1983
St. Lawrence	10	17	1.7	110	2×10^4	Yeats and Bewers, 1982
Hudson	10–20	70–80	4–8	200–300	$2 \times 10^4 - 3 \times 10^4$	Klinkhammer and Bender, 1981
Orinoco				2		Grant et al., 1982
Yangtse				2		Grant et al., 1982
Gota	25–100	14–64	0.3–0.6	9–25	$2 \times 10^4 - 3 \times 10^4$	Danielsson et al., 1983

^{a}Calculated value assuming a suspended matter concentration of 75 mg/L.

Table 2. Cadmium in Waters of the St. Lawrence Basin

Location	Dissolved Cadmium (ng/L)	Reference	Dissolved Cadmium (ng/L)	Reference
Lake Superior	23	Poldoski and Glass, 1978		
Lake Michigan	15–41 (avg 28)	Muhlbaier et al., 1982	43	Rossman, 1984
Lake Huron	34	IJC, 1977	41	Rossman, 1984
Lake Erie	40	Lum and Leslie, 1983	58	Rossman, 1984
Lake Ontario	45–53	Nriagu et al., 1981	68	Rossman, 1984
St. Lawrence River (Quebec City)	110	Yeats and Bewers, 1982		

Shield, and this may explain the relatively high value for cadmium, compared to that in some rivers like the Amazon, even at the head of the system. Attributing the trend in the cadmium distribution in the system to anthropogenic activity is compatible with the conclusions drawn from sediment–stratigraphy studies. Kemp et al. (1976), for example, assessed the degree of enrichment of certain metals in the recent sediments of Lake Erie by relating the surficial element concentrations (normalized to aluminum) to those below the *Ambrosia* pollen horizon and concluded that several metals, including cadmium, were significantly enriched in recent sediments. From the perspective of natural mobilization, the Great Lakes/St. Lawrence Basin should be in steady state, but this may not be the case in relation to anthropogenic inputs. If one was to assume that approximately 25 ng/L was the concentration of cadmium in the waters of this basin solely under natural, predominantly primary weathering conditions, it would be concluded that the natural concentration was augmented by about a factor of 3 due to the anthropogenic emissions to the system. This corresponds to an increased discharge of cadmium, in fresh water leaving the system, of 2×10^5 kg/yr as compared with about 7×10^4 kg/yr under natural conditions. The St. Lawrence is, however, an atypical world river in that it has a substantially lower suspended particulate load (~ 10 mg/L) than the world average (~ 550 mg/L). Thus, the discharge of cadmium in the dissolved form from the St. Lawrence River dominates the particulate cadmium discharge even at a distribution coefficient of 10^5. It would, therefore, be useful to examine the situation in another river having a more typical suspended load (Holeman, 1968) and with equal likelihood of anthropogenic influence. A suitable choice, for which appropriate data exist, is the Rhine River.

The Rhine has been extensively studied from a variety of compositional perspectives during this century because of its importance to the economies of several Western European countries for navigation and agricultural purposes [see Salomons and Foerstner (1984) and references therein]. Rhine River sediment samples collected in the Netherlands section of the river at various times during the 1900s have been analyzed for a number of trace metals. The results show increases in the cadmium concentrations from less than 2 mg/kg before 1920 to about 40 mg/kg in the early 1970s. This is thought to reflect increased contamination* of the Rhine as a result of human and industrial activities. Following the early 1970s, there appears to be a small decline in concentrations. Corresponding results for other metals, such as arsenic, which exhibits increased concentrations to 1960 and a substantial decline thereafter, and mercury, which increases to 1975 and then declines, would seem to reflect reductions in the use of arsenic-containing compounds as pesticides and the increased attention to the reduction of mercury discharges from the chlor-alkali industry, respectively. This tends to give credence to the assumption that the trends predominantly reflect anthropogenic effects. Furthermore, independent measurements of the transport of metals in dissolved and particulate phases at the Dutch–German border during the 1970s show a decline in the rates of transport of cadmium and several other metals. These declines are of much greater magnitude for the dissolved forms of cadmium and zinc than they are for their corresponding particulate forms, whereas the declines in both dissolved and particulate forms of chromium, copper, mercury, and lead are of comparable magnitude. For there to exist a different balance between the dissolved and particulate forms of cadmium and zinc would imply that the nature of the suspended material might have altered, but such a trend is not supported by the data for the other metals. This raises the possibility that some of the apparent decline in the transport of the dissolved forms of cadmium and zinc results from improvements in sampling and analytical capability during the same decade in which substantial declines in the concentrations of metals in seawater were being reported.

Other, spatial, evidence of the effect of industrial activities on cadmium in the Rhine basin can also be gleaned from Salomons and Foerstner (1984). These authors reported the distribution of cadmium, zinc, lead, and mercury in the pelitic (<2 μm) fraction of river sediments of the Rhine basin. The selection of the pelitic fraction for analysis will compensate for grain size variations in the sediments of the system but may not wholly normalize for sediment composition changes. Cadmium concentrations lie in the range 5–80 mg/kg range with the lowest values in the head of the Rhine, downstream of Bodensee (Lake Constance), the highest values in the lower Neckar River, and concentrations of about 20 mg/kg in the lower Rhine. The high values in the sediments of the Neckar were subsequently shown to be correlated with extremely high cadmium levels in water, particularly in the Enz River, a tributary of the Neckar. Nevertheless, downstream of the point of inflow of the Enz, high concentrations of dissolved cadmium persisted to the

---

*The term *contamination* is used throughout to indicate the existence of additional amounts or concentrations of an element resulting from anthropogenic activity. The term *pollution* would only be used in instances where deleterious effects, for example, on human health, of the increased environmental incidence of elements are evident.

confluence of the Neckar and the Rhine. Clearly, some time is required to reestablish an equilibrium balance between cadmium in the dissolved, suspended particulate, and sedimentary phases, and this shows up as a decrease in the dissolved concentration and increased particulate concentrations if the supply of excess cadmium is largely in the dissolved form. Salomons and Foerstner (1984) provided several other examples of the incidence of excess cadmium in river sediments that were attributed to contamination from industrial sources. Overall, dissolved cadmium increases from the source (4–12 ng/L, Nuernberg et al., 1976; Mart et al., 1984; Sigg, 1985) to the mouth of the Rhine (400 ng/L, Duinker and Nolting, 1978) consistent with the addition of cadmium from industrial sources within the drainage basin.

If we now turn to examinations of the presence of cadmium in coastal zone sediments, a similar, if equally tentative, picture emerges. Table 3 provides data on the concentrations of cadmium in coastal sediments of several areas that might be thought to reflect increased rates of cadmium influx to marine sediments during recent times. However, the higher concentrations in surficial shelf sediments than the subsurface sediments may reflect the consequences of remobilization of cadmium from material being oxidized or otherwise regenerated during early diagenesis in sediments rather than simply some increased rate of deposition. It seems likely that some cadmium, particularly that associated with organic matter, may be remobilized during the oxidation of organic material in surficial sediments, and it is not known to what extent this cadmium may be able to escape into the overlying water rather than being retained in association with the more refractory components of the sediments such as clay particles. Nevertheless, the ratio of surficial to subsurface sediment cadmium concentrations is higher than similar ratios for other metals, and this may reflect the increased introduction of cadmium to the coastal marine environment from anthropogenic sources.

Table 3. Instances of Elevated Levels of Cadmium in Coastal Zone Sediments

Location	Concentration of Cadmium in Subsurface Sediments (mg/kg)	Concentration of Cadmium in Surficial Sediments (mg/kg)
North Sea	0.3^a	2^a
Baltic	0.25^a	2^a
United States, West Coast	—	$1–2^b$
New Bedford and Buzzards Bay	$0.3–0.4^c$	$2–50^c$

[a] From Salomons and Foerstner, 1984.
[b] From Bruland et al., 1974.
[c] From Nriagu, 1981.

2.3. NATURAL MASS BALANCE FOR CADMIUM IN OCEAN

Having presented the evidence for some change in the rate of mobilization in environmental processes, it seems appropriate to make some attempt at quantifying the extent of this change. We will first attempt to construct a budget for cadmium introduction and removal from the ocean under natural (preindustrial) conditions. In doing this, we can take advantage of an assumption that the ocean is in steady-state balance for purely natural cadmium fluxes. We will start by making choices as to the concentrations of cadmium in transport pathways into and out of the ocean, construct a budget, and then determine how close it is to a balance. It is becoming increasingly clear, from recent measurements, that the concentration of dissolved cadmium in uncontaminated rivers is about 10 ng/L. Riverine suspended particles contain approximately 0.25–0.30 mg/kg cadmium, equal to or slightly greater than the cadmium content of weathered rock material. We base this on the assumption that the distribution coefficient between dissolved and suspended particulate phases in rivers is between 1.5×10^4 and 2×10^4, which at 10 ng/L gives 0.15–0.20 mg/kg for the exchangeable-particulate-phase cadmium concentration. To this concentration, however, we need to add the cadmium associated with particles that is not in exchangeable-phase equilibrium with cadmium in solution (i.e., the so-called detrital, or crystalline, cadmium), which amounts to about 0.1 mg/kg based on crustal rock abundances. This implies that the average concentration of cadmium in river-borne particles is in the range 0.25–0.30 mg/kg of which about 60–70% is in the exchangeable phase and about 30–40% is detrital and incorporated into the mineral lattices of the aluminosilicate particles.

Atmospheric deposition into the ocean is the most difficult to quantify with a known degree of certainty. The literature is replete with measurements of cadmium in atmospheric precipitation, but because of the severity of contamination, use of unsuitable analytical methods, lack of representative sampling, and other artifacts, most of these data are questionable (D. J. MacGregor, personal communication). The most reliable data in the recent literature are those of Mart (1983), who made measurements of cadmium in snowfall in the arctic. He found that fresh snow contained 0.4 ng/L cadmium, but after deposition, the snow became rapidly augmented to 4 ng/L, presumably through dry fallout. It should, however, be remembered that precipitation rates in the arctic are low but that dry snow has a poorer scavenging capacity than rain and therefore is likely to have a lower proportion of washout-derived material than rain. We will assume that Mart's (1983) value for the cadmium content of fresh arctic precipitation is equal to the concentration of cadmium in the rainout component of wet precipitation over the global ocean under wholly natural conditions. We will further assume that the detrital content of cadmium in particulate deposition is similar to the detrital crustal abundance of 0.1 mg/kg. In the absence of more information on the likely deposition of cadmium on land under natural conditions, we will assume that the cadmium content of wet deposition is double

Mart's (1983) value of 0.4 ng/L in remote arctic areas and that aggregate particle deposition on land surface is equal to that over the world ocean. Finally, in order to obtain estimates of the exchangeable-phase cadmium in atmospheric deposition, we will use Nriagu's (1981) estimate of the natural emission of cadmium to the atmosphere of 8×10^5 kg/yr and assume that the difference between the sum of wet and detrital particulate depositional fluxes and the emission figure corresponds to the depositional flux of cadmium in the exchangeable particulate phase. It should be noted that the exchangeable-phase particulate concentration on precipitating atmospheric particles calculated in this way corresponds to approximately 0.5 mg/kg, which compares well with the range of cadmium in soils listed by Ure and Berrow (1982).

We wish to include in the mass balance both continental shelf and pelagic sediments as sinks for oceanic cadmium and deal with these two removal routes separately. We will characterize basement shelf sediments as containing 0.25 mg/kg, which corresponds to basal values in near-shore sediment cores. For the pelagic routes of cadmium removal, we will deal separately with the clay, carbonate, siliceous, and ferromanganese concretion phases. We have had to characterize these various rates of removal as products of

Table 4. Carrier Phase Data for Construction of Oceanic Cadmium Mass Balance

Carrier Phase	Value	Reference
River water discharge	3.2×10^{16} kg/yr	Livingstone, 1963
River-suspended particulate discharge	1.8×10^{13} kg/yr	Holeman, 1968
Atmospheric precipitation into ocean	4.0×10^{17} kg/yr	Chorley, 1969
Atmospheric particle deposition into pelagic ocean	5×10^{11} kg/yr	Bewers and Haysom, 1974
Pelagic clay accumulation rate	1.1×10^{12} kg/yr	Ku et al., 1968
Pelagic silicate accumulation rate	3.6×10^{11} kg/yr	Wollast, 1974
Pelagic carbonate accumulation rate	1.1×10^{12} kg/yr	Milliman, 1974
Polymetallic nodule accretion rate	2.5×10^{10} kg/yr	Krishnaswamy and Lal, 1972
Proportion of river-borne particulate material that survives near-shore and shelf sedimentation	4%	Figure based on amount of residual particulate flux required to account for pelagic clay sedimentation plus a proportion of carbonate sedimentation

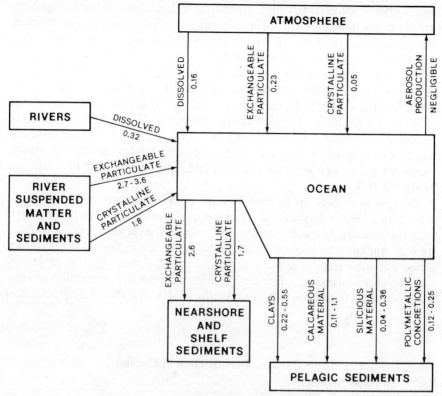

Figure 1. Oceanic cadmium budget (10^6 kg/yr).

carrier-phase deposition (see Table 4) and by ranges of concentration for cadmium in these materials because of uncertainties in the true means. Pelagic clays should have cadmium concentrations in the range 0.2–0.5 mg/kg, carbonates 0.1–1 mg/kg, silicates 0.1–1 mg/kg, and polymetallic nodules 5–10 mg/kg (Aston et al., 1972; Boyle, 1981; Cronan, 1976; Heinrichs et al., 1980; Mullin and Riley, 1956). Combining these concentration ranges with the carrier-phase fluxes given in Table 4 provides the natural cadmium budget depicted in Figure 1. Ranges of influx and efflux for cadmium have been provided for riverine input and pelagic sedimentation based on two different riverine k_d values ($1.5 \times 10^4 - 2 \times 10^4$) and the ranges for cadmium concentrations in pelagic sediment phases. The budget results in a balance between oceanic influxes and effluxes of cadmium with the overall throughput of cadmium being between 5×10^6 and 6×10^6 kg/yr. It also indicates that some postdepositional remobilization of cadmium from near-shore and shelf sediments is required, especially for the high riverine k_d value, in order to account for the additional cadmium entering pelagic sediment phases over that supplied by dissolved cadmium and cadmium associated with particles that survive near-shore and shelf sedimentation.

2.4. POTENTIAL MAGNITUDE OF ANTHROPOGENICALLY MOBILIZED CADMIUM

Having established a hypothetical oceanic budget for cadmium under natural conditions, we will now attempt to assess the extent to which the influx of cadmium to the ocean might have been augmented by cadmium derived from human and industrial activities. Simpson (1981) estimated that the current anthropogenically derived cadmium influx to the ocean is 3.7×10^6 kg/yr. Nriagu (1981) concluded that anthropogenic sources give rise to emissions of 7.3×10^6 kg/yr of cadmium to the atmosphere of which approximately 2×10^6 kg/yr enters the ocean in atmospheric deposition. The much higher proportion of anthropogenically derived cadmium deposition on land, relative to the partitioning of natural cadmium, can be explained on the basis that cadmium released from industrial sources is removed from the atmosphere over comparatively smaller distance scales due to the association of cadmium with particles of greater size than those involved in purely natural scavenging processes in the atmosphere. If these independent estimates are about right, we would expect between 1.5×10^6 and 2×10^6 kg/yr to enter the ocean by way of river runoff. This corresponds to a global increase in the total cadmium content of rivers of between 46 and 62 ng/L, as compared with the natural (aggregate dissolved and particulate) concentration of between 90 and 120 ng/L (corresponding to the range of river discharge fluxes given in Fig. 1). While such increases are unlikely to be distributed uniformly throughout world rivers, it might be possible that 10% of world river runoff might be substantially augmented by cadmium from anthropogenic sources (i.e., runoff from industrialized areas of the world might directly affect 10% of the global runoff). In such circumstances, increased cadmium discharge of between 1.5×10^6 and 2×10^6 kg/yr would result in increased aggregate cadmium concentrations in the affected rivers of between 460 and 620 ng/L. Thus, we can ask whether there is evidence for a significant number of rivers draining industrialized areas having aggregate cadmium concentrations in dissolved and exchangeable particulate phases as high as this. An examination of the data shown in Table 1 suggests that such an aggregate concentration is achievable in several rivers such as the Rhine, Scheldt, and Meuse. The Hudson and St. Lawrence have somewhat lower aggregate values because, despite the substantial augmentation of suspended matter with cadmium over natural levels, their suspended loads are very small. Thus, it would seem that sufficiently elevated cadmium concentrations in rivers are not unusual, but far more data on the composition of other rivers in industrialized areas would be needed to determine how extensive these high concentrations are.

2.5. CONSEQUENCES OF ANTHROPOGENICALLY AUGMENTED INFLUX OF CADMIUM TO OCEAN

It seems likely that the cadmium influx to the ocean might have been augmented by about 60% over the natural influx and that further increases in the influx will continue to occur over the next two decades as a result of anthropogenic activities (Simpson, 1981). It now seems reasonable to examine what the consequences of these increased influxes will be and what effects might already be evident.

If all the additional cadmium from anthropogenic sources were to become incorporated in coastal and shelf sediments, it would correspond to an increase in average shelf sediment cadmium content of 0.24 mg/kg or, alternatively, an increase in the cadmium content of 10% of the shelf sediments to 2.5 mg/kg. There exists some evidence of augmentation of cadmium in regional shelf sediments by comparable amounts but no evidence of any significant augmentation of shelf sediments on a global scale.

Examination of the distribution of dissolved (P. A. Yeats, unpublished data) and total dissolvable (Bruland and Franks, 1983) cadmium in surface waters of the open North Atlantic Ocean shows that cadmium bears a relationship to salinity that extrapolates to relatively high concentrations at zero salinity (Fig. 2). This implies that there exists a significant signal associated with sources of oceanic cadmium, although it is not possible to discriminate between atmospheric and riverine sources in this respect. Such relationships are likely to be common for conditions pertaining to the Atlantic Ocean, which receives proportionately higher river runoff and lower atmospheric precipitation than other oceans. The zero-salinity intercept of these relationships is approximately 140 ng/L, considerably higher than the assumed natural cadmium concentration in rivers and atmospheric precipitation. While it can be hypothesized that the in-shore end member of this relationship is defined by cadmium regenerated from near-shore sediments, any such regeneration should be in balance with the rates of influx of cadmium in the exchangeable particulate phase. Thus, we can argue that this relationship is a consequence of increased influxes of cadmium to the North Atlantic relative to the natural rate of influx and represents transport of at least some of the anthropogenically derived cadmium to the pelagic ocean.

In the ocean as a whole, it is unlikely that any effects of increased influxes will be evident because of the dominance of internal transport fluxes of the element over the influxes. For example, global oceanic convection results in the upwelling of cadmium in deep water in the amount of about 5×10^8 kg/yr as compared with the total influx from the continents of about 1×10^7 kg/yr.

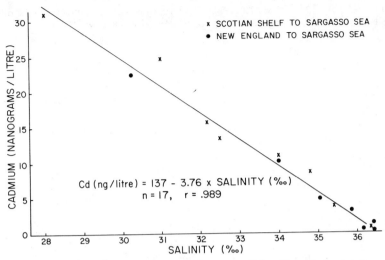

Figure 2. Plot of surface cadmium concentrations vs. salinity. New England data are from Bruland and Franks (1983).

This upwelling flux is balanced primarily by the downward transport of cadmium in rapidly regenerating biogenic material. The net result is a vertical distribution of cadmium characterized by surface depletion and subsurface maximum that closely follows the phosphate distribution. Only a few percent of the material settling out of the surface layer needs to survive intact to the ocean floor in order to account for the pelagic sedimentation of cadmium. Thus, almost all the internal transport of cadmium is in phases with which phosphate is associated, and the rate of internal cycling is so much greater than the rate of cadmium introduction to the ocean that no distributional changes, resulting from increased influxes of cadmium to the ocean, would yet be evident. If the rate of cadmium supply to the ocean continues to increase, as predicted by Simpson (1981), there will be an eventual change in the cadmium–phosphate ratio and a buildup of cadmium at the depths of net phosphate regeneration, but it will be several decades before such a change is detectable. Over much longer time scales, approaching the residence time of cadmium in the ocean (i.e., 15,000 yr based on the natural oceanic budget depicted in Fig. 1), some small cadmium augmentation of pelagic sediments might be expected to occur.

REFERENCES

Aston, S. R., Chester, R., Griffiths, A., and Riley, J. P. (1972). Distribution of cadmium in North Atlantic deep sea sediments. *Nature* (Lond.), **239**, 393.

Bewers, J. M., and Haysom, H. H. (1974). The terrigenous dust contribution to fluoride and iodide in atmospheric precipitation. *J. Rech. Atmospher.,* **8,** 689–697.

Bewers, J. M., and Yeats, P. A. (1977). Oceanic residence times of trace metals. *Nature (Lond.),* **268,** 595–598.

Boyle, E. A. (1981). Cadmium, zinc, copper and barium in foraminifera tests. *Earth Planet. Sci. Lett.,* **53,** 11–35.

Boyle, E. A., Sclater, F. R., and Edmond, J. M. (1976). On the marine geochemistry of cadmium. *Nature* (Lond.), **263,** 42–44.

Boyle, E. A., Huested, S. S., and Grant, B. (1982). The chemical mass balance of the Amazon plume II: Copper, nickel and cadmium. *Deep Sea Res.,* **29,** 1355–1364.

Bruland, K. W. (1980). Oceanographic distributions of cadmium, zinc, nickel and copper in the North Pacific. *Earth Planet. Sci. Lett.,* **47,** 176–198.

Bruland, K. W., and Franks, R. P. (1983). Mn, Ni, Cu, Zn and Cd in the western North Atlantic. In *Trace Metals in Seawater,* Wong, C. S., Boyle, E., Bruland, K. W., Burton, J. D., and Goldberg, E. D. (eds.). Plenum, New York, pp. 395–414.

Bruland, K. W., Bertine, K., Koide, M., and Goldberg, E. D. (1974). History of metal pollution in southern California coastal zone. *Environ. Sci. Technol.,* **8,** 425–432.

Chorley, R. J. (ed.). 1969. *Water, Earth, and Man.* Methuen, London.

Cronan, D. S. (1976). Basal metalliferous sediments from the eastern Pacific. *Geol. Soc. Am. Bull.,* **87,** 928–934.

Danielsson, L. G., Magnusson, B., Westerlund, S., and Zhang, K. (1983). Trace metals in the Gota River estuary. *Est. Coast. Shelf Sci.,* **17,** 73–85.

Duinker, J. C., and Nolting, R. F. (1978). Mixing, removal and mobilization of trace metals in the Rhine estuary. *Neth. J. Sea Res.,* **12,** 205–223.

Duinker, J. C., Hillebrand, M. T. J., Nolting, R. F., Wellershaus, S., and Jacobsen, N. R. (1980). The River Varde Å: Processes affecting the behaviour of metals and organochlorines during estuarine mixing. *Neth. J. Sea Res.,* **14,** 237–267.

Duinker, J. C., Hillebrand, M. T. J., Nolting, R. F., and Wellershaus, S. (1982a). The River Elbe: Processes affecting the behaviour of metals and organochlorines during estuarine mixing. *Neth. J. Sea Res.,* **15,** 141–169.

Duinker, J. C., Hillebrand, M. T. J., Nolting, R. F., and Wellerhaus, S. (1982b). The River Weser: Processses affecting the behaviour of metals and organochlorines during estuarine mixing. *Neth. J. Sea Res.* **15,** 170–195.

Edmond, J. M. (1980). Pathways of nutrients and organic matter from land to ocean through rivers. In *Proceedings of a SCOR/ACMR/ECOR/IAHS/UNESCO/CMG/IABO/IAPSO Workshop on River Input to Ocean Systems, Rome 26–30 March, 1979,* Martin, J.-M., Burton, J. D., and Eisma, D. (eds.). UNESCO Paris, pp. 31–32.

Grant, B,., Hu, M. H., Boyle, E., and Edmond, J. (1982). Comparison of the trace metal chemistry in the Amazon, Orinoco and Yangtse plumes. *EOS,* **63,** 48 (Abstract).

Heinrichs, H., Schulz-Dobrick, B., and Wedepohl, K. H. (1980. Terrestrial geochemistry of Cd, Bi, Tl, Pb, Zn and Rb. *Geochim. Cosmochim. Acta,* **44,** 1519–1533.

Holeman, J. N. (1968). The sediment yield of major rivers of the world. *Wat. Resourc. Res.,* **4,** 737–747.

IJC (1977). *The Waters of Lake Huron and Lake Superior,* Vol. II, Upper Lakes Reference Group, International Joint Commission.

Kemp, A. L. W., Thomas, R. L., Dell, C. I., and Jaquet, J. M. (1976). Cultural impact on the geochemistry of sediments in Lake Erie. *J. Fish. Res. Bd. Can.,* **33,** 440–462.

Klinkhammer, G. P., and Bender, M. L. (1981). Trace metal distributions in the Hudson River estuary. *Est. Coast. Shelf Sci.,* **12,** 629–643.

References

Krishnaswamy, S., and Lal, D. (1972). Manganese nodules and budget of trace solubles in the oceans. In *The Changing Chemistry of the Oceans,* Dyrssen, D., and Jagner, D. (eds.). Wiley, New York, pp. 307–320.

Ku, T. L., Broecker, W. S., and Opdyke, N. (1968). Comparison of sedimentation rates measured by paleomagnetic and ionium methods of age determination. *Earth Planet. Sci. Lett.,* **4,** 1–16.

Li, Y-H., Burkhardt, L., and Teraoka, H. (1984). Desorption and coagulation of trace elements during estuarine mixing. *Geochim. Cosmochim. Acta,* **48,** 1879–1884.

Livingstone, D. A. (1963). Chemical composition of rivers and lakes. In *Data of Geochemistry,* 6th ed., Fleischer, M. (ed.), Geological Survey Professional paper 440-G, U.S. Government Printing Office, Washington, DC, vii + 64 pp.

Lum, K. R., and Leslie, J. K. (1983). Dissolved and particulate metal chemistry of the central and eastern basin of Lake Erie. *Sci. Tot. Environ.,* **30,** 99–109.

Mart. L. (1983). Seasonal variations of Cd, Pb, Cu and Ni levels in snow from the eastern Arctic Ocean. *Tellus.* **35B,** 131–141.

Mart, L., Nuernberg, H. W., and Rutzel, H. (1984). Comparative studies on cadmium levels in the North Sea, Norwegian Sea, Barents Sea and the eastern Arctic Ocean. *Fres. Z. Anal. Chem.,* **317,** 201–209.

Mart, L., Nuernberg, H. W., and Rutzel, H. (1985). Levels of heavy metals in the tidal Elbe and its estuary and the heavy metal input into the sea. *Sci. Tot. Environ.,* **44,** 35–49.

Milliman, J. D. (1974). *Marine Carbonates.* Springer-Verlag, New York.

Muhlbaier, J., Stevens, C., Graczyk, D., Tissue, T. (1982). Determination of cadmium in Lake Michigan by mass spectrometric isotope dilution analysis or atomic absorption spectrometry following electrodeposition. *Anal. Chem.,* **54,** 496–499.

Mullin, J. B., and Riley, J. P. (1956). The occurrence of cadmium in seawater and in marine organisms and sediments. *J. Mar. Res.,* **15,** 103–122.

Nriagu, J. O. (1981). Human influence on the global cadmium cycle. In *Cadmium in the Environment,* Part I, *Ecological Cycling,* Nriagu, J. O. (ed.). Wiley, New York, pp. 2–12.

Nriagu, J. O., Wong, H. K. T., and Coker, R. D. (1981). Particulate and dissolved trace metals in Lake Ontario. *Wat. Res.,* **15,** 91–96.

Nuernberg, H. W., Valenta, P., Mart, L., Raspor, B., and Sipos, L. (1976). Applications of polarography and voltammetry to marine and aquatic chemistry. *Zeits. Anal. Chem.,* **282,** 357–367.

Poldoski, J. E., and Glass, G. E. (1978). Anodic stripping voltammetry at a mercury film electrode: Baseline concentrations of cadmium, lead and copper in selected natural waters. *Anal. Chim. Acta,* **101,** 79–88.

Rossman, R. (1984). Trace metal concentrations in the offshore waters of Lakes Erie and Michigan. Special Report No. 8, Great Lakes Research Division, Great Lakes and Marine Waters Center, University of Michigan, Ann Arbor, MI.

Salomons, W., and Foerstner, U. (1984). *Metals in the Hydrocycle.* Springer-Verlag, Berlin.

Schaule, B. K., and Patterson, C. C. (1983). Perturbation of the natural lead depth profile in the Sargasso Sea by industrial lead. In *Trace Metals in Seawater,* Wong, C. S., Boyle, E., Bruland, K. W., Burton, J. D., and Goldberg, E. D. (eds.). Plenum, New York, pp. 487–503.

Sigg, L. (1985). Metal transfer mechanisms in lakes; the role of settling particles. In *Chemical Processes in Lakes,* Stumm, W. (ed.). Wiley, New York, pp. 283–310.

Simpson, W. R. (1981). A critical review of cadmium in the marine environment. *Prog. Oceanogr.,* **10,** 1–70.

Taylor, D. (1984). Cadmium: A case of mistaken identity? *Mar. Pollut. Bull.,* **15,** 168–170.

Trefry, J. H., and Presley, B. J. (1976). Heavy metal transport from the Mississippi River to the Gulf of Mexico. In *Marine Pollutant Transfer,* Windom, H. L., and Duce, R. A. (eds.). Lexington Books, Lexington, MA, pp. 39–76.

Trefry, J. H., Metz, S., and Trocine, R. P. (1983). Decreased inputs of lead and cadmium to the Gulf of Mexico from the Mississippi River. *EOS,* **64,** 244 (Abstract).

Ure, A. M., and Berrow, M. L. (1982).The elemental constituents of soils. In *Environmental Chemistry,* Vol. 2, Bowen, H. J. M. (ed.), Royal Society of Chemistry, London, pp. 94–204.

Wollast, R. (1974). The silica problem. In *The Sea,* Vol. 5, Goldberg, E. D. (ed.). Wiley, New York, pp. 359–392.

Yeats, P. A., and Bewers, J. M. (1982). Discharge of metals from the St. Lawrence River. *Can. J. Earth Sci.,* **19,** 982–992.

Yeats, P. A., and Bewers, J. M. (1983). Potential anthropogenic influences on trace metal distributions in the North Atlantic. *Can. J. Fish. Aquat. Sci.,* **40** (Supplement 2), 124–131.

3

CADMIUM IN FRESH WATERS: THE GREAT LAKES AND ST. LAWRENCE RIVER

K. R. Lum

Environmental Contaminants Division
National Water Research Institute
Burlington, Ontario, Canada

3.1. Introduction
3.2. Cadmium Concentrations in the Great Lakes
3.3. Cadmium in Bottom Sediments in the Great Lakes
3.4. Cadmium in Suspended Particulate Material
3.5. Partitioning of Cadmium
3.6. Mass Balance Budget for Cadmium
References

3.1. INTRODUCTION

In spite of continuing international interest in the distribution, cycling, and environmental impact of cadmium and its compounds, the data base pertaining to the Great Lakes has not been summarized and evaluated. Such an overview would be valuable because the Great Lakes are the largest freshwater reservoir in the world, containing about 20% of the world's freshwater resources. The Great Lakes Basin itself extends over about 785,450 km^2 (Nriagu, 1986), and it is the home of some 40 million North Americans whose urban and industrial activity have resulted in environmental problems, particularly in Lakes Michigan, Erie, and Ontario. As population growth and further development of the urban industrial complex within the basin are inevitable, continued conflict between the use of the Great Lakes as a waste sink and as a source of drinking water would be expected to occur.

Clearly, therefore, it is important to evaluate the environmental behavior of the toxic metal cadmium in the largest freshwater ecosystem in the world. In this chapter, the recent data on cadmium in water, suspended and bottom sediments, are summarized from the literature. The available information is supplemented with data obtained in the author's field studies and an attempt made to provide an overview of the transport and fate of water-borne cadmium in the Great Lakes and its outflow to the Gulf of St. Lawrence. The atmosphere as a source and a major perturbation on the fluxes of cadmium in the Great Lakes Basin has been recently reviewed (Tisue and Fingleton, 1984; Schmidt and Andren, 1984), and hence this aspect will not be addressed here.

3.2. CADMIUM CONCENTRATIONS IN THE GREAT LAKES

The data base for the concentrations of cadmium forms in the Great Lakes is quite limited in contrast to the greater abundance of reports for the marine environment. Table 1 is a summary of cadmium determinations on Great Lakes waters that have been reported in the past eight years. The first good measurements were made by Poldoski and Glass (1978). Using anodic stripping voltammetry, they obtained an average value of 19 ng/L (s.d. = 8 ng/L) for samples from western Lake Superior at Duluth and streams in wilderness areas of northwestern Minnesota. The total cadmium concentrations ranged from 6 to 34 ng/L. Offshore waters of Lake Superior have recently been analyzed (Rossmann, 1986), and the mean value for surface samples from 20 stations was 27 ng/L (s.d. = 10). The range of total cadmium concentrations in this study was 7–44 ng/L. Rossmann also determined dissolved cadmium and obtained an average concentration of 6.6 (s.d. = 3.0) ng/L with a range from 1–12 ng/L. The latter results are comparable to those recently reported for open-ocean waters (e.g., Mart et al., 1984; Statham et al., 1985; Kremling, 1985) and are consistent with the pristine nature of Lake Superior's waters.

In a study of cadmium in the southern basin of Lake Michigan, total cadmium was determined by isotope dilution mass spectrometry and graphite furnace atomic absorption spectrometry (Muhlbaier and Tisue, 1981). The mean value for 17 samples was 26.6 (s.d. = 9.3) ng/L, with a range of 12–45.6 ng/L. Subsequent work by this group (Muhlbaier et al., 1982) gave a range of total cadmium concentrations of 15–41 ng/L. The eight samples analyzed in the latter study averaged 28 (s.d. = 8) ng/L. Somewhat higher concentrations were obtained for 11 offshore samples by Rossmann

Table 1. Average Cadmium Concentrations in the Great Lakes and St. Lawrence River[a]

Location	Cadmium Concentration (ng/L)			Reference
	Dissolved	Particulate	Total	
Lake Superior	6.6 (3.0)	21 (10)	27 (10)	Rossman, 1986
	—	—	19 (8)	Poldoski and Glass, 1978
Lake Michigan	—	—	27 (9)	Muhlbaier and Tisue, 1981
	—	—	28 (8)	Muhlbaier et al., 1982
	56 (44)	—	—	Rossmann, 1984
Lake Huron	—	—	16 (13)	Rossmann, 1982
Lake Erie	71 (27)	—	98 (81)	Rossmann, 1984
	40 (20)	6 (5)	—	Lum and Leslie, 1983
	29 (15)	—	—	Lum and Callaghan, 1986
Lake Ontario	—	21 (19)	50 (10)	Nriagu et al., 1981
	10 (5)			Lum and Callaghan, 1986
St. Lawrence River,	110 (290)			Yeats and Bewers, 1982
St. Lawrence River May 1985	14 (8)			Lum et al., 1986
St. Lawrence River October 1985	16 (12)			Lum et al., 1986
St. Lawrence River At Quebec	15 (7)			Lum et al., 1986

[a] Numbers in parentheses are standard deviations.

(1984), with an average value of 42 (s.d. = 18) ng/L and a range of 19–87 ng/L.

Total cadmium in Lake Huron and Georgian Bay have been reported to average 16 (s.d. = 13) ng/L with a range of 1.7–61 ng/L (Rossmann, 1982). There are no other results with which to compare these data as Rossmann's measurements are the only reported ones to date.

Total cadmium concentrations in the offshore waters of Lake Erie have been reported to average 98 (s.d. = 81) ng/L, with a range of 39–320 ng/L (Rossmann, 1984). Dissolved cadmium was also determined, and Rossmann obtained an average value for 11 measurements of 71 (s.d. = 27) ng/L, with a range of 41–120 ng/L. The latter results are appreciably higher than the average dissolved values (40; s.d. = 20 and a range of 10–90 ng/L) found for 21 samples from offshore waters of the central and eastern basins of Lake Erie that were collected in September 1978 (Lum and Leslie, 1983). More recent measurements in the author's laboratory have produced an average concentration of dissolved cadmium in Lake Erie that is remarkably similar to the average concentration for southern Lake Michigan obtained by Muhlbaier and colleagues (1981, 1982). In May 1984, 60 surface water samples taken from the near-shore zone of the three basins, and also from a transect across the mouth of the Detroit River and from the river itself, gave an average of 32 (s.d. = 30) ng/L, with a range of 3–170 ng/L. If one excludes the four samples taken above, at, and below the confluence of the Detroit and the Rouge Rivers (concentrations of 34, 170, 162, and 60 ng/L), the standard deviation of the measurements drops to 15 ng/L, but the average value is not significantly changed, that is, 29 ng/L. These recent data suggest that the results for samples collected in 1978 are on average about 25% higher. There are two possible explanations for this. Contamination of the earlier samples is possible because the determinations were done by a preconcentration procedure using Chelex-100 chelating ion exchange resin. The detection limit of the analytical procedure was 10 ng/L, in contrast to the less than 2 ng/L of the present procedure, which is based on direct injection graphite furnace AAS analysis of acidified samples (Lum and Callaghan, 1986). Thus it is possible that the difference in average concentrations may simply reflect the "analytical noise" associated with the less sensitive procedure used for the 1978 samples. Note that the concentration difference between the two average values is at the detection limit of the older procedure. The alternative scenario assumes that the data are reliable, and hence the difference could be caused by a decrease in the loadings of cadmium to Lake Erie, principally from the Detroit River watershed, the largest single source of contaminants in general to western Lake Erie. The decrease in iron loadings in the period 1968–1980 has been documented (IJC, 1981), and this would also apply to other metals associated with the steel and automotive complexes in the Detroit, Michigan, area and related metallurgical operations. Furthermore, the recession of the early 1980s would have resulted in major decreases in industrial production with consequent reductions in both effluents and atmospheric emissions. There is insufficient evidence on which to unequivocally reject either scenario as whole lake measurements need to be carried out for several years before the trends, or patterns, discussed above can be clearly delineated.

In Lake Ontario, acid-labile (raw water acidified to pH 2) cadmium averaged 50 (s.d. = 10) ng/L for 16 samples collected at various depths (Nriagu et al., 1981). The range of concentrations was 45–53 ng/L. The acid used to preserve the samples would be expected to leach cadmium from particulate surfaces, and hence these results can be regarded as an estimate of near-total cadmium. Recent measurements (Lum and Callaghan, 1986) indicate that the concentration of dissolved cadmium is substantially lower. The range of concentrations determined for 58 surface water samples collected during the spring runoff in the first week of April 1984 was 3–34 ng/L. The average value was 10 (s.d. = 5) ng/L. The relatively uniform concentrations throughout Lake Ontario reflect the well-mixed isothermal conditions prevailing at the time of sampling.

The St. Lawrence River serves as the channel by which dissolved and particulate-bound constituents from the lower Great Lakes and their connecting channels are discharged to the Gulf of St. Lawrence and the marine environment (Allan, 1986). Previous measurements of cadmium in this system (Yeats and Bewers, 1982) indicate an average dissolved concentration of 110 ng/L for 17 samples collected at different times of the year at Quebec

City. Recent work by the author using the procedure described in Lum and Callaghan (1986) yields an average dissolved cadmium concentration of 25 (s.d. = 41) ng/L for surface water (1 m) taken at 26 stations distributed from the outflow of Lake Ontario to Quebec City in May 1985. If one deletes two stations at which the highest concentrations were measured, 200 and 111 ng/L (both at Montreal), the average value and standard deviation are 14 and 8 ng/L, respectively. Two stations within a kilometer of each other at Ile d'Orleans (Quebec City) showed values of 9 and 11 ng/L. In October 1985, 32 samples were collected throughout the river, and the average dissolved cadmium concentration was 20 (s.d. = 19) ng/L. Again, if one deletes two stations at Montreal showing the highest concentrations, the average value and standard deviation are 16 and 12 ng/L. During the October cruise, 10 samples were collected at 1 m and 2 m off bottom at two-hourly intervals at an anchor station off Levis opposite Quebec City. The average concentration here was 15 (s.d. = 7) ng/L. The recent data on the St. Lawrence River and the Great Lakes in general indicate that the concentration of dissolved cadmium reported by Yeats and Bewers (1982) is very likely too high. The reason for the discrepancy can partly be explained by the relatively high detection limit (30 ng/L) of the method used by them (Bewers et al., 1976). On the basis of the above recent data, the annual flux of dissolved cadmium based on a mean dissolved cadmium concentration of 15 ng/L and a mean discharge of 12,000 m^3/s at Quebec City is 5675 kg. This is an order of magnitude less than that reported by Yeats and Bewers (1982). The difference can be attributed mainly to the average dissolved concentration used by these workers, that is, 110 ng/L.

3.3. CADMIUM IN BOTTOM SEDIMENTS IN THE GREAT LAKES

The distribution of cadmium in the surficial sediments of the Great Lakes reflects degrees of contamination related to proximity to major industrial/urban centers and the role of the depositional basins as a sink for sedimenting particles (Fig. 1). For example, the average concentration of cadmium is three times greater in southern Lake Michigan compared with the average for the lake as a whole (Table 2). Predictably, local variations are important. Cline and Chambers (1977) found that the cadmium concentration in 38 sediment samples from northeastern Lake Michigan near Sleeping Bear Point averaged 0.47 µg/g (s.d. = 0.49). Whereas in Green Bay (Lake Michigan), the average of five samples was 1.3 µg/g (s.d. = 0.5) (Christensen and Chien, 1981). For Lake Erie, the decreasing concentrations (west to east) progressing from the shallow, eutrophic western basin to the deeper eastern basin with relatively little industry in its watershed, has been demonstrated by Nriagu et al. (1979). In addition, the large amounts of eroding bluff material from the north shore of the lake would be expected to be an effective diluent reducing the total concentration of cadmium measured in the

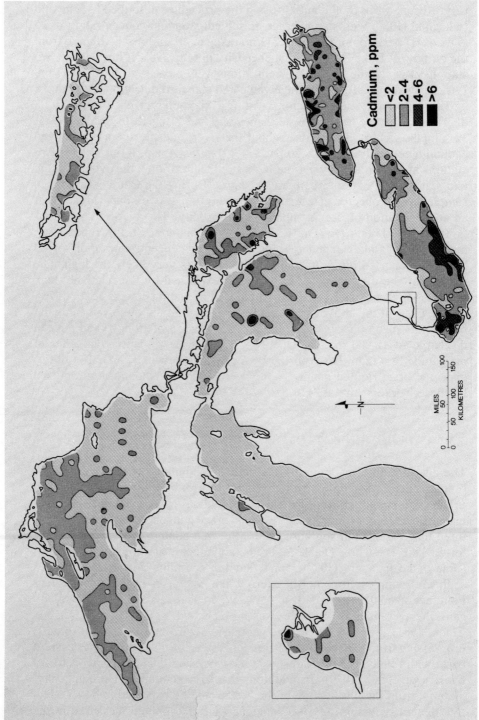

Figure 1. Distribution of cadmium in the sediments of the Great Lakes (from Thomas and Mudroch, 1979).

Table 2. Cadmium Concentration in Surficial Sediments of the Great Lakes[a]

Location	Cadmium Concentration (μg/g)	Reference
Lake Superior ($n = 5$)	2.2 (0.5)	Kemp et al., 1978
Lake Huron ($n = 6$)	2.0 (0.5)	Kemp et al., 1978
Georgian Bay ($n = 3$)	3.3 (1.1)	Kemp et al., 1978
Southern Michigan ($n = 40$)	3.0	Muhlbaier and Tisue, 1981
Lake Michigan ($n = 286$)	0.9 (0.4)	Cahill and Shimp, 1984
Green Bay, MI ($n = 5$)	1.3 (0.5)	Christensen and Chien, 1981
Sleeping Bear Point, MI ($n = 38$)	0.47 (0.49)	Cline and Chambers, 1977
Lake Erie	3.6	Kemp and Thomas, 1976
Lake Erie ($n = 7$)	3.7 (1.6)	Nriagu et al., 1979
Lake Ontario	5.1	Kemp and Thomas, 1976

[a] Numbers in parentheses are standard deviations.

depositional zone of the eastern basin. The average concentration of cadmium in the surficial sediments of Lake Ontario is 5.6 μg/g (Kemp and Thomas, 1976). This value represents the three main basins and does not include the value reported by these authors for a station close to Hamilton Harbor or for a station at the outflow of Lake Ontario. The latter had an anomalously high concentration of 49 μg/g, which is inconsistent with the fact that the area is a transitional zone for deposition and the surrounding watershed does not have any major industrial or municipal sources of cadmium that would contribute the high levels of cadmium necessary to contaminate the bottom sediment to such an elevated level.

Comparison of the average levels of cadmium in the surficial sediments of the depositional basins of the five Great Lakes shows that Lake Ontario is the most contaminated (Table 2). The average level is about twice that for southern Lake Michigan and Lake Erie. In fact, the average value is comparable to that reported by Nriagu et al. (1979) for western Lake Erie. The latter measurements are not surprising because of the extensive industrial activity in the watershed, particularly in the Detroit River area. Inputs of particle-bound cadmium from the Niagara River (outflow of Lake Erie plus industrial discharges on the river) cannot account for the enrichment of the sediments (see next section). The high cadmium levels in the surficial sediments appears to be consistent with an efficient scavenging process operating in the lake combined with effective trapping of sedimented particles in the depositional zones. The eastern Rochester basin has a maximum depth of 244 m, and the west-to-east transport of particles may be viewed as being focused into this deep hole. The fact that the average concentration of dissolved cadmium is one-third that of Lake Erie and the general increase in cadmium content in the sediments from west to east (away from sources such as Hamilton, Toronto, and the Niagara River) provides support for these processes operating together. Kemp and Thomas (1976) reported concentrations of 3.7, 5.1, 5.6, and 6.2 µg/g along this west-to-east transect.

3.4. CADMIUM IN SUSPENDED PARTICULATE MATERIAL

The available data on the cadmium content of suspended particulate material in the Great Lakes and the St. Lawrence River are summarized in Table 3. Information on the Mississippi and the Mackenzie rivers are included for comparative purposes.

The Detroit River particulates are from 19 stations distributed throughout the river and at the river–lake interface in the western basin of Lake Erie. Results from five stations close to industrial/municipal outfalls, namely, the mouth of the Rouge River at which the Detroit STP discharges adjacent to the Great Lakes Steel complex and the lower Trenton Channel, are excluded from the average concentration given in Table 3. The disparity between the average concentration of cadmium in 29 samples from the central and eastern basins of Lake Erie (Lum and Leslie, 1983) and the value for 25 samples collected in 1984 is almost entirely caused by the small samples (200 mg) used in the former study, which resulted in a higher detection limit. Half a gram of sample was used in the 1984 set and the final volume was 25 mL instead of 50 mL. The average value for the 1984 samples also includes

Table 3. Concentration of Cadmium in Suspended Particulate Material from the Great Lakes and St. Lawrence River[a]

Location	Cadmium Concentration (µg/g)	Reference
Southern Michigan	0.5 –2.3	Parker et al., 1982
Detroit River	2.6 (1.2)	Lum, unpublished data
Lake Erie	4 (2)	Lum and Leslie, 1983
Lake Erie	2.4 (0.9)	Lum, unpublished data
Lake Ontario	9.7 (2.2)	Nriagu et al., 1981
Western Lake Ontario	3.2 (1.7)	Lum, unpublished data
Lake Ontario outflow	2.3 (1.0)	Lum, unpublished data
Niagara River at Fort Erie	1.7 (0.70)	Lum, unpublished data
Niagara River at Lake Ontario	1.8 (1.0)	Lum, unpublished data
St. Lawrence River[b]	2.8 (0.3)	Yeats and Bewers, 1982
St. Lawrence River[b]	1.8 (0.5)	Cossa and Poulet, 1978
St. Lawrence River[b]	1.09 (0.37)	Lum et al., 1986
St. Lawrence River	1.88 (0.76)	Lum et al., 1986
Mississippi River	0.60 (0.12)	Trefry et al., 1983
Mackenzie River	0.81	Martin and Meybeck, 1979

[a] Numbers in parentheses are standard deviations.
[b] Samples taken at Quebec City.

stations in the western basin and thus can be regarded as representative of the whole lake. Comparison of the lakewide average, 2.4 μg/g with the average concentration for 14 samples taken at the outflow of Lake Erie on the Niagara River at Fort Erie, 1.7 μg/g indicates that there is some retention and/or removal of particulate cadmium, most likely in the deep hole of the eastern basin. Discharges in the Niagara River do not appear to contribute significantly to the concentrations of particulate-bound cadmium as the average for 56 samples obtained at the mouth of the Niagara River at Niagara-on-the-Lake is not significantly different from that at the Lake Erie end. Significant enrichment of particulate cadmium in the western basin of Lake Ontario is evident from the average concentration (43 samples) for this basin. The most probable source of the particulate cadmium is Hamilton Harbor, the location of the largest steel companies in Canada. The data offered by Nriagu et al. (1981) seem high in light of the results summarized in Table 3. These workers do not provide details on the analysis, and hence it is not possible to infer a likely cause for their high values. The average concentration for 16 samples taken from the outflow of Lake Ontario at Wolfe Island suggests that removal and/or retention processes are responsible for the decrease in particulate-bound cadmium. As previously discussed, there is a discernible west-to-east increase in the surficial bottom sediment cadmium concentrations, the highest concentration occurring in the deep eastern basin, which is evidently performing a useful role as the last sedimentary sink before Lake Ontario discharges into the St. Lawrence River.

The concentration of cadmium in suspended particulate material from the St. Lawrence River is relatively uniform and remarkably similar to the average concentrations in the Niagara River. The value shown in Table 3 is the mean of 52 samples collected from the outflow of Lake Ontario to Quebec City during May and October 1985. This is in excellent agreement with the average of 11 measurements for samples from three stations in the vicinity of Quebec City (Cossa and Poulet, 1978). In contrast, the geometric mean for 33 samples collected at Quebec City obtained by Yeats and Bewers (1982) appears to be high. The arithmetic mean for 9 samples collected at two hourly intervals at an anchor station off Quebec City in October 1985 was 1.09 μg/g. This is at variance with the results of Cossa and Poulet, and the discrepancy is likely caused by differences in analytical methods. These workers digested particulates collected on a filter (possibly 5–10 mg) and analyzed the concentrated nitric acid extracts by graphite furnace AAS. Unless the background absorbances are simultaneously measured during atomization, there is no guarantee that accurate background compensation of the signals produced by the high acid content and by major elements such as iron and calcium in the samples would occur. Thus, there is a strong likelihood of background undercompensation. In the author's laboratory, the atomic absorption measurements are carried out with a Zeeman effect AAS system, and the background absorbance signals are also recorded. Thus, the recent data indicate that at Quebec City particulate-bound cadmium concentrations have decreased by approximately 50%. This is not surprising because most of the inputs are along the international section and in the Montreal area, and hence dilution of the suspended load might be expected during transport to Quebec City.

The discharge of particulate-bound cadmium from the St. Lawrence River can be estimated using the average cadmium content of particulates at Quebec City and the annual suspended solids loadings. The latter are on the order of 5.2×10^9 kg/yr at Quebec City (Tremblay, 1985). Thus, an estimated 5670 kg of cadmium are discharged annually by the St. Lawrence River. Alternatively, the particulate cadmium flux may be calculated from the average concentration of suspended particulate material at the time of sampling, namely, 9.67 mg/L (s.d. = 1.16) and their average cadmium content (in nanograms per liter). The particulate cadmium discharged in October 1985 is therefore 3980 kg. These estimates are about one-third to a half less than the figure of 9000 kg/yr calculated by Yeats and Bewers (1982). The discrepancy is likely caused by the uncertainty discussed by Yeats and Bewers resulting from the determination of particulate cadmium as the difference between total and dissolved forms. As for the particulate flux calculated by the author, it is worth noting that more analytical data on the cadmium content of particulates as a function of season are needed in order to derive an acceptable annual discharge by the St. Lawrence River.

3.5. PARTITIONING OF CADMIUM

The speciation of cadmium is generally regarded as being dominated by dissolved forms except for situations where the concentration of suspended particulate material is high (e.g., Yeats and Bewers, 1982; Trefry et al., 1983; Lum and Leslie, 1983; Li et al., 1984; Balls, 1985a; Mart et al., 1985). Dissolved forms in the St. Lawrence River account for about 60% of the total cadmium transported by the river, which is characterized by a particulate matter concentration that averages 10 mg/L (Table 4). In contrast, the 200–300 mg/L suspended load of the Mississippi River presents much greater amounts of sorption sites, and this is evident in the low proportion (10%) of dissolved forms in this system. For Lake Erie, with an average suspended load of 1.7 mg/L during calm isothermal conditions, dissolved forms predominate at an estimated 87% of the total cadmium in the central and eastern basins (Lum and Leslie, 1983). Recent measurements on the tidal Elbe show that about 63% of the cadmium input to the estuary is as dissolved forms (Mart et al., 1985). In coastal marine waters, dissolved forms also dominate the partitioning of cadmium (Balls, 1985a,b). In the context of the above findings, it is surprising to note the dominance of particulate-bound cadmium in Lakes Huron and Superior, which contain levels of suspended material comparable to the open ocean (Rossmann, 1982, 1986). Examination of the latter study clearly indicates that the dissolved cadmium results are within the accepted range (Table 1). However, the discrepancy in the distribution coefficient is very likely an artifact resulting from the determination of cadmium in several hundred micrograms of particulate material collected on a membrane filter.

The distribution coefficient K_d summarized in Table 4 has been expressed in liters per nanogram as described by Tisue and Fingleton (1984). Li et al. (1984) use units of milliliters per gram, and their results have been converted by multiplying by 10^{-12}. Where the K_d is not given, it has been calculated using either of the following two relationships:

$$K_d = \frac{\text{particulate cadmium (ng/g)}}{\text{dissolved cadmium (ng/L)}} \times \frac{1 \text{g}}{10^9 \text{ ng}}$$

or

$$K_d = \frac{\text{particulate cadmium (ng/L)}}{1 \times 10^{-3} \text{ g/L}} \times \frac{1}{\text{dissolved cadmium (ng/L)}} \times \frac{1}{10^9 \text{ ng}}.$$

Note that the latter relationship will yield a K_d for a 1-mg/L suspended particulate matter concentration.

In general, the distribution coefficient ranges from 0.50×10^{-7} to 2.0×10^{-7}. The relative constancy of the value for a wide range of natural waters

Table 4. Partitioning of Cadmium in the Great Lakes and the St. Lawrence River[a]

Location	Percentage Dissolved	K_d ($\times 10^7$ L/ng)	Reference
St. Lawrence River, May 1985	60 (22)	2.01 (1.74)	Lum et al., 1986
St. Lawrence River, October 1985	69 (17)	1.86 (2.11)	Lum et al., 1986
St. Lawrence River, Quebec	56 (15)	0.92 (0.53)	Lum et al., 1986
Lake Erie	87	1.0	Lum and Leslie, 1983
Lake Michigan	—	1.0	Tisue and Fingleton, 1984
Lake Ontario	—	2.0	Nriagu et al., 1981
Lake Huron	0	—	Rossman, 1982
Lake Superior	22	32	Rossmann, 1986
Greifensee	—	0.65	Imboden et al., 1980
Mississippi	10	0.41	Trefry et al., 1983
Hudson River	—	~0.15	Li et al., 1984
Mississippi	—	~0.45	Li et al., 1984
Northern North Sea	94	0.63	Balls, 1985a

[a] Numbers in parentheses are standard deviations.

of varying trophic levels and suspended particulate matter concentrations is quite remarkable. The slightly lower K_d in the Mississippi River is simply a reflection of its higher suspended load and greater scavenging of cadmium by particulates.

It is tempting to speculate on the use of an average K_d for a given aquatic system for predicting the particulate elemental concentration, for example, from measurements of the concentrations of suspended particulate matter and dissolved elemental forms. The former is a straightforward measurement that is most often simply the operational phase separation step for the determination of dissolved elemental forms. Such predictive capability would be especially valuable in fluvial and estuarine environments where suspended particulate matter concentrations vary over a wide range and also in the lower Great Lakes, which have seasonal maxima in phytoplankton production superimposed on storm-induced shoreline erosion yielding large suspended loadings to open waters. Using the equation for calculating K_d that incorporates the suspended matter concentration, predicted concentrations of particulate cadmium were obtained for the 26 stations sampled during May 1985 from the St. Lawrence River. The range of dissolved cadmium concentrations was 3–200 ng/L, and the particulate matter concentrations ranged from 1.42 to 13.5 mg/L. The particulate cadmium concentrations obtained from the sum of the cadmium contents in sequential 1 N HCl and residual extracts of 0.500 g of sample and determined by Zeeman flame AAS (Lum and Edgar, 1983) ranged from 2.2 to 23.1 ng/L. Of the 26 stations, there was agreement (for this exercise, assumed to be within 30% of each other) for only 10 pairs of measured and predicted values. This lack of agreement is perhaps not surprising for a number of reasons. First, the distribution coefficient assumes an equilibrium distribution between the dissolved and particulate phases. In a nonpristine river having many tributaries and industrial and municipal discharges such as the St. Lawrence, equilibrium may be the exception rather than the rule, in comparison with, for example, laboratory experiments such as those carried out by Li et al. (1984). The variation in the K_d values calculated at each station supports this argument as the range was 0.23×10^{-7} to 7.2×10^{-7}. Second, the error in the determination of the suspended load concentration is significant. For the St. Lawrence River at Quebec City, the precision of this determination was ± 4 mg/L for duplicate measurements on eight samples (Tremblay, 1985). Third, random errors in the determination of the dissolved cadmium concentrations, which are inevitable in any large field investigation, will be another source of variation, although the data set (Table 1) for samples collected in May and October suggests that this is not a major factor.

Although the use of analytical results and an average K_d for specific aquatic systems in a predictive fashion cannot be ruled out on the basis of the above discussion, further investigation is clearly required to test the limits of this potential tool.

3.6. MASS BALANCE BUDGET FOR CADMIUM

The anthropogenic inputs of cadmium to the Great Lakes has recently been summarized (Table 5). A useful comparison can therefore be made using the data presented in this review to determine whether there are any inconsistencies in the mass balance budgets that may have resulted from deficiencies in the cadmium data base.

Taking Lake Ontario first, it was estimated above that about 6 t each of particulate and dissolved cadmium were exported via the St. Lawrence River. Using the sediment accumulation rate given by Nriagu (1986), 4.4×10^6 t/yr and the average concentration of cadmium in particulate material, 3.2 µg/g (Table 3), about 14 t of cadmium are retained in the bottom sediments on an annual basis. The quantity of cadmium in the waters of Lake Ontario may be estimated from the average concentration in the dissolved phase and the volume of the lake, 1.223×10^{16} L. The concentration of suspended particulate matter in Lake Ontario is generally <2 mg/L, and in the open lake it is often <1 mg/L. Thus, dissolved forms of cadmium would be expected to dominate. This calculation indicates that there are 16 t of cadmium dissolved in the lake. The total of 47 t agrees well with the total input of 47 t reported by Nriagu (1986).

In the case of Lake Erie, the export of particulate-bound cadmium can be calculated from the average concentration at the outflow at Fort Erie (Table 3), the average flow of the Niagara River, 6400 m^3/s, and the average suspended particulate matter concentration, 8.4 mg/L. This works out to be about 3 t/y. Similarly, taking the average concentration of dissolved cadmium in Lake Erie, the export of dissolved forms is about 6 t. With a lake volume of 4.835×10^{14} L, the estimated quantity of dissolved cadmium in Lake Erie is 14 t. Finally, at an accumulation rate of 13 t/yr, the sediments retain 31 t Cd/y. Total output is therefore about 54 t. This is half of the input estimated by Nriagu (1986), who indicated that atmospheric inputs contributed 75 t of cadmium. Consideration of the water surface area of Lake Erie (Table 5) suggests that the figure of 75 t is anomalously high when compared with the other lakes. Assuming that the atmospheric loading for Lake On-

Table 5. Cadmium Inputs to the Great Lakes (t/yr)[a]

Location	Atmosphere	Total	Water Surface Area ($\times 10^{10}$/m^2)
Lake Superior	82	84	8.21
Lake Michigan	58	94	5.78
Lake Huron	60	62	5.96
Lake Erie	75	108	2.56
Lake Ontario	28	47	1.97

[a] From Nriagu (1986).

tario can be used to represent the situation for Lake Erie and correcting for the difference in their surface areas, the input to Lake Erie would be 36 t. Total inputs now work out to be 69 t with fair agreement with the output of 54 t. However, because the atmospheric loadings of cadmium and other trace elements have been evaluated most thoroughly for Lake Michigan (Schmidt and Andren, 1984; Tisue and Fingleton, 1984), it is more logical to use this lake as the basis on which to do a comparative calculation. Interestingly, this calculation shows that the atmospheric loadings to Lakes Superior and Huron as a function of their surface areas to be 82 and 60 t, identical to the figures in Nriagu (1986). For Lake Erie, the calculated loading is 26 t, giving a total of 59 t, which is in good agreement with the total output.

There is insufficient information for the remaining lakes to evaluate their mass balance budgets for cadmium. At least for Lakes Erie and Ontario, inputs are almost completely balanced by outputs. Calculations for the southern basin of Lake Michigan (Muhlbaier and Tisue, 1981) indicate that this basin is a net source of cadmium to the rest of the lake. This review clearly shows the need for measurements of cadmium in suspended particulate matter and in the dissolved phase in the upper lakes and in their connecting channels before such questions can be answered as whether a potential imbalance between influx and efflux of cadmium exists in the upper lakes, which may lead to increases in cadmium concentrations to deleterious levels (Marshall et al., 1981; Tisue and Fingleton, 1984).

REFERENCES

Allan, R. J. (1986). The limnological units of the Lower Great Lakes–St. Lawrence River corridor and their role in the source and aquatic fate of toxic contaminants. *Can. J. Wat. Pollut. Res.*, **21**, 168–186.

Balls, P. W. (1985a). Trace metals in the northern North Sea. *Mar. Pollut. Bull.*, **16**, 203–207.

Balls, P. W. (1985b). Copper, lead and cadmium in coastal waters of the western North Sea. *Mar. Chem.*, **15**, 363–378.

Bewers, J. M., Sundby, B., and Yeats, P. A. (1976). The distribution of trace metals in the western North Atlantic off Nova Scotia. *Geochim. Cosmochim. Acta*, **40**, 687–696.

Cahill, R. A., and Shimp, N. F. (1984). Inorganic contaminants in Lake Michigan sediments. In *Toxic Contaminants in the Great Lakes*, Nriagu, J. O., and Simmons, M. S. (eds.). Wiley. New York, pp. 393–423.

Christensen, M. R., and Chien, N-K. (1981). Fluxes of arsenic, lead, zinc and cadmium to Green Bay and Lake Michigan. *Environ. Sci. Technol.*, **15**, 553–558.

Cline, J. T., and Chambers, R. L. (1977). Spatial and temporal distribution of heavy metals in lake sediments near Sleeping Bear Point, Michigan. *J. Sed. Petrol.*, **47**, 716–727.

Cossa, D., and Poulet, S. A. (1978). Survey of trace metal contents of suspended matter in the St. Lawrence Estuary and Saguenay Fjord. *J. Fish. Res. Bd. Can.*, **35**, 338–345.

IJC (International Joint Commission). (1981). Report on Great Lakes Water Quality. Appendix: Great Lakes Surveillance, IJC Regional Office, Windsor, Ontario, Canada.

References

Imboden, D. M., Tschopp, J., and Stumm, W. (1980). Die rekonstruktion frueherer Stoffrachten in einem See mittels. Sedimentuntersuchungen. *Schweiz. Z. Hydrol.*, **42**, 1–14.

Kemp, A. L. W., and Thomas, R. L. (1976). Impact of man's activities on the chemical composition in the sediments of Lakes Ontario, Erie and Huron. *Wat. Air Soil Pollut.*, **5**, 469–490.

Kemp, A. L. W., Williams, J. D. H., Thomas, R. L., and Gregory, M. L. (1978). Impact of man's activities on the chemical composition of the sediments of Lakes Superior and Huron. *Wat. Air Soil Pollut.*, **10**, 381–402.

Kremling, K. (1985). The distribution of cadmium, copper, nickel, manganese, and aluminum in surface waters of the open Atlantic and European shelf area. *Deep-sea Res.*, **32**, 531–555.

Li, Y-H., Burkhardt, L., and Teraoka, H. (1984). Desorption and coagulation of trace elements during estuarine mixing. *Geochim. Cosmochim. Acta*, **48**, 1879–1884.

Lum, K. R., Callaghan, M. and P. R. Youakim (1986). Dissolved and particulate cadmium in the St. Lawrence River. In *Proceedings of the Second International Conference on Environmental Contamination, Amsterdam.*

Lum, K. R., and Callaghan, M. (1986). Direct determination of cadmium in natural waters by graphite furnace AAS without matrix modification. *Anal. Chim. Acta*, **187**, 157–162.

Lum, K. R., and Edgar, D. (1983). Determination of cadmium and silver in sediments by Zeeman Effect Flame Atomic-Absorption Spectometry. *Analyst*, **108**, 918–924.

Lum, K. R., and Leslie, J. K. (1983). Dissolved and particulate metal chemistry of the central and eastern basins of Lake Erie. *Sci. Tot. Environ.*, **30**, 99–109.

Marshall, J. S., Mellinger, D. L., and Parker, J. I. (1981). Combined effects of cadmium and zinc on a Lake Michigan zooplankton community. *J. Great Lakes Res.*, **7**, 215–223.

Mart, L., Nürnberg, H. W., and Dyrssen, D. (1984). Trace metal levels in the eastern Arctic Ocean. *Sci. Tot. Environ.*, **39**, 1–14.

Mart, L., Nürnberg, H. W., and Rutzel, H. (1985). Levels of heavy metals in the tidal Elbe and its estuary and the heavy metal input into the sea. *Sci. Tot. Environ.*, **44**, 35–49.

Martin, J-M., and Meybeck, M. (1979). Elemental mass-balance of material carried by major world rivers. *Mar. Chem.*, **7**, 173–206.

Muhlbaier, J., and Tisue, G. T. (1981). Cadmium in the southern basin of Lake Michigan. *Wat. Air Soil Pollut.*, **15**, 45–59.

Muhlbaier, J. C., Stevens, D., and Tisue, G. T. (1982). Determination of cadmium in Lake Michigan by mass spectrometric isotope dilution analysis or atomic absorption spectrometry following electrodeposition. *Anal. Chem.*, **54**, 496–499.

Nriagu, J. O. (1986). Metal pollution in the Great Lakes in relation to their carrying capacity. In *The Role of the Oceans as a Waste Disposal Option,* Kullenberg, G. (ed.). Reidel, Dordrecht, pp. 441–468.

Nriagu, J. O., Kemp, A. L. W., Wong, H. K. T., and Harper, N. (1979). Sedimentary record of heavy metal pollution in Lake Erie. *Geochim. Cosmochim. Acta.*, **43**, 247–258.

Nriagu, J. O., Wong, H. K. T., and Coker, R. D. (1981). Particulate and dissolved trace metals in Lake Ontario. *Water Res.*, **15**, 91–96.

Parker, J. I., Stanlaw, K. A., Marshall, J. S., and Kennedy, C. W. (1982). Sorption and sedimentation of Zn and Cd by seston in southern Lake Michigan. *J. Great Lakes Res.*, **8**, 520–531.

Poldoski, J. E., and Glass, G. E. (1978). Anodic stripping voltammetry at a mercury film electrode: Baseline concentrations of cadmium, lead and copper in selected natural waters. *Anal. Chim. Acta*, **101**, 79–88.

Rossmann, R. (1982). Trace metal chemistry of the waters of Lake Huron. Publication 21, Great Lakes Research Division, University of Michigan, Ann Arbor, MI.

Rossmann, R. (1984). Trace metal concentrations in the offshore waters of Lakes Erie and Michigan. Special Report No. 108, Great Lakes Research Division, University of Michigan, Ann Arbor, MI.

Rossmann, R. (1986). Trace metal concentrations in the offshore waters of Lake Superior. Draft Report, Great Lakes Research Division, University of Michigan, Ann Arbor, MI.

Schmidt, J. A., and Andren, A. W. (1984). Deposition of airborne metals into the Great Lakes: An evaluation of past and present estimates. In *Toxic Contaminants in the Great Lakes*, Nriagu, J. O., and Simmons, M. S. (eds.). Wiley, New York, pp. 81–103.

Statham, P.·J., Burton, J. D., and Hydes, D. J. (1985). Cd and Mn in the Alboran Sea and the adjacent North Atlantic: Geochemical implications for the mediterranean. *Nature,* **313**, 565–567.

Thomas, R. L., and Mudroch, A. (1979). Small craft harbours: Sediment survey, Lakes Ontario, Erie and Lake St. Clair. Report to Small Craft Harbours, Ontario Region from Great Lakes Biolimnology Laboratory, Burlington, Ontario.

Tisue, G. T., and Fingleton, D. (1984). Atmospheric inputs and the dynamics of trace elements in Lake Michigan. In *Toxic Contaminants in the Great Lakes,* Nriagu, J. O., and Simmons, M. S., (eds.). Wiley, New York, pp. 105–125.

Trefry, J. H., Metz, S., and Trocine, R. P. (1983). Decreased inputs of cadmium and lead to the Gulf of Mexico from the Mississippi River. *EOS,* **64**, 244 (Abstract).

Tremblay, G. H. (1985). Variations temporelles des concentrations en ions majeurs du fleuve Saint-Laurent et evaluation de l'apport fluvial dans la zone estuarienne. Rapport technique canadien sur l'hydrographie et les sciences oceaniques, No. 64, Department of Fisheries and Oceans, Ottawa, Canada.

Yeats, P. A., and Bewers, J. M. (1982). Discharge of metals from the St. Lawrence River. *Can. J. Earth Sci.,* **19**, 982–992.

4

CADMIUM ASSOCIATIONS IN FRESHWATER AND MARINE SEDIMENT

M. Kersten
U. Förstner

Technische Universität Hamburg-Harburg
Hamburg, West Germany

4.1. Introduction
4.2 Sampling and Sample Handling
4.3. Chemical Extraction Methods for Partitioning of Cadmium in Sediments
    4.3.1. Single-Leaching Methods
    4.3.2. Sequential Extraction Procedures
    4.3.3. Specificity of Extraction Procedures
4.4. Changes in Sedimentary Cadmium Associations under Different Physicochemical Conditions
4.5. Release of Cadmium at Sediment–Water Interface
References

4.1. INTRODUCTION

The major uses of sediment analyses, especially on toxic trace metals such as cadmium but also on artificial radionuclides and on persistent organic chemicals such as PAHs and halogenated hydrocarbons, can be defined as follows:

the identification, surveillance, monitoring, and control of sources and distribution of pollutants and

the evaluation of the environmental impact of polluted sediments.

Sediments influence the current quality of the water systems and preserve the historical development of certain hydrologic and pollution conditions. Comparative analysis of the total concentrations of longitudinal profiles and sediment cores is performed to determine metal anomalies in zones of mineralization and from polluted areas. The study of dated sediment cores has proven especially useful as it provides a historical record of the various influences on the aquatic system by indicating both the natural background levels and the man-induced accumulation of elements over an extended period of time. Sediment analysis is used to estimate point sources of trace metals that upon being discharged to surface waters do not remain in solution but are rapidly adsorbed by particulate matter, thereby escaping detection by water monitoring. Thus, sediment data play an increasing role within the framework of environmental forensic investigations (Meiggs, 1980).

In addition, sediments are increasingly recognized as a pollutant source itself. Metals, for example, are not necessarily fixed ultimately by the sediment and soil but may be recycled via biological and chemical processes both within the sedimentary compartment and also back to the water column or the biosphere. This has been demonstrated especially in the course of "dredged materials," which threatened not only the quality of water but also organisms. Human activity promote the accumulation of polluted sediments in the waterways, and the increased maintenance dredging activities result in a large amount of contaminated solids for which safe disposal sites have to be found. A detailed discussion of these problems, which have had at least an important stimulating effect on the development of our knowledge about the chemical mobility of cadmium in sediment–water systems, has been given in the review of Khalid (1980).

In the discussion of environmental problems with cadmium, it is also increasingly recognized that the ecotoxicological significance of its impact is determined rather by the specific form and reactivity of its particulate association than by its accumulation rate. This effect typically refers to the proportions of anthropogenic source, which may be primarily associated with labile surface-binding sites and become partly mobilized in the aquatic milieu by repetitive changes of physicochemical conditions or accumulation of organic chelators. Cadmium can be expected to partition into several chemical forms or associations in relation to the solid phases of soil and sediment: water-soluble fraction in adsorption equilibrium with solid surfaces and fractions closely associated with or residing in carbonates, organic matter, manganese oxides, amorphous as well as crystalline iron oxides, sulfides, silicates, and other resistant minerals. It is a general experience from sediment studies that the surplus of cadmium introduced into the aquatic system by man's activities usually exists in relatively unstable chemical forms and should, therefore, be more accessible for short- and middle-term geochemical processes—including biological uptake—than the detrital, predominantly natural metal compounds that had undergone the geological cycle (Salomons and Förstner, 1980). Thus, relevant environmental studies should be related to the chemistry of the particle's surface and/or to the metal species highly enriched in autochthonous phases and surface coatings.

Basically, there are three methodological concepts for determining the distribution of a trace metal within or among small particles, and a discussion of these follows.

1. Relatively new physical instrumental techniques may be utilized to directly determine the chemical partitioning of cadmium in polluted sediments. Analysis of single particles by X-ray fluorescence using either electron or proton microprobes can identify differences in the matrix composition between individual cadmium-rich particles (Lee and Kittrick, 1984). Other techniques such as X-ray photoelectron spectroscopy (ESCA), X-ray transmission diffractometry, and (laser- or secondary-ion-induced) ion mass spectroscopy (LAMMA, SIMS) have been used on many types of geologic materials, including sediments, but seem to be limited to major elements due to their high detection limits. Various physical preconcentration techniques may be useful to obtain valuable information on the particulate associations of cadmium. Some physical separation techniques used in speciation studies of trace metals in natural waters have been reviewed by DeMora and Harrison (1983). The total concentration of cadmium can be determined as a function of particle size (e.g., by sequential filtration) (Hoffmann et al., 1981). Other physical fractionation techniques include magnetic and especially density separations (Linton et al., 1980; Pilkington and Warren, 1979; Mattigot and Ervin, 1983). Direct species determination of cadmium in polluted sediments, however, seems not to be possible as yet by either of these techniques. In addition, the rigorous conditions under which these analyses have to be performed may be effective in changing the initial cadmium associations, especially in anoxic sediments (Kersten and Förstner, 1986).

2. The composition of interstitial waters in sediments is perhaps the most sensitive indicator of the types and the extent of reactions that take place between trace-metal loaded sediment particles and the aqueous phase that contacts them. The large surface area of fine-grained sediment with respect to the small volume of its trapped interstitial water ensures that minor reactions with the solid phases will be indicated by major changes in the composition of the aqueous phase. Thus, the theoretical assessment of the nature of trace-metal phases via the equilibrium solution composition method has been often used in soil science, which relates the equilibrium activities of metals and ligands to the formation of complexes in solution and to the stability of certain solid associations. Data obtained from this type of analysis refer to the closed steady-state system whose chemistry is con-

trolled by thermodynamic considerations and by the kinetics of adsorption–desorption and precipitation–dissolution reactions. Although the results for well-characterized systems are impressive, the value of the predicted chemical speciation in natural sediment–water systems obtained from these models is limited (Florence and Batley, 1980). Some of the important sources of errors include the limited thermodynamic data file (Baham, 1984). In addition, trace-metal solution chemistry involving experimental measurement techniques such as dialysis, ultrafiltration, anodic stripping voltametry, ion-selective electrodes, ion exchange resins, and liquid chromatography has not generally kept pace with the remarkable progress in the sophistication of computer-aided chemical modeling. Some of the dissolved cadmium species predicted by computer speciation cannot be measured yet by direct techniques. Most natural sediment–water systems probably never reach complete equilibrium with respect to the constituent chemical components and corresponding solid materials. Other complications are that the interstitial water chemistry of the upper horizon of shallow-water sediments (unlike those published from deep-sea basins) can respond to changes in the composition of the overlying water. Moreover, it is subject to the effects of tidal action and bioturbation. Just as aquatic systems are chemically dynamic, they are also open to physical and biological dynamics. Even when such dynamics directly affect only selected chemical equilibria, these equilibria are likely to be coupled to others, which will then be disturbed in turn.

3. One of the oldest and most commonly used methods of operational fractionating polluted sediments involves the use of partial chemical leaching. Chemical leaching—apart from the characterization of the reactivity of specific metals—can provide information on the behavior of metal pollutants under typical environmental conditions. Common single-reagent leachate tests such as introduced by, for example, the U.S. Environmental Protection Agency (USEPA), American Society for Testing and Materials (ASTM),and German Water Chemistry Group ("Deutsche Einheitsverfahren") use either distilled/sampled water or acetic acid. A highly sophisticated single-step method of the USEPA designed for studies on the leachability of waste products consists of a mixture of sodium acetate, acetic acid, glycine, pyrogallol, and iron sulfate (Ham et al., 1979). A number of test protocols have been designed initially for the assessment of plant-available soil nutrients and speciation of trace metals in sewage-sludge-amended soils (Jackson, 1958; Lake et al., 1984). The purposes of selective extractions in geochemical exploration are, according to Chao (1984), (1) to elucidate the mode of occurrence of trace metals in soils and sediments; (2) to enhance the "geochemical contrast" between mineralized and background areas above what can be accomplished by the bulk analysis; and (3) to differentiate between effects on metal distribution caused by mineralization and those resulting from lithological and environmental factors. In the sediment-petrographic field, interest was focused initially on the differentiation between authigenic and detrital phases in ferromanganese concretions from

deep-sea deposits (e.g., Chester and Hughes, 1967; Lyle et al., 1984). In general, these studies serve for the determination of the major accumulative phases for metals in sedimentary ore deposits and the diagenetic accumulation mechanisms (Förstner, 1982). The early procedures involved two- or, at most, three-step multiple extractions used in a sequence of increasing strength, which attempted to partition trace metals into different chemical components of the materials sampled. This concept is based on the idea that a particular solvent is either phase or mechanistic specific (Horowitz, 1984; e.g., acetic acid will only attack and dissolve carbonates, and ammonium acetate at pH 7 will only remove adsorbates). The sequences became more and more complex in an attempt to differentiate more clearly between anthropogenic and natural inorganic pollutants and to try and predict or even estimate bioavailability. Although the major problem of these techniques is that the extraction reagents are not as selective as many users claim, and extraction efficiencies vary due to length of treatment and sediment to extractant ratio (Förstner, 1985), most studies conducted on cadmium speciation in freshwater and marine sediments refer to their potential means of providing the requisite chemical data for the assessment of the speciation of cadmium.

There are only few studies reported in the literature where more than one of these approaches have been used simultaneously for determining the cadmium species present in sediments. McGrath et al. (1984) measured the proportions of cadmium in contaminated soil solutions present in its divalent ion using an ion exchange equilibrium technique and cadmium-ion-selective electrodes. The results of both methods were compared with predictions made by various computer speciation programs, but agreement was not good. The authors explained their findings with the fact that soil organic matter contains a heterogeneous mixture of complexing sites. Thus, it is impossible to specify a unique stability constant with cadmium. Current experiments aim to improve the thermodynamic data files of computer speciation programs specifying two or more stability constants for each metal to consider, which are characteristic of its reactions with strong and weak complexing sites in soil solutions (Jones et al., 1985). Lee and Kittrick (1984) examined an anoxic harbor sediment heavily polluted with cadmium using the electron beam microprobe to obtain direct information about the elements associated with cadmium-rich particles. For these particles, cadmium was found to be most frequently associated with sulfur (in ~90% of the particles). They additionally predicted the formation of cadmium sulfides in these sediments using thermodynamic calculations. In contrast, exchangeable (~34%), carbonate (~36%), and manganese–iron oxide-bound (~22%) cadmium represent the most important chemical extraction fractions from the same samples. This disagreement is probably due to an improper sample handling because the authors have dried their anoxic sediment samples prior to applying the sequential extraction procedure. This important problem, which has been frequently neglected in soil sciences, will be considered in more detail in the next section.

4.2. SAMPLING AND SAMPLE HANDLING

Sampling procedure for source reconnaissance analysis have been reviewed by Förstner and Salomons (1980). In this chapter, we will concentrate on cadmium speciation by the aid of chemical sequential extraction. The appropriate equipment for sediment collection and processing has been described by Jenne et al. (1980).

Bottom sediments are usually taken by a grab sampler of the Van Veen or Ekman–Birge type. Material from the upper, fluid, light brown, oxidized layer is generally dissimilar to the layers below it (Luoma and Bryan, 1981). It is suggested that the chiefly dark layers directly underneath are more representative of the pollution situation over the last few years, especially in estuarine deposits exhibiting rapidly fluctuating sedimentation rates, and should be given priority for subsequent investigations. To complement this, surface sediment marked by current contamination should be examined separately. Concentrations of extractable trace metals in oxidized estuarine sediments are influenced by the manner and time of storage between collection and extraction (Thomson et al., 1980). Development of anoxic conditions during any short storage should be avoided by deep freezing the samples. In anaerobic environments with a relatively uniform sedimentation rate and without any significant biological activity (e.g., from eutrophic lakes and anoxic marine basins with specific geomorphological features), a more favorable procedure involves taking of vertical profiles with a piston corer. A core profile of approximately 1 m covers a historical period of up to 200 yr, and its development can be traced by virtue of the metal content in the individual layers. Dating of the layers [e.g., by radioactive isotopes of lead (^{210}Pb) and cesium (^{137}Cs)] may be useful for such studies. The interpretation of dated trace-metal profiles as historical records of deposition, however, assumes the absence of any significant postdepositional remobilization within the sediment column, such as introduced by acidification (Carignan and Nriagu, 1985) and oxidation (Kersten and Kerner, 1985). Core sediments with a well-developed redox potential discontinuity may be useful in studying general changes of cadmium speciation in sediments with changing Eh–pH conditions. Anoxic sediment samples require different sampling preservation techniques such as oxygen exclusion. Drying (also freeze drying) can seriously disturb the initial chemical speciation of cadmium, as will be shown in Figure 4 (see Section 4.4).

Extraction of interstitial waters from sediments is usually performed by either centrifugation or squeezing; oxidation must be prevented during these procedures (Lyons et al., 1979). Watson et al. (1985) showed that sediment stored prior to the separation of interstitial water yield significant changes in chemical composition compared to samples processed within 24 h of collection. In addition, the authors described spatial and seasonal variations in the chemistry of estuarine sediment interstitial waters, which need to be carefully examined before to define a proper sampling strategy. Kinniburgh and Miles (1983) described a new centrifugation technique using immiscible displacement with a dense, inert fluorocarbon liquid. Yields of interstitial water from soils at field capacity are typically 20–50% of the total water present. The authors showed that the difference in the chemical composition of the extracted and the remaining water in their samples was negligible.

In situ methods are considered more promising because of their inherent simplicity and appear well adapted to the study of trace metals at the sediment–water interface under field conditions. A technique described by Mayer (1976) consists of a dialysis bag filled with distilled water, which is displaced into the sediment allowing equilibrium to take place between the inserted and ambient water over a period of some days to weeks. An improved sampler of this type has been described by Bottomley and Bayly (1984). Another *in situ* sampler for close interval pore water studies as presented by Hesslein (1976) can be made from a clear acrylic plastic panel with small compartments predrilled in 1-cm-steps or less. This panel can be covered by a nondegradable dialysis membrane or by a polysulfone membrane filter sheet (Carignan, 1984). Yoshida (1984) reported that the measured values for nurients from centrifugal separation are much larger than those obtained by the dialysis sampler. He stressed that nutrients are released dynamically during the centrifugal separation and thus does not indicate the true values, which are obtained statically by the dialysis technique. More recently, Carignan et al. (1985) compared both techniques and found that centrifugation yields interstitial water cadmium values equivalent to dialysis when high-speed centrifuges were used and the supernatant was filtered through 0.2-μm membranes. The chemical data obtained by a multicomponent analytical approach should be checked for internal consistency using the electroneutrality criteria or even more sophisticated methods (Merino, 1979).

Suspended sediments are recovered by filtration, continuous-flow centrifugation, or sediment traps (Etcheber and Jouanneau, 1980). Because metal content evolves extensively in time and space, especially in estuarine waters, many samples are required to gain a representative pattern. Speciation studies need a sufficient quantity of particulate matter (at least some grams, to carry out all the analysis: particle size distribution, mineralogy, total and sequential chemical extractions). To separate the appropriate amounts of sampled water within, at most, the limited time on station required on research vessels, most researchers use continuous-flow centrifugation. Although particles are separated by density rather than by size, this technique is chosen due to its rapid functioning and good recovery rate (Etcheber and Jouanneau, 1980; Van Der Sloot and Duinker, 1981). The development of *in situ* water-sampling systems with filtration devices provide the capability of extraction particulate and dissolved cadmium from large volumes of seawater (Winget, 1982; Green, 1984). It should be noted, however, that both separation procedures are operationally defined and still remain subject to a tremendous amount of interpretation and argument as universally accepted methods in this field have not as yet been devised. Various authors have shown that taking suspended matter by the centrifugation technique results in material different from that obtained by filtration (Duinker et al., 1979; Ongley and Blanchford, 1982; DeMora and Harrison, 1983). The most significant distinctions were found with respect to small low-density particles, whereas the deviations in mineral composition of particles with grain sizes on the order of micrometers seem to be limited to 10–30% (Salomons and Mook, 1980).

4.3. CHEMICAL EXTRACTION METHODS FOR PARTITIONING OF CADMIUM IN SEDIMENTS

The key to appreciate the fate of cadmium in the sediment–water system lies in identifying and quantifying the various forms of cadmium present in the sedimentary components, the various mechanisms affecting the availability of sediment-bound cadmium, and the influence of physicochemical parameters on the relative partition of cadmium between the sedimentary components. On the basis of a literature review on chemical partitioning, Horowitz (1984) classified two approaches that have been widely used. The first attempts to determine how trace metals are retained on or by sediments, the so-called *mechanistic approach*. The second approach determines where inorganic constituents such as cadmium are retained on or by sediments (phase or site), the so-called *phase approach*. This term is used in the thermodynamic sense and incorporates such components as interstitial water, carbonates, oxides, sulfides, organic materials, and silicates. Despite the simplicity of these two approaches, most attempts of chemically partitioned complex sediment samples combine aspects of both. Table 1 gives a compilation of the various mechanisms, products, and substrates of metal enrichment and their estimated significance in aquatic sediments. This scheme includes the contribution of both the mechanistic and the phase approach.

It is suggested that the inert bonding of metals (e.g., in silicates and heavy minerals) is the dominant effect of metal accumulation in natural lithogenic sediments. With respect to surface-associated mechanisms, the accumulation on organic substances and hydrous iron–manganese oxyhydroxides seems to be particularly important. Jenne and Luoma reviewed the state of knowledge in 1977 on the physicochemical partitioning of trace metals, including cadmium, into various forms and suggested that the most general particulate sinks for this metal are both oxides and organic substances. They concluded from their compilation of the published experimental evidence that the equilibrium concentrations of cadmium in the sediment–water inter-

Table 1. Mechanisms and Substrates of Metal Enrichment in Aquatic Systems

	Rock Debris and Solid Waste Material[a]	Metal-Hydroxide, Metal-Carbonate, and Metal-Sulfide	Organic Substance (reactive)	Hydrous Fe–Mn Oxides[b]	Calcium Carbonate
Inert bonding	× × ×			×	×
Precipitation		× × ×		× ×	× ×
Adsorption	×	×	×	×	×
Coprecipitation		× ×		× × ×	× ×
Coatings			× × ×	× × ×	×
Flocculation			×	× ×	

[a] Including organics.
[b] Mainly authigenic.

face and hence its biological availability was regulated via sorption–desorption and dissolution–precipitation reactions. Thus, application of chemical extraction techniques on the basis of the above-mentioned mechanistic approach can be useful to assess the mobility of cadmium in the aquatic milieu.

4.3.1. Single-Leaching Methods

Sediment-leaching techniques can be classified as single-leaching steps and sequential extraction procedures. The application of an extractant may be either as a "nonselective" extraction—the application of an individual extractant on a selected material—or as a "selective" extraction step in successive multiple sequences. The terms *nonselective* and *selective,* as used here, are strictly relative, as will be discussed later. A compilation of single-leaching steps involved in common procedures is given in Table 2 and in Figure 1. The readily water-soluble and exchangeable fractions of cadmium in sediments are the first portion to be brought into solution by any chemical extraction technique. The amount of exchangeable cadmium proportion, which is suggested to be in equilibrium with the aqueous phase and thus represents the readily bioavailable fraction, is usually determined through percolation or extraction with a solution containing a displacing cation such as neutral ammonium acetate, chloride, or nitrate salts (Lake et al., 1984). Exchangeable forms of trace metals normally constitute only a small proportion of the total amount of metals in uncontaminated soils and sediments. While in polluted sediments at most <1% of total copper and lead are exchangeable, exchangeable portions of cadmium can reach values as much as >20%. The ion-exchangeable fraction of trace metals, however, is still ill-

Table 2. Methods for Extraction of Metals from Major Chemical Components in Sediments[a]

Method	Reference
Adsorption and Cation Exchange	
BaCl$_2$–triethanolamine, pH 8.1	Jackson, 1958
MgCl$_2$, pH 7	Gibbs, 1973
Ammonium acetate, pH 7	Jackson, 1958
Carbonate Phases	
CO$_2$ treatment of suspension	Patchineelam, 1975
Acidic cation exchange resin	Deurer et al., 1978
NaOAc/HOAc buffer, pH 5	Tessier et al., 1979
HOAc	Chester and Hughes, 1967
Reducible Phases (in approximate order of release of iron)	
Acidified hydroxylamine–hydrochloride	Chao, 1972; Chao and Zhou, 1983
EDTA, pH 8	Borggaard, 1976
Ammonium oxalate buffer (Tamm's reagent)	Schwertmann, 1973
Hydroxylamine–acetic acid	Chester and Hughes, 1967
Dithionite–citrate buffer	Holmgren, 1967
Oxidizable Phases (including sulfides and organic matter)	
Acidified hydrogen peroxide, extracted with ammonium acetate in 6% HNO$_3$	Gupta and Chen, 1975
NaOH, extracted with H$_2$SO$_4$	Volkov and Fomina, 1974
Organic solvents	Cooper and Harris, 1974
Sodium hypochlorite	Gibbs, 1973
Alkali pyrophosphate	Eaton, 1979
0.1 N HCl bubbled with air (sulfides)	Kitano and Fujiyoshi, 1980

[a] After Salomons and Förstner (1980).

defined, as is indicated by significant metal releases in replicate experiments (Salomons and Förstner, 1980). It is clear that attempts to measure the levels of exchangeable cadmium in saline, sodic, and calcareous material, such as marine and salt marsh sediments, have encountered serious difficulties. Displacement of adsorbed cadmium using chlorides such as MgCl$_2$ causes considerable complexation effects. Neutral NH$_4$OAc causes dissolution of manganese–oxyhydroxides (Jenne, 1977) and reactive metal oxide coatings in general (Gibbs, 1973). Dissolution of organic matter via the formation of ammoniated organic complexes may also result from the use of ammonium salts (Frink, 1964). Other less known reagents were discussed by Robbins et al. (1984) NaOAc, pH 8.2, has been suggested as an extractant for exchangeable cations as it causes little dissolution of carbonates (Bascomb, 1964). However, Rhoades and Krueger (1968) have shown that this extractant can

EXTRACTANT TYPE	RETENTION MODE						
	Ion Exchange Sites	Surface adsorption	Precipitated (CO_3, S, OH)	Co-ppted. (amorphous hydrous oxides)	Co-ordinated to organics	Occluded (crystalline hydrous oxides)	Lattice component (mineral)
Electrolyte	$MgCl_2$	------▶					
Acetic Acid (buffer)	HOAc	HOAc/OAc⁻	------▶				
(reducing)	HOAc +	NH_2OH		------▶			
Oxalic Acid (buffer)	HOx +	NH_4Ox				Light (UV) ------▶	
dil. Acid (cold)		0.4 m	HCl	------	------	------	------▶
Acid (hot)	HCl +	HNO_3;	HNO_3 +	$HClO_4$		------▶	
Mixtures (+HF)		HCl +	HNO_3 +	HF			
Chelating Agents	EDTA,	DTPA	------	------	------	------▶	
	$Na_4P_2O_7$	------		------▶			
	$Na_4P_2O_7$ + $Na_2S_2O_7$	------			------▶		
	$Na_2S_2O_7$ + citrate +		HCO_3^-			------▶	
Basic Solns.			(alk.ppte)	------	NaOH ------▶		
					NaF		
Fusion (+ Acid leach)			Na_2CO_3				

Figure 1. Schematic representation of the ability of different extractant agents to release metal ion retained in different modes or associated with specific soil fractions. Dashed segments indicate areas of uncertainty. (From Pickering, 1981.)

dissolve structural cations from clay minerals. Polemio and Rhoades (1977) have proposed the use of ethanolic NaOAc–NaCl, pH 8.2, to remove exchangeable cations but concede that sodium-acetate may not completely saturate exchange sites in competition with divalent cations. NaOAc extraction releases interlayer cations in dioctaedral layer silicates into solution, in contrast to NH_4^+, which collapses the edges of these layer silicates, especially vermiculite (Bower, 1950). To release cadmium ions bound to carbonates, 1 M NaOAc buffered with acetic acid to pH 5.0 is most widely used. The preferential carbonate binding of cadmium often described (e.g., Span and Gaillard, 1986) can be attributed to the relatively high stability of $CdCO_3$ under the physicochemical conditions of common fresh waters (Hem, 1972).

From Figure 1, it follows clearly that extractants releasing cadmium ions bound to organic and other specific inorganic phases should be less specific. They would release at least all the exchangeable and soluble metal additionally. Dissolution of amorphous oxyhydroxides can be prompted by reducing the iron and manganese to lower valency states, for example, through weakly reducing agents like hydroxylamine, acidified oxalate (Tamm's reagent), or dithionite (Chao, 1972; Chao and Zhou, 1983; Pfeiffer et al., 1982). Oxalate treatment did not appear to attack crystalline material (Schoer, 1985), but it has been shown that the degree of interaction varies with illumination (Schwertmann, 1973). The acidified hydroxylamine is extremely sensitive to the presence of any carbonates in sediments due to the poor buffering capacity of 0.01 N nitric acid (Pfeiffer et al., 1982). Neutralization dramatically affected extraction of cadmium (Kersten, unpublished data). Thus, the hydroxylamine reagent is meaningful only in sequence to carbonate dissolving steps or for sediment samples where carbonates are present in very low concentrations. For routine investigations, an extraction with diluted HCl is widely accepted because this procedure requires fewer manipulations (Malo, 1977). The cadmium fraction associated with organic matter has been extracted through oxidation of this component with NaOCl or H_2O_2. The disadvantage in this method lies in the concomitant oxidation of the sulfidic cadmium association, which can dominate the partition of total cadmium (Lee and Kittrick, 1984). In a comparative study on chemical dissolution of nine common crystalline sulfide minerals, Chao and Sanzalone (1977) found that galena, chalcopyrite cinnabar, molybdenite, pyrite, sphalerite, and tetrahydrite were decomposed to the extent of 43–66% of the total sulfur content by a combined hydrogen peroxide–ascorbic acid leach. Other methods have since been applied using dissolving agents such as sodium hydroxide, pyrophosphate, or surfactants (e.g., hot sodium dodecyl sulfate solution, pH 8.8, with $NaHCO_3$; Robbins et al., 1984). The apparent ability of both EDTA and HCl to release metal ions associated with both inorganic and organic components has led to these being used as extractants for a rapid, inexpensive way to evaluate the total nondetrital fraction in trace-metal-polluted sediments.

Hydrochloric acid can dissolve iron oxides of varying degrees of crystallinity, depending on the concentration of the acid and the temperature of reaction (Chao and Zhou, 1983). At elevated temperatures, HCl can attack even primary silicate minerals as well as clays. Also, this acid will dissolve carbonate minerals and replace metal ions adsorbed on inorganic and organic surface sites. Partial single extraction of cadmium from sediments using both EDTA and HCl has been reported by, for example, Farrah and Pickering (1978) and Luoma and Bryan (1981). Kitano and Fujiyoshi (1980) reported partitioning of cadmium into sulfidic and organic fractions in core sediments from Osaka Bay using 0.1 N HCl under oxidized conditions and 30% hydrogen peroxide. In case of anoxic sediments, this partition scheme can be useful, as will be shown in Section 4.4.

Determination of the extracted metals is in principle possible either by analyzing the residue by, for example, INAA or XRF or at most using the separated effluent, analyzing by, for example, AAS or ICP and comparing the results with the appropriate bulk concentration. For the complete dissolution of detrital or lattice-bound cadmium, which is usually a minor fraction in polluted sediments, a combination of acids is concededly useful. Nitric acid promotes destruction of organic matter, but when used alone or in conjunction with hydrochloric or perchloric acid, this digest does not completely dissolve some types of silicate minerals (Malo, 1977).

4.3.2. Sequential Extraction Procedures

In connection with the problems arising from the disposal of contaminated dredged material, sequential extraction procedures have been developed, which include the successive leaching of metals from interstitial waters and from ion-exchangeable, easily reducible organic and residual sediment components (Brannon et al., 1976). Based on these early experiments and experiences in our own laboratory, an extraction procedure consisting of four steps was suggested for the differentiation of cadmium in oxic sediments (Salomons and Förstner, 1980):

1. Extraction 1 M NH$_4$OAc at pH 7 for exchangeable cations,
2. Extraction with acidified hydroxylamine hydrochloride at pH 2 for easily reducible phases,
3. Digestion with acidified hot hydrogen peroxide followed by an extraction with ammonium acetate to remove any readsorbed metal ions, and
4. Dissolution of the residual solid with HF–HClO$_4$–HNO$_3$.

This scheme was used for the study of cadmium partition in sediments from 10 different European river systems (Salomons and Förstner, 1980). In Figure 2, the rivers are arranged according to their approximate geographical position from north to south.

Throughout, the results show relatively low amounts of cadmium present in the residual (lithogenous) fraction in these more or less polluted river

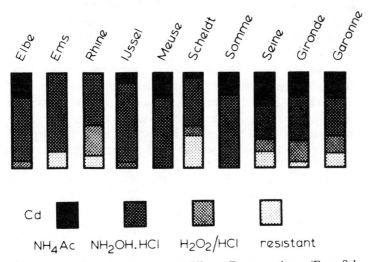

Figure 2. Partition pattern of cadmium in 10 different European rivers. (From Salomons and Förstner, 1980.)

systems. In nearly all these oxic examples, the "reducible phase" is an important sink for cadmium. Compared to elements such as lead and copper, cadmium is characteristically enriched in the more mobile fraction "cation exchange" and "easily reducible phases" (of which the latter includes the fraction bound by carbonates). It is, therefore, more mobile than most of the other toxic heavy metals.

In extension of this sequence and using the experience of the work from Tessier et al. (1979), a six-step sequential extraction protocol was developed for the assessment of the environmental compatibility of dredged material that avoids some of the inadequacies of earlier experiments (Förstner and Calmano, 1982):

1. Exchangeable cations: The sediment sample is extracted for 2 h on a mechanical shaker using 1 M ammonium acetate, pH 7, with a solid–solution ratio of 1:20 (Tessier et al., 1979: 1 N $MgCl_2$, pH 7, 10 min).
2. Carbonatic fraction: The centrifuged residue from step 1 is leached for 5 h with 1 M sodium acetate adjusted to pH 5.0 with acetic acid.
3. Easily reducible fraction: The residue from the preceding step is leached for 12 h (overnight) with 0.1 M hydroxylamine hydrochloride + 0.01 N nitric acid. The solid–solution ratio is now increased to 1:100.
4. Moderately reducible fraction: The residue from step 3 is extracted for 24 h with 0.1 M ammonium oxalate buffer, pH 3.
5. Oxidizable metal bound to organic matter and sulfides: The residue is digested twice with acidified H_2O_2 at 85° C and then extracted with 1 M ammonium acetate in 6% HNO_3 overnight.
6. Residual fraction: The residue is dissolved in concentrated nitric acid and/or other acids.

This scheme has been applied, apart from the major studies on river sediments and dredged materials, to demonstrate the relative mobility of particulate cadmium in lake sediments affected by acid rain (Reuther et al., 1981), in sand filter column experiments (Förstner et al., 1979), and in fresh (Förstner et al., 1981) and incinerated sewage sludge (Fraser and Lum, 1982). Some of these data are summarized in Table 3. A study made by Harrison et al. (1981) with six street dust and four roadside soil samples indicates that cadmium is there to a substantial proportion in the exchangeable fraction while zinc and lead is predominantly associated with carbonates and Fe–Mn oxides (copper is predominantly found in organic association). The high cation and chloride concentration (1 M $MgCl_2$) used in the exchange solution may reflect conditions in soils contaminated with deicing salt (Harrison et al., 1981). In comparison to the street-dust-leaching data, the experiments on roadside soils indicated significant lower extraction rates

Table 3. Chemical Extraction Results (%) of Cadmium, Zinc, and Lead from Different Fractions of Solid Waste Materials[a]

	Cadmium				Zinc				Lead			
	A	B	C	D	A	B	C	D	A	B	C	D
Exchangeable	20	26	17	—	2	3	4	—	2	1	—	—
Carbonate	38	24	61	—	44	31	69	—	43	26	11	1
Reducible	28	25	6	1	43	34	14	4	38	44	29	1
Organic sulfide	8	8	11	10	7	9	7	6	7	12	16	1
Residual fraction	6	18	5	89	4	23	6	90	10	17	44	98

[a] A = street dust; B = roadside soil (Harrison et al., 1981); C = fresh sewage sludge (Förstner et al., 1981); D = ash samples from incinerated sewage sludge (Fraser and Lum, 1982).

in the more mobile fractions and an increase of the residual metal proportions. This was connected with the lower total metal contents in the roadside soil samples compared to the dust specimen. Despite the relatively low concentration of cadmium in the sewage sludge sample (C, sample from a treatment plant at Landau/Pfalz), there is a significant shift to higher percentages in the carbonate fraction. This is also valid for zinc, whereas lead is enriched in the organic and residual forms, compared to the atmospherically derived samples (A and B). The general experience that the more enriched cadmium is also the more reactive one seems to be valid only for waste material, which was not treated by high-temperature processes. This is exemplified by the data from incinerated sludge ash (sample D, Hamilton Municipal Incinerator), where the remaining potentially toxic metals are highly enriched but are rather immobile with ~90–100% of cadmium and lead bound in "residual fraction" (Fraser and Lum, 1982). A review on the chemical speciation and environmental effects of cadmium and other critical elements in combustion residues is given by Förstner (1986).

4.3.3. Specificity of Extraction Procedures

Although "selectivity" for a specific phase in a strictly thermodynamic sense cannot be expected for these procedures, there are also differences in the specificity between the various extractants and methods used. The compilation given by Pickering (1981) as redrawn in Figure 1 implies that any extractant is only crudely differentiated between the different forms of selected species in sedimentary material. In particular, despite the clear advantages of a differentiated analysis over bulk characterization of sediments, it was pointed out that a single extraction method is not selective by itself. Several problems are encountered in part in our laboratory, which may illustrate this point (Pfeiffer et al., 1982; Calmano and Förstner, 1983), and by the useful evaluation summary given by Robbins et al. (1984):

- Labile metal phases could be transformed during sample preparation, which can occur not only for anoxic but also for oxic surface sediments and suspended matter (Thomson et al., 1980). In both cases, the transition of oxic to anoxic conditions, and vice versa, has to be avoided.
- Processes of readsorption and precipitation have to be considered, particularly during extraction with less acid steps (Rendell et al., 1980). This phenomenon could result in underestimation of the metal released from solid phases and could contribute to "downstream" errors when a succession of extraction steps is carried out (Nirel et al., 1985). To minimize such readsorption errors during extractions, each step could be repeated several short times using fresh solutions, as shown by Robbins et al. (1984). Chelation of metals in leachates of higher pH values using EDTA can be utilized rather than acidification to prevent as much as possible the hydrolysis of the released metals and flocculation of humic acids.
- Reactions are influenced by the duration of the experiment and by the ratio of solid matter to volume of extractant. Some of the extractants are more reactive than others as shown by repeated experiments (Heath and Dymond, 1977). A too high solid content with increased buffer capacity may cause the system to overload. Such an effect is reflected by changes of pH values in time-dependent tests. However, in certain cases, this disadvantage may provide additional data for differentiation of trace-metal binding mechanisms (Bowser et al., 1979).
- There are major objections to the use of barium chloride–triethanolamine or magnesium chloride for the determination of exchangeable cadmium, since there are chelating effects by the organic agent and by formation of soluble-metal–chloro complexes.
- Hydrolyzation of cadmium may be initiated during high-pH conditions, which arise, for example, during humic extraction with sodium hydroxide or sodium pyrophosphate (Burton, 1978).
- Difficulties occur with the carbonate extraction by the sodium acetate–acetic acid buffer treatment. This reagent has the disadvantage of releasing free hydrogen ions to attack other mineral species, especially labile oxyhydrates. It apparently does not attack, however, the detrital iron and manganese oxide minerals to any great extent (Loring, 1976) and leaves the lattice structure of silicate minerals intact (Robbins et al., 1984). The major advantage of the NaOAc step lies in the reduction of the buffer capacity of the sample before applying the following leaching steps. The buffering capacity of the NaOAc–HOAc extractant solution is sufficient for the first repetitive extract to dissolve all the $CaCO_3$ in 0.5 g of a carbonate-rich sample that contains 68% $CaCO_3$, resulting in a final pH of 5.5 (Robbins et al., 1984).

- Borggaard (1982) reported a selective extraction procedure to dissolve amorphous iron oxides from soils by using EDTA at slight alkaline pH conditions, with time of extraction varying from a few days to several months. This time-consuming procedure, however, may serve no purpose to its application in environmental monitoring.
- Problems in the acid-reducing steps are encountered particularly with the utilization of the dithionite–citrate extraction: contamination by zinc and cadmium in the reagent, rapid clogging of the burner during atomic adsorption analysis, decomposition of the extractant, and formation of sulfides. Additionally, it attacks iron-rich layer silicates (Pawluk, 1972).
- Treatment with hydrogen peroxide affects both labile and relatively stable bound metal compounds and also oxidizes sedimentary sulfides to varying degrees (Chao and Sanzalone, 1977). It is catalytically decomposed by oxides, with a resultant suppression of organic matter oxidation. Metals released by H_2O_2 treatment are susceptible to scavenging by oxyhydrates in the sediment (Guy et al., 1978), which need to be removed prior to the H_2O_2 step. Taylor and McKenzie (1966) utilized a prolonged treatment with acidified 10% hydrogen peroxide (pH 3) to remove manganese oxides from soils. Thus, the H_2O_2 treatment as part of a sequential multiple extraction scheme should follow the removal of reducible oxides to avoid an adverse effect on the fractionation pattern of sediments. Poorly buffered, organic-rich or sulfide-rich sediments become very acidic when treated with H_2O_2, leading to decomposition of silicates and oxides (Farmer and Mitchell, 1963). Conditions promoting an expansion of clays during H_2O_2 treatment seem to enhance the ease and extent of structural cation loss (Farmer et al., 1971).
- Treatment with organic solvents seems to be poorly suited for routine application (Salomons and Förstner, 1980). Dispersion and dissolution of the reactive organic phase by alkali pyrophosphate or by surfactants at neutral to slightly alkaline pH values and elevated temperatures (Eaton, 1979; Robbins et al., 1984) seems preferable to chemical oxidation approaches, when differentiation between organic and sulfidic association of cadmium by sequential extraction procedures is achieved. Both pyrophosphate and surfactants do not attack sulfides nor do these reagents dissolve significant amounts of amorphous iron oxides (Chao, 1984; Robbins et al., 1984). Because of the highly dispersive state of the suspension resulting from these reagents, however, strong centrifugation is necessary to separate the supernatant solution. Other complications arising from these extractants are described by Jeanroy and Guillet (1981) and Robbins et al. (1984).

Of course, the degree of fractionation steps required depends on the purposes of the study. In discussions of the extraction techniques where a single "selective" extractant as one step in a protocol of a sequential multiple extraction is considered, the choice of reagent is determined by the component of the sedimentary matrix that is sought by the speciation study. With a careful designed combination of various single steps to a sequential extraction scheme, however, the specificity of some steps has been significantly improved, which led to such sequence formulations as preferred by, for example, Engler et al. (1977), Filipek and Owen (1978), Tessier et al. (1979, 1980), and Khalid et al. (1981).

The procedure given by Förstner and Calmano (1982) has been tested in our laboratory using well-defined natural mineral phases (Rapin and Förstner, 1983). Figure 3 shows that the carbonate phase is undoubtedly attacked during the second step (sodium acetate) of the extraction sequence; however, step 1 (ammonium acetate) also dissolves a significant part of this phase (as has already been mentioned by Jackson, 1958). In the case of deep-sea ferromanganese nodules, cadmium was found in the exchangeable fractions (steps 1 and 2) and is probably strongly adsorbed on the oxyhydroxides of iron and manganese since the latter are known for their high adsorption capacity. Similarly, the predominant proportion of the total leachable cadmium from crystallized iron oxides such as heamatite is extracted by step 2. From an artificial amorphous FeS sample, as expected, the predominant fraction was leached by the oxidizing step 4; however, over 20% was released in step 1. Similarly, using a sample of galena (PbS), nearly all the total content was extracted by step 4. The remaining proportion, however, was exchanged by ammonium acetate. This fact is probably due to a partial oxidation of these samples, which affects the amorphous phase to a stronger degree than the crystalline mineral.

A more complex and less specific behavior with regard to the cadmium results in Figure 3 showed goethite, vivianite, and clay samples. The clay mineral group (illite, chlorite, montmorillonite, and kaolinite) is attacked predominantly by the last "residual" step in the case of iron, manganese, zinc, nickel, and chromium but not in the case of cadmium. This is probably due to the rather rigid digestion by acidified hot H_2O_2 leading to part dissolution of silicates (Van Langeveld et al., 1978). Farrah and Pickering (1978) examined extractability of heavy-metal ions sorbed on clays and found that cadmium preadsorbed on clay (kaolinite, illite, and montmorillonite) at pH 5 or 7 is effectively displaced by neutral ammonium acetate. Iron(II) phases such as vivianite and siderite are not stable in oxidized solutions. Thus, most elements were predominantly extracted in the first three steps. Quantitative extrapolation of the results of this study and a similar one presented recently by Nirel et al. (1985) is difficult. The crystalline substrates used in both studies are not fully representative of their natural equivalents, and the quantitative impact of transfers from a given fraction to the following ones is obviously dependent on the initial composition of the sample analyzed. For this "matrix effect," as shown by Nirel et al. (1985), careful tests are to be undertaken to check the efficiency of this method. Application to natural

Chemical Extraction Methods for Partitioning of Cadmium in Sediments

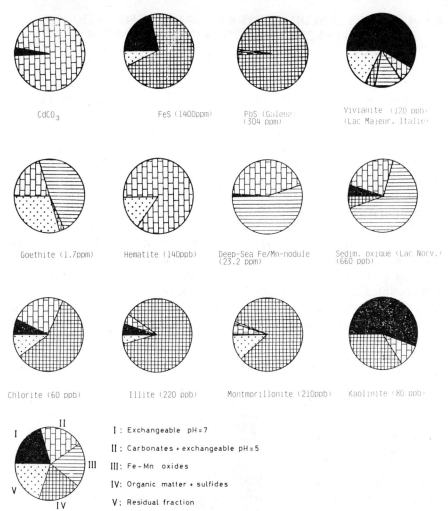

Figure 3. Sequential extraction results of cadmium from different minerals. (From Rapin and Förstner, 1983.)

sediments will give reproducible results as long as the relative abundance of their organic and inorganic components does not show large variations, but a reliable identification of a given particulate cadmium species (even operationally defined: Jenne, 1977) may be questionable in different matrices. These limitations have led us, among others, to the conclusion that the results given by the sequential chemical extraction are rather indicative to certain operationally defined groups of cadmium forms and associations than to direct cadmium speciation in sediments. With this practical obstacle recognized, the application and the potential usefulness of this technique in environmental studies can be logically discussed.

4.4 CHANGES IN SEDIMENTARY CADMIUM ASSOCIATIONS UNDER DIFFERENT PHYSICOCHEMICAL CONDITIONS

The redox potential of a sediment–water system is a controlling factor in regulating the chemical form of cadmium, whereas pH and salinity influence the stability of the various forms. This fact has been confirmed by a number of publications based on sequential extraction studies (reviews given by Jenne and Luoma, 1977; Khalid, 1980; and Salomons et al., 1986) and can be shown by a simple but impressive experiment (Fig. 4). Anoxic sediment samples from Hamburg Harbor were divided into four series under an argon-flushed glove box in order to study the effect of various sample pretreatments, including aeration and dehydration, on the chemical forms of cadmium (Kersten and Förstner, 1986).

Manipulations of the first series were all done under the inert atmosphere to serve as a control. The second series was treated by the elutriate test modified for air bubbling (Lee et al., 1976). This test was initially designed by the USEPA to detect any short-term release of chemical contaminants from polluted material during dredging manipulations and disposal. This test involves the mixing of one volume of the harbor sludge with four volumes of the dredging or disposal site water for 30-min shaking period. If the soluble chemical constituent in the water exceeds 1.5 times the ambient concentration in the original water, special conditions will govern the remedial mea-

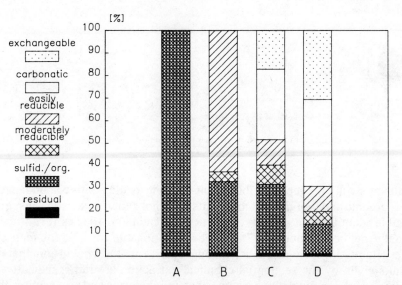

Figure 4. Partition of cadmium in anoxic mud from Hamburg Harbor in relation to the pretreatment procedures : (*A*) control extracted as received under oxygen-free conditions; (*B*) after treatment with elutriate test; (*C*) freeze dried; and (*D*) oven dried (60 °C).

sures to be undertaken (Engler, 1980). A study conducted on the factors influencing the results of the elutriate test has shown that this test as originally developed cannot yield a reliable estimate of the potential release, especially of cadmium, since it did not define the conditions of mixing to enable a well-defined, reproducible oxygen status existing during the test period (Lee et al., 1976). The result was that a modified elutriate test has been proposed in which compressed air agitation is utilized during the mixing period.

The third subsample series was freeze dried, and the fourth series was dried under air in a convection oven at 60° C. Subsequent to the preservation and pretreatment measures, respectively, the subsamples were extracted by the sequential technique in order to trace the changes of chemical forms of cadmium. The significant differences shown in Figure 4 can be ascribed to the contact of the sediment with air and by dehydration rather than to experimental artifacts such as inhomogeneity of the sediments or variations in the extraction protocol. Indeed, no differences were obtained for oxic suspended matter (Kersten and Förstner, 1985), and the sum of the metal concentrations in the individual fractions of all four series of each sample agreed within 10%. Extraction of the residual fraction was the only step not affected significantly by the sample pretreatment. This fact supports the observation that residual metal binding forms are not involved in controlling cadmium availability (Salomons and Förstner, 1984). In the original subsample, the sulfidic–organic fractions of cadmium dominate, with proportions up to 98%. The remainder of the cadmium was associated with residual sediment components or oxyhydrates. Similar data have been reported from San Francisco Bay sediments, which indicate that 90–94% of total particulate cadmium was complexed by insoluble organic matter or precipitated as sulfides (Serne and Mercer, 1975). The distribution of cadmium in the Ashtabula and Bridgeport Harbor sediments as reported by Brannon et al. (1976) also shows a dominating association with the sulfide–organic fraction.

Following the application of the elutriate test, the oxidizable sulfidic–organic portion of cadmium decreases drastically and is now found in the easily reducible fraction. This drastic change of cadmium forms in the short aeration experiment is likely caused by rapid oxidation of the sulfidic form. Coprecipitation and adsorption of cadmium with the precipitated oxyhydrates may have removed the liberated metal from solution. Freshly precipitated oxyhydrates are much more effective in scavenging high concentrations of trace metals because of greater reactive surface area than aged crystalline materials (Jenne, 1977; Lee, 1975). After freeze and oven drying of the initially anoxic samples, cadmium proportions were found even in the most mobile operationally defined carbonatic and exchangeable fractions. The formation of carbonatic association of cadmium is probably promoted by the decomposition of $MnCO_3$ (rhodochrosite), which is also indicated from the extraction results (Kersten and Förstner, 1986). The high concentration of cadmium present in these fractions may have a hazardous impact on water quality during dredging and disposal operations as well as upland disposal of these sediments (Gambrell et al., 1978; Khalid, 1980).

The published experimental evidence confirms that equilibrium concentrations of cadmium at the sediment–water interface and its bioavailability are regulated via adsorption–desorption and precipitation–dissolution mechanisms, according to the suggestions made by Jenne and Luoma (1977) and by Salomons (1985). The latter author has stressed indeed that it is important to know whether cadmium concentrations are controlled by precipitation–dissolution reactions or by solid-phase adsorption–desorption processes in a distinct environment. If precipitation–dissolution processes control the cadmium availability, the dissolved concentrations will not depend on the total metal content, and an increased input to the system will not affect its concentration in the interfacial water. On the other hand, when adsorption–desorption is the main process for binding metals to the sediments, an increased input will cause an increase in soluble-metal concentrations. This is especially true with changing physicochemical conditions, as will be discussed in the next paragraph.

Sediment samples collected on four campaigns as summarized in Table 4 represent different early diagenetic environments according to the classification of Berner (1981). The Hamburg Harbor sediment represent methanogenic muds of an eutrophic freshwater environment. The Piraeus Harbor samples consist of sulfide-rich muds of a marine euxinic environment. The NRA Heuckenlock samples were taken from the subsurface of a frequently water-logged tidal high-flat soil with a slightly reducing environment, where denitrification takes place. These samples represent a postoxic milieu. The

Table 4. Samples Selected for Comparison of Chemical Fractionation Patterns of Cadmium from Different Environments

Sample	Location	Comments and References
1–4	Hamburg Harbor basins at river Elbe, FRG	Muddy sediments rich in organic matter and strongly affected by industrial and municipal waste (Kersten and Förstner, 1986)
5–7	Piraeus Harbor area, Greece	Muddy sediments rich in iron minerals and free sulfide ions, highly polluted by Cd (Boboti et al., 1985)
8–11	Tidal floodplain "Heukenlock" of river Elbe near Hamburg	Subsamples from an undisturbed, well-characterized sediment core (Kersten and Kerner, 1985)
12–20	Elbe and Weser Estuary	Suspended matter and oxic, unconsolidated surface sediments (Förstner et al., 1981; Kersten and Förstner, 1985)

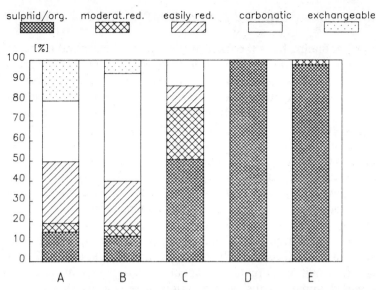

Figure 5. Fractionation results of cadmium from samples of different early diagenetic stages: (A) estuarine suspended matter; (B) oxic surface sediments from the Elbe estuary: (C) postoxic subsurface core sediments; (D) marine sulfidic; and (E) freshwater methanic mud samples.

suspended matter samples were found in oxygenated marine and fresh waters containing both detrital and secondary minerals representing the earliest stage of diagenesis. The fractionation patterns of cadmium as determined by the sequential extraction technique described above are presented in Figure 5 for these samples. The sum of the individual proportions extracted by the first five steps is normalized to 100%. This normalizing procedure provides a discernible comparison of the metal compounds characteristic to the respective environment. Interferences by various amounts of residual fractions due to fluctuations in residual metal content not involved in recent geochemical cycles are avoided by this means. In addition, from the varying cadmium content of 1.4–31.8 mg/kg (dry wt.), the sediment samples were chosen from similar mineral and size characteristics to avoid matrix effects on the results of sequential extraction.

The general trend in changing the metal fractionation pattern from oxic to anoxic environments is similar to that encountered in the course of sample aeration prior to extraction. The suspended freshwater sample shows the highest portions of cadmium recovered by the first three steps and suggests that in this environment as well as in the oxic one cadmium is bound chiefly to exchangeable sites and carbonatic phases. No characteristic influence is found on the percentage of oxidizable cadmium in both the marine and freshwater environments.

4.5. RELEASE OF CADMIUM AT SEDIMENT–WATER INTERFACE

An important finding from the fractionation studies described in the preceding section relates to the dominant role of the nonresidual sulfidic fraction of cadmium in anaerobic sediments. Due to these associations of low solubility, even strong accumulation of anthropogenic cadmium is rendered less bioavailable. This is particularly effective since bottom sediments, especially those contaminated by cadmium and organic wastes, always tend to become more reduced because of the continuous release of electrons to the environment through the respiration processes of microorganisms.

Chemical transformations in organic-rich estuarine sediments are controlled by microbially mediated oxidation of the organic matter, which involves a sequence of electron acceptors ranging from oxygen to nitrate, Mn(IV), Fe(III), sulfate, and bicarbonate. The sequence of these reactions follows the order of free energy that can be gained during bacterial metabolism (Stumm and Baccini, 1978; Fenchel and Blackburn, 1979). Under steady-state conditions, organisms capable of deriving the highest metabolic energy will dominate. Thus, a succession of organisms should exist with increasing depth in the sediments due to the decreasing supply of the appropriate electron acceptors. It has been consistently shown from examining, for example, deep-sea core sediments that individual metabolic processes are effective within vertically isolated zones. However, in estuarine areas, the upper oxic zone is found mostly to reach no more than a few millimeters in depth. The metabolic products of the organic matter oxidation processes (e.g., HCO_3^-, HPO_4^{2-}, volatile organic ligands) and that of coupled inorganic reduction processes [e.g., Fe(II), Mn(II), HS^-] accumulate in the sediment until their concentrations are limited by (biologically or physically mediated) dispersive and advective transport, by subsequent microbial utilization, or by the formation of authigenic minerals like sulfides (Suess, 1979; Berner, 1981a). Thus, each biogeochemical sediment zone reveals its own characteristic secondary mineral assemblage, which is critical both in buffering interstitial water components (Salomons, 1985) and in affecting transfer of chemical species across the sediment–water interface. On the basis of typical secondary iron and manganese mineral associations, a scheme for the recognition of the different early diagenetic zones has been proposed by Berner (1981b).

The excessive supply of electrons accounts for the fact that most natural sedimentary environments remain reduced once these processes are initiated. A reversal of these processes and hence the oxidation of the stable sulfidic cadmium forms needs an energy input such as provided by both natural and technical activities. Examples are given for freshwater, estuarine, and marine environments:

Hoeppel et al. (1978) were among the first to observe changes in cadmium forms during dredged soil disposal on land containment areas. They have

Table 5. Cadmium Forms in Solids from Confined Land Disposal[a]

Chemical Fraction	Percentage of Total Cadmium Content	
	Influent	Effluent
Exchangeable	21.0	18.0
Carbonatic	21.4	56.7
Easily Reducible	9.2	11.8
Remaining Phases	49.3	13.5

[a] From Hoeppel et al., 1978.

compared five influent and effluent samples of suspended matter in such ponding systems (Table 5). From their findings, it is obvious that cadmium concentrations increased significantly in the carbonate phase of the effluent samples, presumably as a direct result of the organic–sulfidic associations present in the influent slurries.

The natural estuarine system, on the other hand, is also characterized by a high-energy input, which dissipates on each tide exchange, varying markedly both with time and locality. This energy source contributes to the degree of sediment remobilization and turbulent mixing as well as to a variety of effects on a much lower energy level. The effects of frictional dissipation of the tidal energy on the sedimentary environment can be characterized in the order of their energy demand. These are (1) flushing of the whole turbidity zone and large bottom deposits by stormy weather events, (2) temporary and permanent resuspension of bottom deposits through currents introduced by tidal action, (3) exchange fluxes between sediments and overlying water through drainage and recycling of tidal water in the floodplains, (4) groundwater table fluctuations within the river wetlands introduced by the tidal range, and (5) advective mixing and replacement of interstitial water in bottom sediments exchanged by denser water. These energy inputs affect cadmium-polluted sediments at any scale by shifting them to a more oxidized environment, transforming the relatively unavailable reduced cadmium to readily bioavailable forms. Two examples from the Elbe River estuary are given here.

Within a 1-yr monitoring program (summer 1979 to summer 1980), water samples were taken by Wilken and Weiler (1984) at Geesthacht Dam, which divides the tidal from the nontidal part of the Elbe River. Their results on the dissolved and particulate cadmium contents showed a pronounced seasonal variation. The first runoffs early in the year 1980 correlated with high cadmium concentrations. The authors suggested that the increased cadmium transport with high water discharges is caused by remobilization of deposits in the river, which can be built up during low water discharges in the summer and autumn. They were able to trace three high water waves transporting contaminated fine-grained material in decreasing amounts and estimated the cadmium freight of these waves to be 7.5t, which was 37% of the total cadmium freight at the dam during the 1-yr period. The authors found additionally that 22% of the total cadmium was dissolved with no variation in the ratio of dissolved to particulate-bound metal in the observed period irrespective of the variations in total content. This result agrees with our observations on the mobility of cadmium in oxidized suspended sediments, which is controlled by adsorption–desorption processes. Thus remobilization of anoxic bottom deposits by strong water currents may cause a remobilization of cadmium by subsequent oxidation.

Another case of oxidative remobilization has been demonstrated within the Heukenlock, a tidal freshwater flat in the upper Elbe estuary near Hamburg (Kersten and Kerner, 1985). Within estuaries, intertidal mudflats provide the sink for pollutants introduced from upstreams. However, at these productive sites, seasonal effects and even diurnal water level fluctuations introduce drastic environmental influences, as has been stressed by Gambrell et al. (1977). This high flat is colonized by monodominant reed stands *(Phragmites communis)*. A strongly elevated cadmium content in the rhizomes of these emerged macrophytes indicates high proportions of bioavailable cadmium species in the rooting zone of this site. In order to study the chemical forms of cadmium, sediment cores have been taken from this site and were subsampled and analyzed applying our sequential extraction technique (Kersten and Kerner, 1985).

Figure 6 shows that the top 4 cm of the silty core sediment densely packed with macrophyte litter revealed a slightly reducing microhabitat, where probably intensive plant debris degradation involving deamination processes take place. Below 4 cm the Eh level increases to a plateau at about 300 mV prevailing down to 8–12 cm depth. Here a well-developed redox potential discontinuity indicates changing environment. A second, less well developed redox plateau at about -50 mV occurs from 16 to 20 $\pm$ 2 cm depth, finally followed by a strongly anoxic sediment zone. This zonation is obviously controlled by the more or less water-logged conditions in the sediment during the studied period, since the groundwater table was found to alternate between 8 and 22 $\pm$ 2 cm below the sediment surface immediately after and before inundation, respectively.

The slower initiation of the reducing conditions within the top 12 cm of the sediment column is likely due to the presence of nitrate in the pore water (Kersten and Kerner, 1985). The nutrient accumulates in the top horizon as a result of pore water drainage and aerobic nitrification. Nitrate has been shown to prevent a rapid lowering of redox potential (Bailey and Beauchamp, 1971). The decline of NH_4–N in the upper 10 cm reflects the aerobic nitrification during the relatively long aeration periods of the sediment surface (Fig. 6). The rapid transition to lower Eh values in the fre-

---

Figure 6. Geochemical characterization of core sediments from the Heukenlock intertidal flat contaminated by cadmium. (From Kersten and Kerner, 1985).

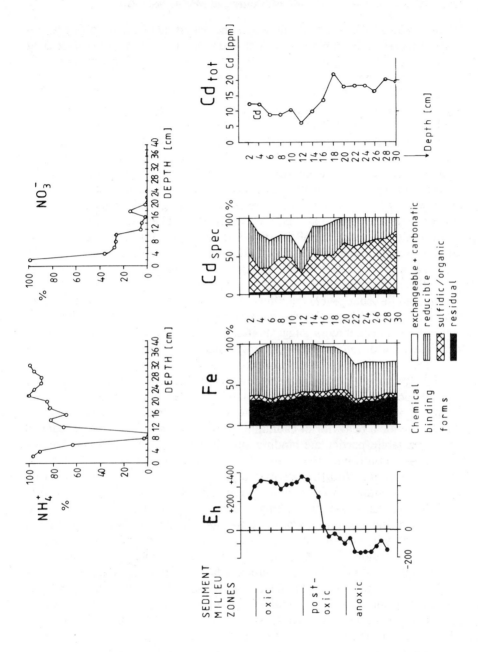

quently water-logged horizon indicates the occurrence of anaerobic respiratory processses, beginning with denitrification. The nitrate level decreases near to the detection limit, whereas the dissolved ammonium concentration is again raised below 12 cm. In the anoxic zone below 18 cm, only a slight depletion in the pore water sulfate concentration was noticed, although a hydrogen sulfide odor was noticed. This was probably due to a constantly replenished supply of SO_4^{2-} from the tidal water drainage.

The recorded changes in total cadmium concentrations with depth are shown in Figure 6. In the oxic top 10-cm sediment column, particulate cadmium contents decrease from 12 to 6 mg/kg (dry wt.) with depth. In the transition zone from 12 to 18 cm, the cadmium concentrations increase again downwardly to a maximum of more than 20 mg/kg at the beginning of the anoxic zone and remain at high levels thereafter. The results of the sequential cadmium extractions of the core sediments separated at 2-cm intervals are also summarized in Figure 6. Cadmium is associated by 60–80% to the sulfidic–organic fraction in the anoxic zone of the profile; in the oxic and transition zone, these associations become successively transformed to carbonatic and exchangeable fractions. Thus, high proportions of mobile cadmium forms correlate with the marked depletion in the total cadmium concentration. This behavior suggests that variation in the chemical form is an important factor in controlling the metal fixation in sediments. The unusual subsurface peak in cadmium concentration found below the oxic zone is obviously caused by the frequent downward flux of oxidized surface water by tidal action. This reverse pore water drainage from oxic to anoxic sediment horizons implies an oxidative pore-water-pumping effect on the biogeochemical behavior of cadmium. In the oxic zone, cadmium is leached from the labile particulate-binding sites because the predominant binding mechanism controlling the availability of cadmium in this zone is adsorption–desorption. With the downward pore water flux, the mobilized metal gets into the anoxic environment, where cadmium is precipitated again with sulfidic–organic associations. From the recorded concentrations, it is expected that long-term removal of up to 50% of cadmium from the sediment subsurface will take place at the anoxic sedimentary sink located a few centimeters below the sediment–water interface, which gives a flux maximum of 0.4 g/m^2 yr in the Heukenlock area. This effect of "oxidative pumping" on the biogeochemical behavior of pollutants in estuarine intertidal flats depends on the level of the tidal energy input, that is, on the physical morphometry of the site, on vertical, spatial, and seasonal hydrological fluctuations, and on the mineralogy of the sediments.

In the *marine environment,* several observations have been made suggesting similar, but significantly slower, oxidation processes on the release of cadmium from polluted sediments. Field evidence for changing cadmium mobilities was reported by Holmes et al. (1974) from Corpus Christi Bay Harbor. During the summer, when the harbor water is stagnant, cadmium is precipitated as CdS at the sediment–water interface. In the winter months, on the other hand, the intrusion of oxygen-rich water into the bay results in the release of some of the precipitated metal. Gendron et al. (1986) found evidence for different release mechanisms of cobalt and cadmium near the sediment–water interface in the St. Lawrence Estuary, in agreement with our results. The hydroxylamine-extractable fraction of the (freeze-dried)sediment surface layer in the Lawrentian trough was enriched in cobalt but depleted in cadmium relative to the subsurface sediment. While in this layer the highest dissolved cadmium concentrations were detected, the concentrations of dissolved cobalt were negligible. The profiles for cobalt resemble those for manganese and iron with increased levels downwards. Suggesting a release in the reduced zone and reprecipitation in the surface layer. On the other hand, cadmium appears to be released in the surface, probably as a result of the aerobic remobilization of organically bound cadmium. The authors found, in addition, that the upward flux of cadmium across the sediment–water interface depends on the flux of oxygen into the sediment and stressed that the cadmium released near the interface is recycled back into the water column. Pore water analyses in shallow-water sediments of Puget Sound estuary indicate higher concentrations of cadmium in the upper oxidized centimeters of sediment covering reduced, H_2S-containing deposits (Emerson et al., 1984). This metal remobilization has been explained by the removal of sulfide from the pore waters via biological-mediated ventilation of the upper sediment layer with oxic overlying water, allowing the enrichment of dissolved cadmium that might otherwise be very low in concentration due to insoluble sulfide formation. The authors suggested a greatly enhanced flux of the metal to the bottom waters by this mechanism. These processes have been shown to be more effective during spring and summer than during the winter months (Hines et al., 1984). From enclosure experiments in Narragansett Bay, it has been estimated that by mechanisms such as oxidation of organic and sulfidic material, the anthropogenic proportion of cadmium in marine sediments is released to the water within approximately 3 yr; for remobilization of copper and lead, approximately 40 and 400 yr, respectively, is needed, according to these extrapolations.

Once the mechanisms controlling availability of particulate cadmium have changed from precipitation–dissolution to adsorption–desorption, changing environmental parameters like pH (acid rain), salinity (in estuaries), or the discharge of complexing agents (NTA) affects remobilization of cadmium from sediments. This fact has been confirmed by a number of laboratory experiments as well as by theoretical considerations. Equilibrium models such as those presented by Sibley and Morgan (1975) can be used to calculate cadmium speciation in quite complicated seawater–freshwater mixtures if the system can be adequately defined. Their mathematical computations included the effect of hydrolysis, complexion, protonation of ligands, oxidation–reduction chains, adsorption, and solid precipitation reactions. The computations showed that solid $CdCO_3$ is an important species in fresh water, while in seawater cadmium is present as a dissolved chloro complex, with maximum concentrations occurring at a seawater–freshwater ratio of 1:1 in their model. Since the negative charge on river-borne suspended colloids is not reversed by adsorption of seawater cations (Hunter, 1983), it can be expected that at least the negatively charged complexes (e.g., $CdCl_3^-$) are preferentially kept in solution. Field investigations by Ahlf (1983) and Calmano et al. (1985) on longitudinal sections of the Elbe and Weser estuaries in northern Germany indicate characteristic mobilization of cadmium (different from other trace metals studied) at the salinity gradient. It should be mentioned that another typical area of cadmium release from particulate matter (derived by atmospheric transport) by chloro complexation is the sea surface microlayer (Lion and Leckie, 1981).

There are still conflicting results with regard to the mechanisms and kinetics of cadmium release from solids to the interfacial water. Both Salomons (1980) and recently Prause et al. (1985) observed in their batch experiments with dredged sediments suspended in estuarine water that a considerable release of cadmium is found only in long-term experiments. No enrichment of cadmium in solution could be observed during resuspension experiments lasting between 12 and 24 h. Even in seawater in most cases, a decrease of the metal concentration was found, probably resulting from adsorption to freshly precipitated Fe/Mn–oxyhydrates. Both authors have extended their experiment to several weeks and have found maximum remobilization rates in weeks 2–4, the maximum release arising after more than 1 month. Prause et al. (1985) stressed that simple cation exchange mechanisms obviously have no importance for the remobilization of cadmium because a new equilibrium at the surface of sorbents will normally be established in less than this time. In fact, our chemical fractionation results have shown that the oxidation of sulfidic cadmium associations and subsequent adsorption to oxyhydroxides lasts at least less than 1 h (see Section 4.4). In order to test the influence of bacteria on the rate of cadmium remobilization, the authors added antibiotics to the system on the day 21. This resulted in an interruption of cadmium release for a period similar to the initial lag phase, as shown in Figure 7. From these results, Prause et al. (1985) suggested that bacterial activity can stimulate cadmium remobilization by a complex web of substrate decomposition and potential organic ligand liberation or formation. In addition, the results of both authors emphasize strongly the fact that the time spans of these release processes are much longer than the typical short-term fluctuations in estuarine properties like tidal advection and associated processes (e.g., sediment mobility). This also shows that hydrodynamic conditions determine the desorption processes. A high particle residence time likely causes a stronger remobilization of the metal from the suspended matter and points to the fact that observations on the quantitative behavior of cadmium as determined for one estuary cannot be extrapolated to other

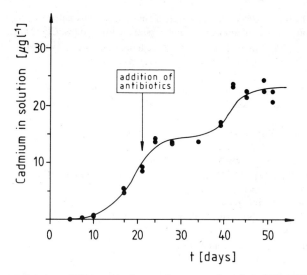

Figure 7. Remobilization of Cd in a batch experiment as a function of time. Sludge from the Europahafen Bremen; bacterial activity suppressed by addition of antibiotics (streptomycine and chloramphenicole, 0.2 g/L each) at day 21 of the experiment. Salinity of the suspension (15 g solids/L): 32%; pH 7.9; $T = 7\,°C$; 9 mg/L = O_2. (From Prause et al., 1985.)

estuaries since chemical and hydrodynamic processes may differ to a large extent.

REFERENCES

Ahlf, W. (1983). The River Elbe: Behaviour of Cd and Zn during estuarine mixing. *Environ. Technol. Lett.*, **4**, 405–410.

Baham, J. (1984) Prediction of ion activities in soil solutions: Computer equilibrium modeling. *Soil Sci Soc. Am. J.*, **48**, 525–531.

Bailey, L. D., and Beauchamp, E. G. (1971). Nitrate reduction and redox potentials measured with permanently and temporal placed platinum electrodes in saturated soils. *Can. J. Soil Sci.*, **51**, 51–58.

Bascomb, C. L. (1964). Rapid method for the determination of cation exchange capacity of calcareous and non-calcareous soils. *J. Sci. Fd. Agric.*, **15**, 821–823.

Berner, R. A. (1981a). A new geochemical classification of sedimentary environment. *J. Sediment. Petrol.*, **51**, 359–365.

Berner, R. A. (1981b). Authigenic mineral formation resulting from organic matter decomposition in modern sediments. *Fortschr. Miner.*, **59**, 117–133.

Boboti, A., Stoffers, P., and Müller, G. (1985). Heavy metal pollution in the harbour area of Piraeus, Greece. In *International Conference on Heavy Metals in the Environment, Athens, September 1985*, Vol. 2, Lekkas, T. D. (ed.). CEP Consultants, Edinburgh, pp. 407–410.

Borggaard, O. K. (1976). The use of EDTA in soil analysis. *Acta Agric. Scand.*, **26**, 144–150.

Borggaard, O. K. (1982). Selective extraction of amorphous iron oxides by EDTA from selected silicates and mixtures of amorphous and crystalline iron oxides. *Clay Miner.*, **17**, 365–368.

Bottomley, E. Z., and Bayly, I. L. (1984). A sediment porewater sampler used in root zone studies of the submerged macrophyte, myriophyllum spicatum. *Limnol. Oceanogr.*, **29**, 671–673.

Bower, C. A. (1950). Fixation of ammonium in difficulty exchangeable form under moist conditions by some soils of semiarid regions. *Soil Sci.*, **70**, 375–383.

Bowser, C. J., Mills, B. A., and Callender, E. J. (1979). Extractive chemistry of equatorial Pacific pelagic sediment and relationship to nodule forming processes. In *Marine Geology and Oceanography of the Pacific Manganese Nodule Province*, Bishoff, J. L., and Piper, D. Z. (eds.). Plenum, New York, pp. 587–619.

Brannon, J. M., Engler, R. M., Rose, J. R., Hunt, P. G., and Smith, I. (1976). Distribution of toxic heavy metals in marine and freshwater sediments. In *Proceedings of the Specialty Conference on Dredging and its Environmental Effects*, Krenkel, P. A., Harrison, J., and Burdick III, J. C. (eds.). American Society of Civil Engineers, New York, pp. 455–495.

Burton, J. D. (1978). The models of association of trace metals with certain compounds in the sedimentary cycle. In *Biogeochemistry of Estuarine Sediments*, Goldberg, E. D. (ed.). UNESCO, Paris, pp. 33–41.

Calmano, W., and Förstner, U. (1983). Chemical extraction of heavy metals in polluted river sediments in central Europe. *Sci. Tot. Environ.*, **28**, 77–90.

Calmano, W., Wellershaus, S., and Liebsch, H. (1985). The Weser Estuary: A study on heavy metal behaviour under hydrographic and water quality conditions. *Veröffent. Inst. Meeres-Forsch. Bremerh.*, **20**, 151–182.

Carignan, R. (1984). Interstitial water sampling by dialysis: Methodological notes. *Limnol. Oceanogr.*, **29**, 667–670.

Carignan, R., and Nriagu, J. O. (1985). Trace metal deposition and mobility in the sediments of two lakes near Sudbury, Ontario. *Geochim. Cosmochim. Acta*, **49**, 1753–1764.

Carignan, R., Rapin, F., and Tessier, A. (1985). Sediment porewater sampling for metal analysis: A comparison of techniques. *Geochim. Cosmochim. Acta*, **49**, 2493–2497.

Chao, T. T. (1972). Selective dissolution of manganese oxides from soils and sediments with acidified hydroxylamine hydrochloride. *Soil Sci. Soc. Am. Proc.*, **36**, 764–768.

Chao, T. T. (1984). Use of partial dissolution technqiues in geochemical exploration. *J. Geochem. Explor.*, **20**, 101–135.

Chao, T. T., and Liyi Zhou. (1983). Extraction techniques for selective dissolution of amorphous iron oxides from soils and sediments. *Soil Sci. Soc. Am. J.*, **47**, 225–232.

Chao, T. T., and Sanzalone, R. F. (1977). Chemical dissolution of sulfide minerals. *J. Res. U. S. Geol. Surv.*, **5**, 409–412.

Chester, R., and Hughes, M. J. (1967). A chemical technique for the separation of ferromanganese minerals, carbonate minerals and adsorbed trace elements from pelagic sediments. *Chem. Geol.*, **2**, 249–262.

Cooper, B. S., and Harris, R. C. (1974). Heavy metals in organic phases of river and estuarine sediments. *Mar. Pollut. Bull.*, **5**, 24–26.

DeMora, S. J., and Harrison, R. M. (1983). The use of physical separation techniques in trace metal speciation studies. *Wat. Res.*, **27**, 723–733.

Deurer, R., Förstner, U., and Schmoll, G. (1978). Selective chemical extraction of carbonate associated trace metal in recent lacustrine sediments. *Geochim. Cosmochim. Acta*, **42**, 425–427.

Duinker, J. C., Nolting, R. F., and Van Der Sloot, H. A. (1979). The determination of suspended metals in coastal waters by different sampling and processing methods (filtration, centrifugation). *Neth. J. Sea Res.*, **13**, 282–297.

Eaton, A. (1979). Leachable trace elements in San Francisco Bay sediments: Indicators of sources and estuarine processes. *Environ. Geol.*, **2**, 333–339.

Emerson, S., Jahnke, R., and Heggie, D. (1984). Sediment–water exchange in shallow water estuarine sediments. *J. Mar. Res.*, **42**, 709–730.

Engler, R. M. (1980). Prediction of pollution potential through geochemical and biological procedures: Development of regulatory guidelines and criteria for the discharge of dredged fill material. In *Contaminants and Sediments*, Vol. 1, Baker, R. A. (ed.). Ann Arbor Science, Ann Arbor, MI, pp. 143–169.

Engler, R. M., Brannon, J. M., Rose, J., and Bigham, G. (1977). A practical selective extraction procedure for sediment characterization. In *Chemistry of Marine Sediments*, Yen, T. F. (ed.). Ann Arbor Science, Ann Arbor, MI, pp. 163–171.

Etcheber, H., and Jouanneau, J. M. (1980). Comparison of the different methods for the recovery of suspended matter from estuarine waters: Deposition, filtration and centrifugation; consequences for the determination of some heavy metals. *Est. Coast. Mar. Sci.*, **11**, 701–707.

Farmer, V. C., and Mitchell B. (1963). Occurrence of oxalates in soil clays following hydrogen peroxide treatment. *Soil Sci.*, **96**, 221–229.

Farmer, V. C., Russel, J. D., McHardy, W. J., Newman, A. C. D., Ahlrichs, J. L., and Rimsaite, J. Y. H. (1971). Evidence for loss of protons and octahedral iron from oxidized biotites and vermiculites. *Mineral. Mag.*, **38**, 121–137.

Farrah, H., and Pickering, W. F. (1978). Extraction of heavy metal ions sorbed on clays. *Wat. Air Soil Pollut.*, **9**, 491–498.

Fenchel, T., and Blackburn, T. H. (1979). *Bacteria and Mineral Cycling*. Academic, New York.

Filipek, L. H., and Owen, R. M. (1978). Analysis of heavy metal distributions among different mineralogical states in sediments. *Can. J. Spectro.*, **23**, 31–34.

Florence, T. M., and Batley, G. E. (1980). Chemical speciation in natural waters. *Crit. Rev. Anal. Chem.*, **9**, 219–296.

Förstner, U. (1982). Chemical forms of metal enrichment in recent sediments. In *Ore Genesis: The state of The Art*, Amstutz, G. C. (ed.). Springer, Berlin, pp. 191–199.

Förstner, U. (1985). Chemical forms and reactivities of metals in sediments. In *Chemical Methods for Assessing Bio-available Metals in Sludges and Soils*, Leschber, R., Davis, R. D., and l'Hermite, P. L. (eds.). Elsevier Applied Science, London, pp. 1–31.

Förstner, U. (1986). Chemical speciation and environmental effects of critical elements in solid waste materials—combustion residues. *TheImportance of Chemical Speciation in Environmental Processes*, Bernhard, M., Brinkman, S. E., and Sadler, P. J. (eds.). Springer, Berlin, Heidelberg, New York, pp. 465–492.

Förstner, U., and Salomons, W. (1980). Trace metal analysis on polluted sediments. I. Assessment of sources and intensities. Environ. Technol. Lett., **1**, 494–505.

Förstner, U., and Calmano, W. (1982). Bindungsformen von Schwermetallen in Baggerschlämmen. *Vom Wasser*, **59**, 83–92.

Förstner, U., Nähle, C., and Schöttler, U. (1979). Sorption of metals in sand filters in the presence of humic acids. *DVWK Bull.* **13**, 95–125.

Förstner, U., Calmano, W., Conradt, K., Jaksch, H., Schimkus, C., and Schoer, J. (1981). Chemical speciation of heavy metals in solid waste materials (sewage sludge, mining wastes, dredged materials, polluted sediments) by sequential extraction. In *International Conference on Heavy Metals in the Environment*, Ernst, W. H. O. (ed.). Amsterdam, September 1981, CEP Consultants, Edinburgh, pp. 698–704.

Fraser, J. L., and Lum, K. R. (1982). Availability of elements of environmental importance in incinerated sludge ash. *Environ. Sci. Technol.*, **17**, 52–54.

Frink, C. R. (1964). The effects of wash solvents on cation-exchange capacity measurements. *Soil Sci. Soc. Am. Proc.*, **28**, 506–511.

Gambrell, R. P., Collard, V. R., Reddy, C. N., and Patrick, W. H. Jr. (1977). Trace and toxic metal uptake by marsh plants as affected by Eh, pH, and salinity. U.S. Army Engineer Waterways Experiment Station, Technical Report D-77-40, CE, Vicksburg, MS.

Gambrell, R. P., Khalid, R. A., and Patrick, W. H. Jr. (1978). Disposal alternatives for contaminated dredged material as a management tool to minimize environmental effects. U.S. Army Engineers Waterways Experiment Station, Technical Report DS-78-8, CE, Vicksburg, MS.

Gendron, A., Silverberg, N., Sundby, B., and Lebel, J. (1986). Early diagenesis of cadmium and cobalt in sediments of the Laurentian Trough. *Geochim. Cosmochim. Acta*, **50**, 741–747.

Gibbs, R. J. (1973). Mechanisms of trace metal transport in rivers. *Science*, **180**, 71–73.

Green, D. R. (1984). *The Development of an In-Situ Water Sampling System*. Seastar Instruments, Sidney, B.C., Canada. Canada V8L3S7.

Gupta, S. K., and Chen, Y. K. (1975). Partitioning of trace metals in selective fractions on nearshore sediments. *Environ. Lett.*, **10**, 129–158.

Guy, R. D., Chakrabarti, C. L., and McBain, D. C. (1978). An evaluation of extraction techniques for the fractionation of copper and lead in model sediment systems. *Wat. Res.*, **12**, 21–24.

Ham, R. K., Anderson, M. A., Stanforth, R., and Stegmann, R. (1979). Background study on the development of a standard leaching test. *U.S. EPA-600/2-79–109*.

Harrison, R. M., Laxen, D. P. H., and Wilson, S. J. (1981). Chemical associations of lead, cadmium, copper, and zinc in street dusts and roadside soils. *Environ. Sci. Technol.*, **15**, 1378–1383.

Heath, G. R., and Dymond, J. (1977). Genesis and transformations of metalliferous sediments from the East Pacific Rise, Bauer Deep and Central Basin, Nazca Plat. *Geol. Soc. Am. Bull.*, **88**, 723–733.

Hem, J. D. (1972). Chemistry and occurrence of cadmium and zinc in surface water and groundwater. *Wat. Resour. Res.*, **8**, 661–679.

Hesslein, R. (1976). An in-situ sampler for close interval pore water studies. *Limnol. Oceanogr.*, **21**, 912–914.

Hines, M. E., Berry Lyons, W. M., Armstrong, P. B., Orem, W. H., Spencer, M. J., and Gaudette, H. E. (1984). Seasonal metal remobilization in the sediments of Great Bay, New Hampshire. *Mar. Chem.*, **15**, 173–187.

Hoeppel, R. E., Meyers, T. E., and Engler, R. M.(1978). Physical and Chemical Characterization of Dredged Material Influents and Effluents in Confined Land Disposal Areas. U.S. Army Engineer Waterways Experiment Station, Technical Report D-78-24, CE, Vicksburg, MS.

Hoffmann, M. R., Yost, E. C., Eisenreich, S. J., and Maier, W. J. (1981). Characterization of soluble and colloidal-phase metal complexes in river water by ultrafiltration. A mass-balance approach. *Environ. Sci. Technol.*, **15**, 655–661.

Holmes, C. W., Slade, E. A., and McLerran, C. J. (1974). Migration and redistribution of zinc and cadmium in marine estuarine systems. *Environ. Sci. Technol.*, **8**, 255–259.

Holmgren, G. S. (1967). A rapid citrate-dithionite extractable iron procedure. *Soil Sci. Soc. Am. Proc.*, **31**, 210–211.

Horowitz, A. J. (1984). A Primer on Trace Metal-Sediment Chemistry. U.S. Geological Survey, Open-File Report 84-709, Doraville, GA.

Hunt, C. D., and Smith, D. L. (1983). Remobilization of metals from polluted marine sediments. *Can. J. Fish. Aquat. Sci.*, **40**, 132–142.

Hunter, K. A. (1983). On the estuarine mixing of dissolved substances in relation to colloid stability and surface properties. *Geochim. Cosmochim. Acta*, **47**, 467–473.

Jackson, M. L. (1958). *Soil Chemical Analysis.* Prentice-Hall, Englewood Cliffs, NJ, pp. 183–204.

Jeanroy, E., and Guillet, B. (1981). The occurrence of suspended ferruginous particles in pyrophosphate extracts of some soil horizons. *Geoderma,* **26,** 95–105.

Jenne, E. A. (1977). Trace element sorption by sediments and soils—sites and processes. In *Symposium on Molybdenum,* Chappell, W., and Petersen, K. (eds.). Marcel Dekker, New York, pp. 425–553.

Jenne, E. A., and Luoma, S. N. (1977). Forms of trace elements in soils, sediments, and associated waters: An overview of their determination and biological availability. In *Biological Implications of Metals in the Environment,* Wildung, R., and Drucker, H. (eds.). ERDA Symposium Series 42, Energy Research and Development Administration, Oak Ridge, TN, pp. 110–143.

Jenne, E. A., Kennedy, V. C., Burchard, J. M., and Ball, J. W. (1980). Sediment collection and processing for selective extraction and for total trace element analysis. In *Sediments and Contaminants,* Vol. 2, Baker, R. A. (ed.). Ann Arbor Science, Ann Arbor, MI, pp. 169–191.

Jones, M. J., Daly, H., and Hart, B. T. (1985). Trace metal complexing capacity in natural waters: Measurement and use in estimating toxic concentrations. In *International Conference on Heavy Metals in the Environment, Athens 1985,* Vol. 2. Lekkas, T. D. (ed.). CEP Consultants, Edinburgh, pp. 481–483.

Kersten, M., and Förstner, U. (1985). Trace metal partitioning in suspended matter with special reference to pollution in the southeastern North Sea. In *Transport of Carbon and Minerals in Major World Rivers, Part 3,* Degens, E. T., Kempe, S., and Herrera, R. (eds.). Mitt. Geol.-Paläontol. Inst. Univ. Hamburg, SCOPE/UNEP Sonderband 58, pp. 631–645.

Kersten, M., and Förstner, U. (1986). Chemical fractionation of heavy metals in anoxic estuarine and coastal sediments. *Wat. Sci. Tech.,* **18,** 121–130.

Kersten, M., and Kerner, M. (1985). Transformations of heavy metals and plant nutrients in a tidal freshwater flat sediment of the Elbe Estuary as affected by Eh and tidal cycle. In *International Conference on Heavy Metals in the Environment. Athens 1985,* Vol. 1, Lekkas, T. D. (ed.). CEP Consultants, Edinburgh, pp. 533–535.

Khalid, R. A. (1980). Chemical mobility of cadmium in sediment–water systems. In *Cadmium in the Environment,* Part I, *Ecological Cycling,* Nriagu, J. O. (ed.). Wiley, New York, pp. 257–304.

Khalid, R. A., Gambrell, R. P., and Patrick, W. H., Jr. (1981). Chemical availability of cadmium in Mississippi River sediment. *J. Environ. Qual.,* **10,** 523–528.

Kinniburgh, D. G., and Miles, D. L. (1983). Extraction and chemical analysis of interstitial water from soils and rocks. *Environ. Sci. Technol.,* **17,** 362–368.

Kitano, Y., and Fujiyoshi, R. (1980). Partitioning of cadmium, copper, manganese and iron into mineral and organic fractions in core sediments from the Osaka Bay. *Geochem. J.,* **14,** 289–301.

Lake, D. L., Kirk, P. W. W., and Lester, J. N. (1984). Fractionation, characterization, and speciation of heavy metals in sewage sludge and sludge-amended soils: A review. *J. Environ. Qual.,* **13,** 175–183.

Lee, F. Y., and Kittrick, J. A. (1984). Elements associated with the cadmium phase in a harbor sediment as determined with the electron beam microprobe. *J. Environ. Qual.,* **13,** 337–340.

Lee, G. F. (1975). Role of hydrous metal oxides in the transport of heavy metals in the environment. In *Heavy Metals in the Aquatic Environment,* Krenkel, P. A. (ed.). Pergamon, Oxford, pp. 137–147.

Lee, G. F., Lopez, J. M., and Piwoni, M. D. (1976). Evaluation of the factors influencing the results of the elutriate test for dredging material disposal criteria. In *Proceedings of the Specialty Conf. on Dredging and its Environmental Effects,* Krenkel, P. A., Harrison, J., and Burdick III, J. C. (eds.). American Society of Civil Engineers, New York, pp. 253–288.

Linton, R. W., Natusch, D. F. S., Solomon, R. L., and Evans, C. A., Jr. (1980). Physicochemical characterization of lead in urban dusts. A microanalytical approach to lead tracing. *Environ. Sci. Technol.,* **14,** 159–164.

Lion, L. W., and Leckie, J. O. (1981). Chemical speciation of trace metals at the air–sea interface: The application of an equilibrium model. *Environ. Geol.,* **3,** 293–314

Loring, D. H. (1976). The distribution and partition of zinc, copper, and lead in the sediments of Saguenay Fjord. *Can. J. Earth Sci.,* **13,** 960–971.

Luoma, S. N., and Bryan, G. W. (1981). A statistical assessment of the form of trace metals in oxidized estuarine sediments employing chemical extractants. *Sci. Tot. Environ.,* **17,** 165–196.

Lyle, M., Heath, G. R., and Robbins, J. M. (1984). Transport and release of transition elements during early diagenesis: Sequential leaching of sediments from MANOP sites M and H. Part I. pH 5 acetic acid leach. *Geochim. Cosmochim. Acta,* **48,** 1705–1715.

Lyons, W. B., Gaudette, H. E., and Smith, G. M. (1979). Pore water sampling in anoxic carbonate sediments: Oxidation artefacts. *Nature,* **277,** 48–49.

Malo, B. A. (1977). Partial extraction of metals from aquatic sediments. *Environ. Sci. Technol.,* **11,** 277–282.

Mattigot, S. V., and Ervin, J. O. (1983). Scheme for density separation and identification of compound forms in size-fractionated fly ash. *Fuel,* **62,** 927–931.

Mayer, L. M. (1976). Chemical water sampling in lakes and sediments with dialysis bags. *Limnol. Oceanogr.,* **21,** 909–911.

McGrath, S. P., Sanders, J. R., Tanrock, N. P., and Laurie, S. H. (1984). A comparison of experimental methods and computer programs for determining metal ion concentrations. In *Proceedings of the International Conference on Environmental Contamination, London 1984.* CEP Consultants, Edinburgh, pp. 707–712.

McKeague, J. A. (1967). An evaluation of 0.1 *M* pyrophosphate and pyrophosphatedithionite in comparison with oxalate as extractants of the accumulation products in podsols and some other soils. *Can. J. Soil Sci.,* **47,** 95–99.

Meiggs, T. O. (1980). The use of sediment analysis in forensic investigations and procedural requirements for such studies. In *Contaminants and Sediments,* Baker, R. A. (ed.). Ann Arbor Science, Vol. 1, Ann Arbor, MI, pp. 297–308.

Merino, E. (1979). Internal consistency of a water analysis and uncertainty of the calculated distribution of aqueous species at 25°C. *Geochim. Cosmochim. Acta,* **43,** 1533–1542.

Nirel, P., Thomas, A. J., and Martin, J. M. (1985). A critical evaluation of sequential extraction techniques. Paper submitted to the Speciation Seminar, Oxford, 16–19 April 1985.

Ongley, E., and Blanchford, D. (1982). Application of continuous flow centrifugation to contaminant analysis of suspended sediment in fluvial systems. *Envion. Technol. Lett.,* **3,** 219–229.

Patchineelam, R. S. (1975). Untersuchungen über die Hauptbindungsarten und die Mobilisierbarkeit von Schwermetallen in fluviatilen Sedimenten. Ph.D. Dissertation, University of Heidelberg.

Pawluk, S. (1972). Measurement of crystalline and amorphous iron removal in soils. *Can. J. Soil Sci.,* **52,** 119–123.

Pfeiffer, G., Förstner, U., and Stoffers, P. (1982). Speciation of reducible metal compounds in pelagic sediments by chemical extraction. *Senckenbergiana Marit.,* **14,** 23–38.

Pickering, W. F. (1981). Selective chemical extractions of soil components and bound metal species. *CRC Crit. Rev. Anal. Chem.,* **12,** 233–266.

Pilkington, E. S., and Warren, L. J. (1979). Determination of heavy-metal distribution in marine sediments. *Environ. Sci. Technol.*, **13**, 295–299.

Polemio, M., and Rhoades, J. D. (1977). Determining cation exchange capacity: A new procedure for calcareous and gypsiferous soils. *Soil Sci. Soc. Am. J.*, **41**, 524–528.

Prause, B., Rehm, E., and Schulz-Baldes, M. (1985). The remobilization of Pb and Cd from contaminated dredge spoil after dumping in the marine environment. *Environ. Technol. Lett.*, **6**, 261–266.

Rapin, F., and Förstner, U. (1983). On the selectivity of various extractants used in sequential leaching techniques for the particulate metal speciation. In *International Conference on Heavy Metals in the Environment, Heidelberg, 1983*, Müller, G. (ed.). CEP Consultants, Edinburgh, pp. 1074–1077.

Rendell, P. S., Batley, G. E., and Cameron, A. J. (1980). Adsorption as a control of metal concentrations in sediment extracts. *Environ. Sci. Technol.*, **14**, 314–318.

Reuther, R., Wright, R. F., and Förstner, U. (1981). Distribution and chemical forms of heavy metals in sediment cores from two Norwegian lakes affected by acid precipitation. In *International Conference on Heavy Metals in the Environment, Amsterdam 1981*, Ernst, W. H. O. (ed.). CEP Consultants, Edinburgh, pp. 318–321.

Rhoades, J. D., and Krueger, D. B. (1968). Extraction of cations from silicate minerals during the determination of exchangeable cations in soils. *Soil Sci. Soc. Am. Proc.*, **32**, 488–492.

Robbins, J. M., Lyle, M., and Heath, G. R. (1984). *A sequential extraction procedure for partitioning elements among co-existing phases in marine sediments*. College of Oceanography, Oregon State University, Reference 84-3.

Robinson, W. O. (1927). The determination of organic matter in soils by means of hydrogen peroxide. *J. Agric. Res.*, **34**, 339–356.

Rohatgi, N., and Chen, K. Y. (1975). Transport of trace metals by suspended particulates on mixing with seawater. *J. Wat. Pollut. Contr. Fed.*, **47**, 2298–2316.

Salomons, W. (1980). Adsorption processes and hydrodynamic conditions in estuaries. *Environ. Technol. Lett.*, **1**, 356–365.

Salomons, W. (1985). Sediments and water quality. Environ. Technol. Lett., **6**, 315–326.

Salomons, W., and Förstner, U. (1980). Trace metal analysis on polluted sediments. II. Evaluation of environmental impact. *Environ. Technol. Lett.*, **1**, 506–517.

Salomons, W., and Förstner, U. (1984). *Metals in the Hydrocycle*. Springer, Berlin.

Salomons, W., and Mook, W. G. (1980). Biogeochemical processes affecting metal concentrations in lake sediments (Ijsselmeer, The Netherlands). *Sci. Tot. Environ.*, **16**, 217–229.

Schoer, J. (1985). Iron-oxo-hydroxides and their significance to the behaviour of heavy metals in estuaries. *Environ. Technol. Lett.*, **6**, 189–202.

Schwertmann, U. (1973). Use of oxalate for Fe extraction from soils. *Can. J. Soil. Sci.*, **53**, 244–246.

Serne, R. J., and Mercer, B. W. (1975). *Dredge Disposal Study, San Francisco Bay and Estuary. Appendix F: Crystalline Matrix Study*. U. S. Army Engineer District, San Francisco, CA.

Sibley, T. H., and Morgan, J. J. (1975). Equilibrium speciation of trace metals in freshwater/seawater mixtures. In *International Conference on Heavy Metals in the Environment, Toronto, Canada 1975*, Hutchinson, T. C. (ed.). CEP Consultants, Edinburgh, pp. 319–340.

Span, D., and Gaillard, J-F. (1986). An investigation of a procedure for determining carbonate-bound trace metals. *Chem. Geol.*, **56**, 135–141.

Stumm, W., and Baccini, P. (1978). In *Lakes: Chemistry, Geology, Physics*, Lerman, A. (ed.). Springer, New York, pp. 91–126.

Suess, E. (1979). Mineral phases formed in anoxic sediments by microbial decomposition of organic matter. *Geochim. Cosmochim. Acta,* **43,** 339–352.

Taylor, R. M., and McKenzie, R. M. (1966). The association of trace elements with manganese minerals in Australian soils. *Aust. J. Soil Res.,* **4,** 29–39.

Tessier, A., Campbell, P. G. C., and Bisson, M. (1979). Sequential extraction procedure for the speciation of particulate trace metals. *Anal. Chem.,* **51,** 844–851.

Tessier, A., Campbell, P. G. C., and Bisson, M. (1980). Trace metal speciation in the Yamaska and St. Francois River (Quebec). *Can. J. Earth Sci.,* **17,** 90–105.

Thomson, E. A., Luoma, S. N., Cain, D. J., and Johnasson, C. (1980). The effect of sample storage on the extraction of Cu, Zn, Fe, Mn, and organic material from oxidized estuarine sediments. *Wat. Air Soil Pollut.,* **74,** 215–233.

Van Der Sloot, H. A., and Duinker, J. C. (1981). Isolation of different suspended matter fractions and their trace metal contents. *Environ. Technol. Lett.,* **2,** 511–520.

Van Langeveld, A. D., Van Der Gaast, S. J., and Eisma, D. (1978). A comparison of the effectiveness of eight methods for the removal of organic matter from clay. *Clays Clay Min.,* **26,** 361–364.

Volkov, I. I., and Fomina, L. S. (1974). Influence of organic matter and processes of sulfide formation on distribution of some trace elements in deepwater sediments of Black Sea. In *The Black Sea: Geology, Chemistry and Biology,* Degens, E. T., and Ross, R. A. (eds.). American Association Petroleum. Geology Memoir 20, pp. 456–476.

Watson, P. G., Frickers, P. E., and Goodchild, C. M. (1985). Spatial and seasonal variations in the chemistry of sediment interstitial waters in the Tamar Estuary. *Estuar. Coast. Shelf Sci.,* **21,** 105–119.

Wilken, R. D., and Weiler, K. (1984). Evaluation of heavy metal contents in non tidal influenced Elbe River water during 1979/80. In *International Conference on Environmental Contamination, London 1984.* CEP Consultants, Edinburgh pp. 380–384.

Winget, C. L. (1982). A self-powered pumping system for in-situ extraction of particulate and dissolved materials from large volumes of seawater. Woods Hole Oceanographic Institution, Technical Report WHOI-82-8, Woods Hole, Ma. 02573, U.S.A.

Yoshida, T. (1984). On the behavior of nutrients in sediment pore water and on sediment surface. In *International Conference on Environmental Contamination, London 1984.* CEP Consultants, Edinburgh pp. 172–175.

5

BIOLOGICAL CYCLING OF CADMIUM IN FRESH WATER*

I. R. McCracken

Atlantic Research Laboratory
National Research Council of Canada
Halifax, Nova Scotia, Canada

5.1. **Bioavailability of Cadmium**
5.2. **Uptake, Distribution, and Elimination of Cadmium by Aquatic Macrophytes**
5.3. **Uptake, Distribution, and Elimination of Cadmium by Bacteria**
5.4. **Uptake, Distribution, and Elimination of Cadmium by Phytoplankton**
5.5. **Uptake, Distribution, and Elimination of Cadmium by Zooplankton**
5.6. **Uptake, Distribution, and Elimination of Cadmium by Benthic Invertebrates**
    5.6.1. Aquatic Insect Larvae
    5.6.2. Mollusks
    5.6.3. Crayfish
5.7. **Uptake, Distribution, and Elimination of Cadmium by Fish**
    5.7.1. Uptake and Distribution
    5.7.2. Elimination
5.8. **Effects of Environmental Factors on Cadmium Uptake**
References

*NRCC 26723

5.1. BIOAVAILABILITY OF CADMIUM

In fresh water, the total cadmium concentration consists of cadmium in ionic, colloidal, complexed, and particulate forms. However, the bioavailability of cadmium is not related to the total concentration but rather to its individual physicochemical forms (Florence, 1982). By convention, the particulate fraction is separated from the dissolved fraction by filtration through a 0.45-μm membrane (Florence, 1982). Using the highly sensitive technique of anodic stripping voltammetry, it has been determined that the dissolved portion consists of labile forms, including free metal ions as well as cadmium dissociated at the electrode surface from weak complexes and colloids, and nonlabile forms made up of metal bound in inert complexes. The bound metal is considered to be unavailable, and therefore, the labile forms approximate the bioavailable fraction (Florence and Batley, 1977, 1980).

Dissolved cadmium is known to exist in fresh waters primarily in the bioavailable free ionic form (Florence, 1977; Gardiner, 1974; Mantoura et al., 1978; McIntosh and Bishop, 1976; Poldoski, 1979; Poldoski and Glass, 1975; Shephard et al., 1980), although its speciation can be influenced by factors such as water hardness, pH, redox potential, suspended particulates, and organic matter (Rai et al., 1981; Hart, 1981).

Aquatic biota may also be exposed not only to the ionic form but also to cadmium in suspended particulate matter and in the sediments. Since the form of the metal in these phases influences its rate of release into the water as well as its availability from ingested sediment, chemical extraction procedures have been developed to identify the trace-metal sinks in suspended particulates and sediments and to estimate their bioavailability (Jenne and Luoma, 1977). Generally, the extraction procedures distinguish between three types of available cadmium: (1) readily available, that is, cadmium adsorbed on cation exchange complexes in equilibrium with the aqueous phase; (2) potentially available, that is, inorganic solid phases [$CdCO_3$, $Cd(OH)_2$, and CdS], chelated and insoluble organic bound cadmium, and cadmium precipitated or coprecipitated with hydrous oxides of manganese and iron; and (3) unavailable, that is, cadmium bound in crystalline lattices of clay minerals (Khalid, 1980). The readily available fraction is available for immediate uptake, whereas, as the name suggests, the unavailable portion is not. The potentially available fraction can be mobilized by several mechanisms including a lowering of pH, a change in redox potential, the action of microorganisms, and the actions of erosion, dredging, and bioturbation (Khalid, 1980; Förstner and Prosi, 1979). Hence, the release of cadmium from the sediments and its availability to aquatic organisms are largely dependent on physicochemical factors and microbial action. [*Note:* Concentrations in the tissues of organisms are reported in this chapter in terms of dry weight of tissue. Values from two papers concerning fish (Section 5.7) have been transformed from a wet-weight basis to a dry-weight basis by multiplying by a factor of 5. Those transformations have been indicated by an asterisk (*) on the numerical value.]

5.2. UPTAKE, DISTRIBUTION, AND ELIMINATION OF CADMIUM BY AQUATIC MACROPHYTES

A limited amount of information is available on the uptake, distribution, and elimination of cadmium by aquatic macrophytes, some of which is of little value in assessing what occurs in the natural environment. For example, the use of nutrient solutions for exposure (Cooley and Martin, 1979; Ornes and Wildman, 1979) may affect the uptake by complexing the cadmium (Pfister, 1982). Using concentrations of cadmium orders of magnitude higher than ambient levels over short periods of time is unrealistic (e.g., Cooley and Martin, 1979; Chigbo et al., 1982). Similarly, the uptake of cadmium cannot be accurately evaluated using plants that have been cut into sections (e.g., Cearley and Coleman, 1973).

Many rooted aquatic plants have a demonstrated ability to accumulate cadmium, primarily in the roots and to a lesser extent in the stems and leaves (Giesy et al., 1981; Miller et al., 1983; Wolverton and MacDonald, 1978). Both water and sediment may act on sources (Mayes et al., 1977). Wolverton and MacDonald (1978) found that in a 24-h period the roots of the water hyacinth, *Eichornia crassipes,* exhibited a linear uptake that was directly related to the concentration in the water. However, the "tops" of the plant accumulated very little. In a long-term study, Giesy et al. (1981) exposed two species of macrophytes to 5 and 10 µg Cd/L for a year in channel microcosms (Table 1). Cadmium was introduced into the channels in March, 1976, and sampling of the rush, *Juncus diffusissimus,* and the water starwort, *Callitriche heterophylla,* began in September and November, respectively, and continued until the following March, when the experiment was terminated. It appeared that, with the exception of the roots of *C. heterophylla,* the leaves and roots of both species at both concentrations had attained equilibrium before the sampling started. For some unexplained rea-

Table 1. Uptake and Depuration of Cadmium by Aquatic Macrophytes[a]

Species	Time of Exposure (yr)	Tissue Analyzed	Concentration in Water (µg/L)	Concentration in Tissue (µg/g dry wt.) After 1 yr Exposure	Concentration in Tissue (µg/g dry wt.) After 5 months Depuration	Bioconcentration Factor
Rush	1	Leaves	5	60	10	12,000
J. diffusissimus	1	Roots	5	250	80	50,000
	1	Leaves	10	120	10	12,000
	1	Roots	10	400	130	40,000
Water starwort,	1	Leaves	5	65	≤5	13,000
C. heterophylla	1	Roots	5	900	100	180,000
	1	Leaves	10	120	≤5	12,000
	1	Roots	10	1,395	190	140,000

[a] Adapted from Giesy et al. (1981).

son, the levels in the roots of *C. heterophylla* at both 5 and 10 μg/L increased markedly in cadmium content at the last sampling time before the flow of cadmium was stopped. This may have been an artifact that resulted in the calculation of extremely high concentration factors. Nevertheless, before this increase occurred, the concentration in the plants was very high. *Juncus diffusissimus* accumulated 60 μg Cd/g [bioconcentration factor (BCF) = 12,000] in the leaves and 250 μg Cd/g (BCF = 50,000) in the roots following exposure to 5 μg Cd/L (Table 1). Similarly, *C. heterophylla* exposed to the same concentration achieved levels of 65 and 900 μg Cd/g in the leaves (BCF = 13,000) and roots (BCF = 180,000), respectively. These extremely high uptake levels may have been facilitated by the soft (11 mg/L $CaCO_3$) acidic (pH 6.5) water to which they were subjected. Miller et al. (1983) also found, in a survey of 46 lakes in central Ontario that were characterized by acidic pH levels and low alkalinity, that higher plants as well as bryophytes were capable of accumulating substantial concentrations of cadmium and other metals. In general, plants with a high ratio of surface area to biomass tended to take up more cadmium than those with a lesser ratio (Franzin and McFarlane, 1980; van der Werff and Pruyt, 1982).

However, concentrations of cadmium in aquatic plants do not necessarily reflect those in the environment. Miller et al. (1983) determined that concentrations of eight metals, including cadmium, in two aquatic plants (*Eriocaulon septangulare* and *Eleocharis acicularis*) found in 46 lakes in central Ontario were not representative of the degree of contamination in the sediments. For instance, the lowest tissue concentration of cadmium was associated with the highest level in the sediments. No relationship could be established between the total cadmium content of the sediment and the tissue levels in the pipewort, *E. septangulare*. In two sets of lakes with the same sediment concentration of cadmium (0.4 mg/kg), the uptake by the pipewort in Clearwater Lake was 2.7 and 2.2 mg/kg in the roots and shoots, respectively, whereas in Harp and Red Chalk lakes the respective uptake was 8.2 and 3.0 mg/kg dry weight. Moreover, the much lower pH of Clearwater Lake (4.4) than in the other lakes (pH 6.4 and 6.5) should have facilitated uptake and resulted in higher accumulations. Similarly, Franzin and McFarlane (1980) could not correlate cadmium concentrations in the sediment and water with those in the submerged portions of the water milfoil, *Myriophyllum exalbescens,* in a number of lakes receiving aeolian input of metals from a base-metal smelter. Apparently, other factors are responsible for these findings. For example, Giesy et al. (1981) found that the amount of organic material in the sediment influenced the uptake of cadmium by *J. diffusissimus,* that is, the higher the organic content of the sediment, the lower the uptake of cadmium. Another possibility is that antagonistic and synergistic interactions of metals may influence their respective uptake rates. When lower than expected tissue levels are observed, a disruption in accumulation may have occurred as a result of competition between metals, as suggested by Miller et al. (1983). On the other hand, synergistic interactions between metals may result in higher than expected tissue concentrations. Hutchinson and Czyrska (1975) established that the simultaneous presence of cadmium and zinc increased the concentrations of both metals in *Lemna valdiviana*.

Franzin and McFarlane (1980) stated that the calcium levels in the water may have influenced the uptake of cadmium (see Section 5.8). They found that an uncontaminated lake with low calcium levels contained plants with high cadmium concentrations. Conversely, a contaminated lake with higher amounts of calcium had plants with lower cadmium levels. It appears, then, that aquatic plants cannot be used reliably as indicators of cadmium pollution until the factors responsible for these results are identified and explained.

The elimination of accumulated cadmium by aquatic macrophytes has only been rarely examined. After a 1-y exposure to 5 and 10 μg Cd/L in channel microcosms, *J. diffusissimus* and *C. heterophylla* were transferred to clean water and allowed to depurate for 5 months (Giesy et al., 1981) (Table 1). The concentrations in the leaves dropped sharply during the first month in both species and continued to decline so that after 5 months 85–95% of the accumulated metal had been lost. The roots of the two species differed in their release. The roots of *C. heterophylla* from both concentrations showed a rapid decline during the first month, followed by a more gradual decrease until an equilibrium level was reached, representing a clearance of about 85%. For *J. diffusissimus*, the roots exhibited a rapid decrease during the first 2 months followed by a gradual attainment of equilibrium after only 68% elimination. Apparently, clearance is greater from the leaves than from the roots, and some of the accumulated cadmium is retained despite rapid initial losses. Unfortunately, most studies only consider the depuration of a substance once the exposure has stopped. As mentioned earlier, with the exception of the roots of water starwort, the leaves and roots of both this species and the rush appeared to be at equilibrium within 6–8 months of the start of exposure. This implies either that cadmium was prevented from entering the plant or that excretion of accumulated cadmium was occurring during the exposure period. There is some evidence that the latter does occur. Mayes et al. (1977) determined that concentrations of accumulated cadmium in the rooted aquatic macrophyte, *Elodea canadensis*, exposed through the water and sediment of a contaminated lake peaked after 3–6 weeks and actually declined at 9 weeks. Similarly, Mayes and McIntosh (1975) found that the levels of cadmium in the nonrooted aquatic plant *Ceratophyllum demersum* in a lake receiving electroplating effluent dropped at 12 weeks after reaching maximum values at 6 and 9 weeks. Unfortunately, no information is available about the mechanisms involved.

McIntosh et al. (1978) estimated the total amount of cadmium held by the rooted aquatic macrophyte *Potamogeton crispus* that filled a lake contaminated with electroplating effluent. They calculated that if there was a massive die-off of all the plants at the same time, with release of the total metal burden, the levels in the water would only increase 0.3–1.0 μg/L. Since a massive die-off would be unlikely, the release of cadmium would be considerably less and would probably not be different from daily fluctuations in the water concentrations. Localized effects might be of importance if a rapid release of cadmium occurred from highly contaminated plants in an area of little mixing.

5.3. UPTAKE, DISTRIBUTION, AND ELIMINATION OF CADMIUM BY BACTERIA

Information concerning the uptake, distribution, and elimination of cadmium by bacteria is scant. With the exception of work by Pfister (1982), most studies deal with the uptake of cadmium by pure cultures of bacteria in growth media rather than with *in situ* studies in water or sediment. This can give misleading results since growth media often contain cadmium-complexing constituents that inhibit bacterial accumulation (Houba and Remacle, 1982; Pfister, 1982). The following discussion therefore focuses on studies in (or approximating) the natural environment.

Pfister (1982) found that the predominant bacterial genera in the sediment of a river receiving effluent from an electroplating facility were *Bacillus* sp., *Pseudomonas* sp., and *Arthrobacter* sp. In a model aquatic system consisting of water and sediment in aquaria into which 5000 μg/L Cd had been added daily for 10 weeks, *Bacillus* sp. were present at higher concentrations in the sediment than in the water. Although the accumulation of cadmium tended to fluctuate, the highest uptake for both water and sediments occurred under conditions when cadmium was most mobile, that is, low pH and high temperature. Bacteria accumulated 1000–3000 μg Cd/g, which was about 10 times greater than the level in the sediments.

The uptake of cadmium can vary considerably, both within and between species. For instance, three strains of *Bacillus subtilus* exposed to the same concentration of cadmium accumulated 0.15, 1.20, and 1.22% of the bacterial dry weight (Macaskie and Dean, 1982). In a microcosm approximating river conditions, Remacle (1981) established that "free" bacteria accumulated 1260 μg/g from a 1000-μg/L cadmium solution, whereas bacteria adhering to the wall of the vessel took up about five times as much (6100 μg/g). Part of this difference may be related to the mechanism of uptake. Generally, cadmium is taken up either intracellularly or bound on the outside of the cell by exuded polymers (Gadd and Griffiths, 1978; Macaskie and Dean, 1982). The adhering bacteria may adsorb more cadmium as a result of their ability to produce a mucilaginous product composed of polysaccharides and peptides that traps the cadmium (Remacle, 1981; Remacle et al., 1982).

Intracellular uptake has also been demonstrated by Khazaeli and Mitra (1981) for *Escherichia coli*. They observed that cells exposed to 340 μg Cd/L experienced a long lag phase before resuming growth, indicating an accommodation to cadmium. Since about 80% of the accumulated cadmium was located in the cytoplasm, it was thought that it must have been sequestered. In fact, over 60% of the total bound cadmium was associated with a high-molecular-weight protein that was only found in accommodated cells, suggesting that it was responsible for rendering the intracellular cadmium innocuous.

There does not appear to be any information on the elimination of cadmium by bacteria. Remacle (1981) speculated that bacteria may remove cadmium from the water column to the sediment by sedimentation. In addition, bacteria adhering to rocks may bind cadmium in the sediments.

5.4. UPTAKE, DISTRIBUTION, AND ELIMINATION OF CADMIUM BY PHYTOPLANKTON

In general, the uptake of cadmium by phytoplankton is rapid, with equilibrium being attained within 24 h (Conway and Williams, 1979; Khummongkol et al., 1982; Parker et al., 1982). It appears from Table 2 that several species have a distinct ability to concentrate cadmium, although to varying degrees. For example, when the diatoms *Asterionella formosa* and *Fragilaria crotonensis* were tested under the same conditions, *A. formosa* was able to accumulate approximately three times as much cadmium (43 µg/g) as *F. crotonensis* (12 µg/g) (Conway and Williams, 1979). Similarly, in a study of a small eutrophic lake contaminated with electroplating effluent, McIntosh and Bishop (1976) found that algae were able to accumulate large amounts of cadmium; the cadmium content of algae exposed to 1.1 µg Cd/L at five sites averaged 29.6 µg/g.

There are two basic modes of uptake by phytoplankton, passive and active. Passive uptake refers to cadmium adsorbed on the surface of algal cells, whereas active uptake involves the use of energy to actively transport cadmium into the cell (Rai et al., 1981). Khummongkol et al. (1982) devel-

Table 2. Uptake of Cadmium by Phytoplankton in Laboratory Studies

Species	Period of Exposure	Cadmium Concentration		Bioconcentration Factor	Reference
		In Water (µg/L)	In Tissue (µg/g)		
Asterionella formosa	24 h	1.9	43	23,000	Conway and Williams, 1979
	24 h	4.1	100	24,000	Conway and Williams, 1979
	24 h	8.8	179	20,000	Conway and Williams, 1979
Fragilaria crotonensis	24 h	1.8	12	6,600	Conway and Williams, 1979
	24 h	4.5	36	8,000	Conway and Williams, 1979
	24 h	8.5	64	7,500	Conway and Williams, 1979
Scenedesmus obliquus	14 days	10	12	1,200	Cain et al., 1980

oped a model for cadmium accumulation based on surface adsorption only for the green alga *Chlorella vulgaris* and tried to correlate the predicted uptake with the observed levels. They found in a 70-h exposure that the predicted uptake was lower than that which had actually occurred, indicating that a secondary mechanism other than adsorption was contributing to the accumulation, possibly active uptake. However, the uptake of cadmium may depend on the species, for there is evidence of active (Hart, 1977), passive (Sakaguchi et al., 1979), and a combination of both (Conway and Williams, 1979) processes. For example, Hart (1977) exposed *Chlorella pyrenoidosa* to 25–500 μg/L Cd in the dark and found that the uptake was only 25% of that in the light. No uptake occurred in dead cells while live cells exposed at 33 °C accumulated twice as much cadmium as those at 21 °C. All of these results suggested that cadmium was being actively transported into the cells. In contrast, Sakaguchi et al. (1979) found that heat-killed *Chlorella regularis* cells took up more cadmium than live cells. An increase in temperature and the use of metabolic inhibitors had little effect on uptake, indicating that metabolic processes were not involved in the accumulation and that uptake was therefore passive. This was confirmed when about 90% of the accumulated cadmium was found to be adsorbed on the surface of both live and heat-killed cells.

Both active and passive uptake were thought to occur simultaneously in *A. formosa* since cells in the dark and dead cells both accumulated some cadmium but only about half as much as those in the light and those that were alive (Conway and Williams, 1979). Moreover, in the determination of the distribution of cadmium in the cells, roughly 50–75% of the cadmium was found intracellularly while the remainder was present either in the organic coating of the frustule or in the frustule itself (Conway, 1978; Conway and Williams, 1979).

Phytoplankton are apparently able to accumulate high concentrations of cadmium before showing adverse physiological effects (Cain et al., 1980; Hart and Scaife, 1977), suggesting that the cadmium is somehow rendered innocuous. Several mechanisms of sequestering cadmium have been proposed. Conway and Williams (1979) suggested that sulfated polysaccharides in the organic coating of the diatom wall were functioning as an ion exchange resin. There is some evidence that intracellular cadmium may be tightly bound to inducible proteins in some species. The exposure of *C. pyrenoidosa* to cadmium resulted in the synthesis of a cadmium-binding protein with characteristics similar to metallothioneins (Hart, 1977). Silverberg (1976) found that the number of intramitochondrial granules containing cadmium in three species of green algae increased after cadmium exposure and that they represented insoluble metal–protein complexes. Rachlin et al. (1982) proposed that polyphosphate bodies were responsible for sequestering cadmium in *Plectonema boryanum*.

With the exception of the work by Conway and Williams (1979), few data are available concerning the elimination of cadmium from algae. They established that two species of diatoms, *A. formosa* and *F. crotonensis,* desorbed 55 and 70% of their cadmium accumulation, respectively, after 2 h in "clean" water. Thereafter a much slower release was observed, such that even after 2 months in cadmium-free water, some cadmium remained in the cell. The initial rapid desorption was ascribed to the loosely bound adsorbed fraction while the slower release was thought to represent that portion bound within the cell. The cadmium that was not released was apparently bound irreversibly. It should be pointed out that diatoms have a siliceous frustule outside the cell that does absorb some cadmium (Conway, 1978; Conway and Williams, 1979) and that findings might be different for cells lacking a frustule. Hart (1977), for example, found that *C. pyrenoidosa* released cadmium only slowly. In fact, some algal species do not release cadmium but rather carry it to the sediments. Parker et al. (1982) examined several size fractions of particles in Lake Michigan water and found that nonliving detrital material, mineral particles, and some small nanoplankton that predominated between sizes of 0.45 and 28 μm and phytoplankton that predominated between 28 and 253 μm were largely responsible for removing cadmium from the water column to the sediments. At low levels of phytoplankton abundance, virtually all of the sorption of cadmium occurred via the detrital fraction. However, during a plankton bloom, the biomass increased from 5% of the total seston to a maximum of 29% and correspondingly sorbed more cadmium, although the sorption of the detrital fraction still accounted for about 76% of the total seston uptake. Hence, during periods of phytoplankton abundance, maximum sorption of cadmium occurs, with phytoplankton playing a significant role. Parker et al. (1982) estimated that 0.1 μg $Cd/cm^2/yr$ was removed from the water column to the sediments by the settling of the seston, indicating that the phytoplankton and detrital material were important in cycling cadmium and, in so doing, made it less available to other organisms.

5.5. UPTAKE, DISTRIBUTION, AND ELIMINATION OF CADMIUM BY ZOOPLANKTON

Studies of this topic in the published literature are rare, but it appears that the uptake of cadmium occurs in two phases. Bertram (1980) established that the exoskeleton of *Daphnia pulicaria* adsorbed cadmium until an equilibrium was reached with the ambient concentration. After that, cadmium entered the body by simple diffusion and was sequestered, thereby preventing it from being passed on to the offspring or shunted to the exoskeleton.

The uptake of cadmium by *Daphnia galeata mendotae* is appreciable, as shown in Table 3. For example, after 22 weeks of exposure to 1 μg Cd/L the tissue concentration reached 17.6 μg/g, representing a concentration factor

Table 3. Uptake of Cadmium from Water by *D. galeata mendotae*[a]

Period of Exposure (weeks)	Cadmium Concentration		Bioconcentration Factor
	In Water (μg/L)	In Tissue (μg/g dry wt.)	
22	1	17.6	18,000
22	2	28.3	14,000
22	4	42.8	11,000
22	8	51.7	6,500

[a]From Marshall (1978).

of 18,000. Exposure to higher levels of cadmium over the same period of time resulted in higher tissue concentrations but reductions in the bioconcentration factors.

If *D. galeata mendotae* is representative of zooplankton as a whole, it would seem that zooplankton should accumulate significant amounts of cadmium from the water. However, in the natural environment with other particles such as detritus and phytoplankton present, the sorption by zooplankton may be relatively insignificant. Parker et al. (1982) studied the uptake of radiolabeled cadmium from Lake Michigan water by detrital material, phytoplankton, and zooplankton and found that ≤7% of the total amount accumulated was sorbed by zooplankton. This may have been an effect of particle size. For example, the greatest sorption occurred in the smallest size fraction (0.45–28 μm), which was composed mostly by detrital material and mineral particles, followed by the intermediate size fraction (28–253 μm), comprising the phytoplankton. Zooplankton, making up the largest size group (≥253 μm), accumulated the least. Simpson (1981) estimated that fecal pellets of zooplankton were important in cycling of cadmium from surface waters to the deep waters and sediments in the marine environment. Unfortunately, no such data are available for fresh water.

The source of uptake of cadmium by zooplankton appears to be limited to water. Hart (1977) and Parker et al. (1982) both found that the transfer of cadmium from contaminated algae to zooplankton did not occur to any great extent, indicating that cadmium is not subject to biomagnification between these trophic levels.

5.6. UPTAKE, DISTRIBUTION, AND ELIMINATION OF CADMIUM BY BENTHIC INVERTEBRATES

Three main groups of benthic invertebrates are discussed in this section: aquatic insect larvae, bivalve mollusks, and crayfish. They are all exposed to cadmium through the water, food, and sediment. Although the concentrations of metals are generally much higher in the sediment than in the water column, only a small amount is bioavailable (Förstner and Prosi, 1979). However, the release of metals from sediments is in facilitated bioturbation. Förstner and Prosi (1979) have suggested that the burrowing activity of oligochaetes releases large amounts of interstitial water containing high metal concentrations. In addition, sediments in the deeper layers of the reducing zone may be transported to the oxidative surface, where further chemical speciation can be initiated. However, in the limited number of studies available, the source of cadmium uptake is confined principally to water.

5.6.1. Aquatic Insect Larvae

Aquatic insect larvae can accumulate cadmium from water to a significant extent, although considerable variation exists between species. For example, a caddisfly, *Hydropsyche betteni,* exposed to 3 μg Cd/L for 28 days accumlated 90 μg/g, whereas a stonefly, *Pteronacys dorsata,* subjected to the same conditions accumulated only 8 μg/g (Table 4). Neither species had attained an equilibrium concentration. The caddisfly appeared to reach an equilibrium at the extremely high exposure concentration of 238 μg/L.

In a channel microcosm study, aquatic insect larvae were exposed to 5 and 10 μg/L for 1 yr, followed by a 6-month depuration period (Giesy et al., 1981). Uptake of cadmium occurred solely through the water and food since concentrations in the sediment were too low to act as a source. The data in Table 5, reflecting the mean accumulation in different taxa during the period of exposure, also indicate distinct differences in uptake. For example, the Ephemeroptera and Coleoptera exposed to 10 μg Cd/L accumulated average levels of 176 and 13 μg Cd/g, respectively. The amounts taken up by the different groups decreased in the following order: Ephemeroptera >

Table 4. Uptake of Cadmium by Aquatic Insect Larvae After 28 days Exposure[a]

Species	Water Concentration (μg/L)	Tissue Concentration (μg/g dry wt.)
P. dorsata (stonefly)	3.0	8
	8.3	30
	27.5	50
	85.5	85
	238	150
H. betteni (caddisfly)	3.0	90
	8.3	250
	27.5	310
	85.5	385
	238	280

[a] From Spehar et al. (1978a).

Table 5. Accumulation and Depuration of Cadmium by Aquatic Insect Taxa[a]

Taxa	Water Concentration (μg/L)	Concentration in Tissue (μg/g dry wt.)	Mean Concentration during Depuration Period
Ephemeroptera	5	40.7	—
(mayflies)	10	176	—
Odonata Anisoptera	5	18.4	—
(dragonflies)	10	34.3	—
Odonata Zygoptera	5	32.4	26
(damsel flies)	10	46.4	32.4
Coleoptera	5	4.1	7.4
(beetles)	10	13.0	24.6
Chironomidae	5	55.4	31.5
(midges)	10	91.6	52.6
Ceratopogonidae	5	23.4	28.1
(biting midges)	10	33.1	33.8

[a] From Giesy et al. (1981).

Chironomidae > Odonata (Zygoptera) > Ceratopogonidae > Odonata (Anisoptera) > Coleoptera.

The exoskeleton may be important in adsorbing cadmium in some species. For the dragonfly, *Pantala hymenaea,* 68% of the total uptake was associated with the molted exoskeleton (exuvium) as opposed to only 32% in the body of the emerging adult. The large uptake by the exoskeleton, and the process of molting, may protect the larvae from the toxic effects of cadmium and suggests that the exoskeleton is important in cycling this metal by acting initially as a sink and, after molting, as a source for other components of the ecosystem (Giesy et al., 1981).

The mean cadmium concentrations in various insect larvae during a depuration period in Giesy's study are listed in Table 5. The data were insufficient to determine the loss of cadmium from the Ephemeroptera and Odonata (Anisoptera). For the remaining groups, there was only a limited amount of clearance or none at all. For example, Zygoptera exposed to 5 and 10 μg Cd/L were able to eliminate 20 and 30% of the total amount accumulated, respectively, while the Chironomidae cleared about 40% at both concentrations. Part of these losses may have been due to the molting of the exoskeleton. Data for Coleoptera and Ceratopogonidae indicated no depuration, suggesting that the accumulated cadmium was irreversibly bound, perhaps to a cadmium-binding glycoprotein as was observed by Clubb et al. (1975) in the stonefly, *Pteronarcys californica*.

5.6.2. Mollusks

Mollusks can accumulate high concentrations of cadmium from the water. The snail, *Physa integra,* exposed to 3.0, 8.3, and 27.5 µg Cd/L for 28 days accumulated 30, 45, and 225 µg/g, respectively (Spehar et al., 1978a). The snails were still taking up cadmium when the exposure ceased.

In bivalve species, the gills act both as a filtering system for food and as a respiratory organ and hence are the primary site of uptake. Of the soft tissues of bivalves collected from the field (Table 6), the gills contained the highest levels of cadmium followed by the viscera and finally the muscle. However, as pointed out by Manly and George (1977), the concentration in the gills may not represent true uptake but rather adsorption to mucus sheaths, which are constantly being renewed. The largest proportion of cadmium in the viscera, which is derived from material passing from the gills into the digestive tract (Anderson, 1977), is generally found in the digestive gland and the kidneys (Manly and George, 1977). It is also important to note

Table 6. Accumulation of Cadmium by Mollusks in Field Studies

Location	Species	Tissue Analyzed	Tissue Concentration (µg/g dry wt.)	Reference
Fox River, IL	*Lampsilis siliquoidea*	Shell Body	3.66 5.89	Anderson, 1977
Fox River, IL	*Lampsilis ventricosa*	Shell Body	2.69 2.71	Anderson, 1977
Lake George, NY	*Lamsilis radiata*	Body	9	Heit et al., 1980
Fox River, IL	*Strophitis rugosus*	Shell Body	3.36 4.36	Anderson, 1977
Fox River, IL	*Sphaerium* sp.	Shell Body	2.85 2.49	Anderson, 1977
Fox River, IL	*Lasmigona complanata*	Shell Gills Viscera Muscle	1.24 1.11 0.90 0.36	Anderson, 1977
Fox River, IL	*Anodonta marginata*	Shell Gills Viscera Muscle	1.35 4.75 2.19 0.47	Anderson, 1977
Lake George, NY	*Elliptio complanatus*	Body	9	Heit et al., 1980
Lake George, NY	*Anodonta granalis*	Body	10	Heit et al., 1980

that the shell tends to accumulate almost as much cadmium as the body (Table 6).

Thus, it appears that mollusks can cycle cadmium to some extent by binding it to the constantly renewed mucus sheaths of the gills. In addition, the shell and tissues may be acting as a sink for cadmium. However, no conclusions can be drawn regarding the release of cadmium as no data are available.

5.6.3. Crayfish

Crayfish exposed to cadmium in the water are able to accumulate substantial amounts (Table 7). For example, *Cambarus latimanus* accumulated 15 µg/g (BCF = 2980) in the whole body after exposure to 5 µg Cd/L for 150 days. Gillespie et al. (1977) attempted to assess the uptake of cadmium from water by *Orconectes propinquus propinquus,* but the experimental procedure involved removal of the specimens for 2.5 h for radioactive counting while subjecting them to significant temperature changes (4–11 °C), thereby casting doubts on the accuracy of their results.

Different tissues display various affinities for cadmium, as illustrated in Table 7, which shows that cadmium is highly concentrated in the gills but not in the dorsal muscle tissue (Dickson et al., 1982). Similarly, Anderson and Brower (1978) found that in the field *Orconectes virilis* accumulated the greatest amount of cadmium in the gills, followed by the viscera, and to a lesser extent the muscle, as was the case for mollusks (Anderson, 1977). Cadmium was also found adsorbed to the exoskeleton.

The uptake of cadmium from both a food source and the water by the crayfish *Procambarus acutus acutus* over a 55-day period was examined by Giesy et al. (1980). In contrast to most studies considering food as a source of cadmium, they found that cadmium was being accumulated both from the food and directly from the water and that the uptake was independent and additive. In other words, the accumulation of cadmium from the food only,

Table 7. Accumulation of Cadmium from Water by Crayfish

Species	Tissue Analyzed	Length of Exposure (days)	Concentration in Water (µg/L)	Concentration in Tissue (µg/g dry wt.)	Reference
Cambarus latimanus	Whole body	150	5	15	Thorpe et al., 1979
			10	22	
Procambarus acutus acutus	Gills	21	5	200	Dickson et al., 1982
	Dorsal muscle tissue	—	5	7	
	Gills	—	10	180	
	Dorsal muscle tissue	—	10	1	

added to that accumulated from the water only, and water separately was not significantly different from the accumulation from food and water together. After transfer to "clean" water for 22 days, no significant release of cadmium was observed. It was hypothesized that cadmium was sequestered by a cadmium-binding protein.

The process of molting may be important in cycling cadmium in the environment. Initially, a portion of the ambient cadmium is bound on mucus sheaths in the gills and on the exoskeleton surface (Anderson and Brower, 1978; Giesy et al., 1980). During molting, cadmium is sloughed off with the mucus sheaths and the exoskeleton and may be released back to the environment. However, cadmium bound to tissues is not released, indicating that crayfish are also acting as a sink.

5.7. UPTAKE, DISTRIBUTION, AND ELIMINATION OF CADMIUM BY FISH

5.7.1. Uptake and Distribution

The routes of uptake of cadmium by fish are through the food and the water or both. In general, the accumulation of cadmium occurs principally from the water. Hatakeyama and Yasuno (1982) found that guppies (*Poecilia reticulata*) fed with the cladoceran *Moina macrocopa* containing concentrations of 69.5, 125.9, and 170.6 µg Cd/g on a daily basis for 30 days accumulated less than 3% of the total amount of cadmium fed, regardless of the cadmium content of the zooplankter. In contrast, guppies exposed to 5 and 10 µg Cd/L in the water over a 25-day period concentrated 13 and 24.5 µg Cd/g, respectively, and were still accumulating cadmium when the experiment ended. Williams and Giesy (1978) exposed mosquitofish (*Gambusia affinis*) to cadmium in the water (10 µg/L) and the food (1 µg/g) for 8 weeks. They established that there was no significant increase in whole-fish cadmium levels in control water regardless of food ration, whereas fish subjected to 10 µg Cd/L in the water had significantly higher cadmium residues than the control. Furthermore, there was no significant difference between cadmium concentrations in fish fed the cadmium-containing diet and those fed the control diet at either the high or low water concentration. It therefore appears that although cadmium may be accumulated by fish to a slight extent from food, the primary source is from the water. Other studies have confirmed this result (McFarlane and Franzin, 1980; Peterson et al., 1983; Ferard et al., 1983; Rehwoldt and Karimian-Teherani, 1976).

The gills are the primary site of uptake of cadmium from the water (Pärt and Svanberg, 1981). This is reflected in the data of Kumada et al. (1980), where rainbow trout exposed to 4 µg Cd/L for 10 weeks accumulated 11* µg Cd/g in the gills as opposed to only 0.55* µg Cd/g after receiving 50* µg Cd/g in their food on a daily basis over the same time period.

Table 8. Whole-Body Uptake of Cadmium from Water by Fish

Species	Exposure Concentration (μg/L)	Length of Exposure (days)	Whole-body Concentration (μg/g dry wt.)	Reference
Salvelinus fontinalis, brook trout	0.9	84	0.68	Benoit et al., 1976
	1.7	84	0.65	
	3.4	84	1.26	
Pimephales promelas, fathead minnow	48	46	10.7	Sullivan et al., 1978

Whole-body uptake levels for fish exposed to low levels of dissolved cadmium in the laboratory (Table 8) and the field (Table 9) are in good agreement. For example, brook trout subjected to concentrations of 0.9–3.4 μg/L in the laboratory accumulated 0.65–1.26 μg Cd/g, while in the field, exposure of several species to 0.9–4.2 μg Cd/L resulted in levels ranging from 0.10 to 0.80 μg Cd/g. Since background concentrations of cadmium are little different (0.10–0.3 μg Cd/g) (Table 10), whole-body levels are not necessarily a good indicator of cadmium exposure. This is particularly evident in Table 9, where the whole-body concentration (0.36 μg Cd/g) of bluegills exposed to 2.0 μg Cd/L in Little Center Lake tended to mask the much higher level in the liver (3.52 μg Cd/g). In addition, whole-body levels vary considerably depending on the species. For instance, in the severely contaminated west basin of Palestine Lake (Table 9), the body residues varied from a low of 0.66 μg Cd/g for the golden shiner, *Notemigomus crysoleucas,* to a high of 8.00 μg Cd/g in the redear sunfish, *Lepomis microlophus.*

Several studies have shown that cadmium is accumulated in specific sites in the body rather than being distributed evenly throughout. In general, cadmium is found predominantly in the kidney, liver, and gills (Benoit et al., 1976; Eaton, 1974; Edgren and Notter, 1980; Kumada et al., 1980; Mount and Stephan, 1967; Roberts et al., 1979; Sangalang and Freeman, 1979). Table 11 illustrates the tissue concentrations and bioconcentration factors in two species of fish. From the data of Benoit et al. (1976), it appears that cadmium accumulates to its highest concentration in the kidney, followed by the liver and gills. The results of Eaton (1974) indicate a roughly equivalent uptake for the kidney and liver and a much lower accumulation in the gills. However, it should be noted that in Eaton's study, the concentrations in the kidney, for some unexplained reason, did not increase much with increasing exposure levels. This is in contrast to most studies, which show an increase in uptake with higher exposure concentrations (Benoit et al., 1976; Smith et al., 1976; Pascoe and Mattey, 1977) accompanied by a decline in bioconcentration factors (Table 8).

The concentrations of cadmium in the tissues increase with exposure time (Benoit et al., 1976; Roberts et al., 1979; Kito et al., 1982b) and usually, several weeks are required before an equilibrium is attained (Mount and Stephan, 1967; Kumada et al., 1973; Benoit et al., 1976; Spehar, 1976; Spehar et al., 1978b). Tissues in which there is little or no accumulation include the brain (Rowe and Massaro, 1974; Smith et al., 1976) and the skin, muscle, heart, and bone (Mount and Stephan, 1967; Benoit et al., 1976; Sangalang and Freeman, 1979; Roberts et al., 1979; Kumada et al., 1980).

Although it is difficult to relate these laboratory findings to those in the field since most field studies confine themselves to a determination of whole-body residues (Table 9), it appears that specific tissue accumulations in the field agree with those from the laboratory. For instance, 15 species of fish in the Toronto Harbor and Baie du Doré were all found to accumulate cadmium in their tissues in the following order: kidney $\geq$ liver $\geq$ muscle (Brown and Chow, 1977).

Unfortunately, other field studies seem to have ignored uptake in the kidney and have concentrated on the liver, muscle, and gills. McIntosh and Bishop (1976) analyzed the gill, liver, and muscle tissues of the bluegill subjected to an average concentration of 2.0 µg Cd/L in a lake receiving effluent from an electroplating facility (Table 9). They established that the highest accumulation occurred in the liver (3.52 µg Cd/g) followed by the gill (0.11 µg Cd/g) and finally the muscle (0.08 µg Cd/g). Similarly, Wilson et al. (1981) established that in the Sacramento River basin, where the concentration of cadmium had risen from 0.1 to 1 µg Cd/L as one moved downstream due to an influx of acid mine waste, the levels of cadmium in the liver of two species of salmonids increased from less than 1.5* µg/g to 20* µg/g, an increase of over 1200%. Cadmium in the muscle remained low (<0.11* µg/g) regardless of the concentration in the water. The liver was also found to be a site of accumulation in northern pike, *Esox lucius,* and the white sucker, *Catastomus commersoni,* in five lakes in the vicinity of a base-metal smelter in Flin Flon, Manitoba (McFarlane and Franzin, 1980).

Early life stages of salmonids have also been observed to accumulate cadmium. The uptake by eggs is considerably greater than that by newly hatched alevins and increases with exposure concentration in rainbow trout (Beattie and Pascoe, 1978) and Atlantic salmon (*Salmo salar*) (Rombough and Garside, 1982). For example, after 100 h exposure to 10 µg Cd/L, the trout eggs contained 5.82 µg Cd/g as opposed to only 1.04 µg Cd/g in the alevins (Beattie and Pascoe, 1978). The much higher concentration of cadmium in the eggs than in the alevins is apparently due to adsorption or binding at the egg surface (Beattie and Pascoe, 1978).

5.7.2. Elimination

Information regarding the elimination of cadmium by fish is extremely limited. Benoit et al. (1976) found that even 12 weeks after being placed in

Table 9. Concentrations of Cadmium in Tissues of Fish from Contaminated Sites

Species	Location	Tissue Analyzed	Total Cadmium Concentration in H$_2$O (μg/L)	Cadmium Concentration in Tissue (μg/g dry wt.)	Reference
Lepomis macrochirus, bluegill	Little Center Lake, IN	Whole body Gill Liver Muscle	3.2 (2.0)[a] 3.2 (2.0) 3.2 (2.0) 3.2 (2.0)	0.36 0.11 3.52 0.08	McIntosh and Bishop, 1976 McIntosh and Bishop, 1976 McIntosh and Bishop, 1976 McIntosh and Bishop, 1976
Lepomis gibbosus, pumpkinseed	Little Center Lake, IN	Whole body	3.2 (2.0)	0.57	McIntosh and Bishop, 1976
Lepomis macrochirus, bluegill	Trimble Creek (effluent stream from Palestine Lake)	Whole body	6.57 (4.2)	0.80	Adams et al., 1980
Pomoxis nigromaculatus, black crappie	Trimble Creek (effluent stream from Palestine Lake)	Whole body	6.57 (4.2)	0.52	Adams et al., 1980
Perca flavescens, yellow perch	Trimble Creek (effluent stream from Palestine Lake)	Whole body	6.57 (4.2)	0.36	Adams et al., 1980
Minytrema melanops, spotted suckers	Trimble Creek (effluent stream from Palestine Lake)	Whole body	6.57 (4.2)	0.62	Adams et al., 1980

Species	Location	Tissue		Value	Reference
Notemigonus crysoleucas, golden shiners	Trimble Creek (effluent stream from Palestine Lake)	Whole body	6.57 (4.2)	0.56	Adams et al., 1980
Lepomis microlophus, redear sunfish	Palestine Lake East Basin	Whole body	1.2 (0.9)	0.13	Murphy et al., 1978b
	West Basin	Whole body	47.6 (17.3)	8.00	Murphy et al., 1978b
Micropterus salmoides, largemouth bass	Palestine Lake East Basin	Whole body	1.2 (0.9)	0.12	Murphy et al., 1978b
	West Basin	Whole body	47.6 (17.3)	1.11	Murphy et al., 1978b
Notemigonus crysoleucas, golden shiner	Palestine Lake East Basin	Whole body	1.2 (0.9)	0.21	Murphy et al., 1978b
	West Basin	Whole body	47.6 (17.3)	0.66	Murphy et al., 1978b
Catostomus commersoni, white sucker	Palestine Lake East Basin	Whole body	1.2 (0.9)	0.12	Murphy et al., 1978b
	West Basin	Whole body	47.6 (17.3)	1.99	Murphy et al., 1978b
Lepomis macrochirus, bluegill	Palestine Lake East Basin	Whole body	1.2 (0.9)	0.10	Murphy et al., 1978b
		Muscle	1.2 (0.9)	0.08	Murphy et al., 1978a
	West Basin	Whole body	47.6 (17.3)	3.40	Murphy et al., 1978b
		Muscle	47.6 (17.3)	0.43	Murphy et al., 1978a
Micropterus salmoides, largemouth bass	Palestine Lake East Basin	Muscle	1.2 (0.9)	0.01	Murphy et al., 1978a
	West Basin	Muscle	47.6 (17.3)	0.08	Murphy et al., 1978a

[a]Numbers in parentheses reflect dissolved cadmium.

Table 10. Concentrations of Cadmium in Fish Whole Body from Uncontaminated Areas

Location	Species	Concentration in H_2O	Concentration in Tissue ($\mu g/g$ dry wt.)	Reference
Fox River, IL	*Perca flavescens*, yellow perch	—	0.2	Vinikour et al., 1980
Fox River, IL	*Lepomis macrochirus*, bluegill	—	0.2	Vinikour et al., 1980
Fox River, IL	*Pomoxis nigromaculatus*, black crappie	—	0.2	Vinikour et al., 1980
Fox River, IL	*Ictalurus melas*, black bullhead	—	0.3	Vinikour et al., 1980
Jubilee Creek, IL	*Micropterus dolomieui*, smallmouth bass	—	0.10	Enk and Mathis, 1977
Jubilee Creek, IL	*Etheostoma flabellare*, fantail darter	—	0.15	Enk and Mathis, 1977

"clean" water following a 105-week exposure to 0.9, 1.7, and 3.4 μg Cd/L, brook trout were only able to clear cadmium from the gills. Concentrations in the liver remained about the same while those in the kidney actually rose slightly, suggesting translocation from other tissues. Kumuda et al. (1980) observed the same effects in rainbow trout after 10 weeks withdrawal from exposure to 4 μg Cd/L over a 10-week period. These results indicate that cadmium may be irreversibly bound in the liver and kidney, possibly to low-molecular-weight proteins (Marafante, 1976; Kito et al., 1982b; Thomas et al., 1983a,b).

To determine whether cadmium is sequestered by fish, two routes of exposure have been used: intraperitoneal injection and ambient exposure in the water. In both cases, it was established that the presence of cadmium stimulated the production of cadmium-binding proteins, primarily in the liver and kidneys of a number of species (Marafante, 1976; Beattie and Pascoe, 1979; Kito et al., 1982a–c; Woodworth and Pascoe, 1983; Woodworth et al., 1983). In rainbow trout, virtually all of the cadmium accumulated (99%) was found in the liver, kidneys, and gills (Thomas et al., 1983a,b). It was also noted that fish preexposed to sublethal levels of cadmium in the water exhibited a greater tolerance to subsequent lethal exposures than fish that were not preexposed (Pascoe and Beattie, 1979; Dixon and Sprague, 1981; Kito et al., 1982c), suggesting a protective effect, perhaps due to the sequestrations of cadmium.

Kito et al. (1982a) indicated that these cadmium-binding proteins were metallothioneins, which are known to bind cadmium and sequester it (Kägi and Nordberg, 1979). Metallothioneins are low-molecular-weight proteins

Table 11. Uptake and Distribution of Cadmium in Various Tissues of Fish[a]

Species	Tissue	Exposure Concentration (μg/L)	Length of Exposure (days)	Tissue Concentration (μg/g dry wt.)	BCF	Reference
Salvelinus fontinalis, brook trout						
First generation	Kidney	3.4	266	50.0	14,705	Benoit et al., 1976
	Liver	3.4	266	10.0	2,941	
	Gills	3.4	266	8.0	2,353	
Second generation	Kidney	3.4	490	64.8	19,059	Benoit et al., 1976
	Liver	3.4	490	9.0	2,647	
	Gills	3.4	490	5.1	1,500	
Lepomis macrochirus, bluegill sunfish	Kidney	31	330	188	6,065	Eaton, 1974
	Liver	31	330	201	6,484	
	Gills	31	330	34	1,097	

[a]The exposure concentration listed for trout is the lowest level found to cause an effect; that for sunfish is a no-effect level.

characterized as having a high cysteine content, few aromatic amino acids, and a strong affinity for metals. However, recent work has shown that the route of exposure, and hence concentration of cadmium governs the distribution of cadmium and whether it is bound to low-molecular-weight proteins or metallothioneins (Thomas et al., 1983a,b). Rainbow trout exposed to 9 μg Cd/L in the water for 12–36 weeks had the highest proportion of cadmium in the kidneys (65%) followed by the liver (20–30%) and finally the gills (5–15%). In contrast, cadmium administered intraperitoneally at high concentrations was found almost exclusively in the liver (93% of the total body burden) and to a minor extent in the kidneys (5%) and gills (2%). The extent of mortality suffered by trout exposed to cadmium also suggested that different mechanisms of sequestration were operating. For example, all fish injected with cadmium survived the 8-week exposure despite having higher body burdens. Those fish exposed for the same period at 18, 36, and 54 μg Cd/L had mortalities of 20, 30, and 70%, respectively, even though their body burdens were only 19, 26, and 40%, respectively, of those receiving cadmium injections. Using chromatographic separation techniques, Thomas et al. (1983b) determined that at relatively low levels of ambient cadmium (9 μg Cd/L), two low-molecular-weight proteins found largely in the kidney and liver were responsible for sequestering the cadmium. In contrast, very high concentrations administered intraperitoneally were bound to metallothioneins found in the liver. This was confirmed in an *in vitro* study in which cadmium was found to bind to metallothionein only when the cadmium concentration was increased to extremely high levels (e.g., 1000 μg/mL) (Thomas et al., 1983b). Hence, in the natural environment, it appears that two low-molecular-weight nonmetallothionein proteins are involved in binding and detoxifying cadmium.

5.8. EFFECTS OF ENVIRONMENTAL FACTORS ON CADMIUM UPTAKE

Most of the information on this topic concerns the effect of water hardness. In general, as the water hardness increases, the uptake of cadmium decreases; the reason for this is unclear (Kinkade and Erdman, 1975) although two hypotheses have been advanced. The first suggests that as hardness increases, the Cd^{2+} ion (the most bioavailable form) becomes complexed into nontoxic fractions, whereas the second hypothesis proposes that antagonism by Ca^{2+} and Mg^{2+} ions is responsible for the reduced uptake (Davies et al., 1979). Although some of the free Cd^{2+} ions may become complexed, a number of studies have suggested that the Ca^{2+} ion is the driving force behind the reduction in uptake and hence toxicity, and that this is a biological phenomenon. The inverse relationship between the uptake of cadmium and the level of Ca^{2+} present in the water seems to be a general effect since it has been observed in aquatic plants (McFarlane and Franzin, 1980), phytoplankton (Sakaguchi et al., 1979), invertebrates (Poldoski, 1979; Wright and Frain, 1981), and fish (McFarlane and Franzin, 1980; Michibata, 1981).

The mechanism underlying this effect is only poorly understood, although it is apparently related to the closely linked uptake of cadmium and calcium. The results of Wright (1980) suggested that calcium and cadmium were in competition for sites on an active calcium regulatory mechanism and that cadmium was accidentally accumulated in place of calcium. The findings of Roch and Maly (1979), who determined that cadmium disrupted ion regulation in rainbow trout, which consequently led to severe hypocalcemia, support this possibility. Furthermore, Zitko and Carson (1976) proposed that calcium and heavy metals competed for cellular binding sites.

Data concerning other environmental factors is fragmentary. An increase in temperature generally results in a greater uptake in many organisms (Hart and Scaife, 1977; Pfister, 1982; Remacle et al., 1982; Rombough and Garside, 1982). A reduction in pH leads to a greater uptake in bacteria (Pfister, 1982) and algae (Hart and Scaife, 1977; Soeder et al., 1978). The presence of complexing agents usually results in a decrease in accumulation, although this is not always the case (Poldoski, 1979). For instance, humic acids, pyrophosphate (P_2O_7), nitrilotriacetic acid (NTA), ethylenediaminetetraacetic acid (EDTA), and EBDP (a structural analogue of EDTA) all inhibited cadmium uptake. On the other hand, diethyldithiocarbamate (DDC) stimulated the uptake of cadmium so that uptake was greater than for comparable concentrations of free cadmium. It was postulated that since the DDC and cadmium formed a hydrophobic coordination compound, the resulting greater lipophilic character was responsible for the enhanced uptake.

REFERENCES

Adams, T. G., Atchison, G. J., and Vetter, R. J. (1980). The impact of an industrially contaminated lake on heavy metals in its effluent stream. *Hydrobiologia,* **69,** 187–193.

Anderson, R. V. (1977). Concentration of cadmium, copper, lead and zinc in six species of freshwater clams. *Bull. Environ. Contam. Toxicol.,* **18,** 492–496.

Anderson, R. V., and Brower, J. E. (1978). Patterns of trace metal accumulation in crayfish populations. *Bull. Environ. Contam. Toxicol.,* **20,** 120–127.

Beattie, J. H., and Pascoe, D. (1978). Cadmium uptake by rainbow trout, *Salmo gairdneri* eggs and alevins. *J. Fish Biol.,* **13,** 631–637.

Beattie, J. H., and Pascoe, D. (1979). A cadmium-binding protein in rainbow trout. *Toxicol. Lett.,* **4,** 241–246.

Benoit, D. A., Leonard, E. N., Christensen, G. M., and Fiandt, J. T. (1976). Toxic effects of cadmium on three generations of brook trout (*Salvelinus fontinalis*). *Trans. Am. Fish. Soc.,* **105,** 550–560.

Bertram, P. E. (1980). Population responses of *Daphnia* to long-term exposure to cadmium. *Diss. Abst. Int. B.,* **41,** 4008.

Brown, J. R., and Chow, L. Y. (1977). Heavy metal concentrations in Ontario fish. *Bull. Environ. Contam. Toxicol.,* **17,** 190–195.

Cain, J. R., Paschal, D. C., and Hayden, C. M. (1980). Toxicity and bioaccumulation of cadmium in the colonial green alga *Scenedesmus obliquus*. *Arch. Environ. Contam. Toxicol.*, **9**, 9–16.

Cearley, J. E., and Coleman, R. L. (1973). Cadmium toxicity and accumulation in southern naiad. *Bull. Environ. Contam. Toxicol.*, **9**, 100–101.

Chigbo, F. E., Smith, R. W., and Shore, F. L. (1982). Uptake of arsenic, cadmium, lead and mercury from polluted waters by the water hyacinth, *Eichornia crassipes*. *Environ. Pollut. Ser. A.*, **27**, 31–36.

Clubb, R. W., Lords, J. L., and Gaufin, A. R. (1975). Isolation and characterization of a glycoprotein from the stonefly, *Pteronarcys californica*, which binds cadmium. *J. Insect Physiol.*, **21**, 53–60.

Conway, H. L. (1978). Sorption of arsenic and cadmium and their effects on growth, micromitrient utilization and photosynthetic pigment composition of *Asterionella formosa*. *J. Fish. Res. Bd. Can.*, **35**, 286–294.

Conway, H. L., and Williams, S. C. (1979). Sorption of cadmium and its effect on growth and the utilization of inorganic carbon and phosphorus of two freshwater diatoms. *J. Fish. Res. Bd. Can.*, **36**, 579–586.

Cooley, T. N., and Martin, D. F. (1979). Cadmium in naturally-occurring water hyacinths. *Chemosphere*, **2**, 75–78.

Davies, P. H., Calabrese, A., Goettl, J. P., Jr., Hodson, P. V., Kent, J. C., McKim, J. M., Pearce, J. B., Reish, D. J., Sprague, J. B., and Sunda, W. G. (1979). Cadmium. In *A Review of the EPA Redbook: Quality Criteria for Water*, Thurston, R. V., Russo, R. C., Feterholf, C. M., Jr., Edsall, T. A., and Barber, Y. M., Jr. (eds.). Water Quality Section, American Fisheries Society, Bethesda, MD. pp. 51–66.

Dickson, G. W., Giesy, J. P., and Briese, L. A. (1982). The effect of chronic cadmium exposure on phosphoadenylate concentrations and adenylate energy charge of gills and dorsal muscle tissue of crayfish. *Environ. Toxicol. Chem.*, **1**, 147–156.

Dixon, D. G., and Sprague, J. B. (1981). Copper bioaccumulation and hepato-protein synthesis during acclimation to copper by juvenile rainbow trout. *Aquat. Toxicol.*, **1**, 69–82.

Eaton, J. G. (1974). Chronic cadmium toxicity to the blue gill (*Lepomis macrochirus* Rafinesque). *Trans. Am. Fish. Soc.*, **103**, 729–735.

Edgren, M., and Notter, M. (1980). Cadmium uptake by fingerlings of perch (*Perca fluviatilis*) studied by Cd-115m at two different temperatures. *Bull. Environ. Contam. Toxicol.*, **24**, 647–657.

Enk, M. D., and Mathis, B. J. (1977). Distribution of cadmium and lead in a stream ecosystem. *Hydrobiologica*, **52**, 153–158.

Ferard, J. F., Jouany, J. M., Truhaut, R., and Vasseur P. (1983). Accumulation of cadmium in a freshwater food-chain experimental model. *Ecotox. Environ. Safe.*, **7**, 43–52.

Florence, T. M. (1977). Trace metal species in fresh waters. *Wat. Res.*, **11**, 681–687.

Florence, T. M. (1982). The speciation of trace elements in waters. *Talanta*, **29**, 345–364.

Florence, T. M., and Batley, G. E. (1977). Determination of the chemical forms of trace metals in natural waters with special reference to copper, lead, cadmium and zinc. *Talanta*, **24**, 151–158.

Florence, T. M., and Batley, G. E. (1980). Chemical speciation in natural waters. *CRC Crit. Rev. Anal. Chem.*, **9**, 219–296.

Förstner, U., and Prosi, F. (1979). Heavy metal pollution in freshwater ecosystems. In *Biological aspects of freshwater pollution*, Ravera, O. (ed.). Pergamon, New York, pp. 129–161.

Franzin, W. G., and McFarlane, G. A. (1980). An analysis of the aquatic macrophyte *Myriophyllum exalbescens*, as an indicator of metal contamination of aquatic ecosystems near a base metal smelter. *Bull. Environ. Contam. Toxicol.*, **24**, 597–605.

References

Gadd, G. M., and Griffiths, A. J. (1978). Microorganisms and heavy metal toxicity. *Microbiol. Ecol.*, **4**, 303–317.

Gardiner, J. (1974). The chemistry of cadmium in natural water—I. A study of cadmium complex formation using the cadmium specific ion electrode. *Wat. Res.*, **8**, 23–30.

Giesy, J. P., Bowling, J. W., and Kania, H. J. (1980). Cadmium and zinc accumulation and elimination by freshwater crayfish. *Arch. Environ. Contam. Toxicol.*, **9**, 683–697.

Giesy, J. P., Bowling, J. W., Kania, H. J., Knight, R. L., and Mashburn, S. (1981). Fates of cadmium introduced into channels microcosm. *Environ. Int.*, **5**, 159–175.

Gillespie, R., Reisine, T., and Massaro, E. J. (1977). Cadmium uptake by the crayfish, *Orconectes propinquus propinquus* (Girard). *Environ. Res.*, **13**, 364–368.

Hart, B. A. (1977). *The Role of Phytoplankton in Cycling Cadmium in the Environment*. Water Resources Research Center, University of Vermont, Burlington, VT.

Hart, B. A., and Scaife, B. D. (1977). Toxicity and bioaccumulation of cadmium in *Chlorella pyrenoidosa*. *Environ. Res.*, **14**, 401–413.

Hart, B. T. (1981). Trace metal complexing capacity of natural waters: A review. *Environ. Technol. Lett.*, **2**, 95–110.

Hatakeyama, S., and Yasuno, M. (1982). Accumulation and effects of cadmium on guppy (*Poecilia reticulata*) fed cadmium-dosed cladocera (*Moina macrocopa*). *Bull. Environ. Contam. Toxicol.*, **29**, 159–166.

Heit, M., Klusek, C. S., and Miller, K. M. (1980). Trace element, radionuclide and polynuclear aromatic hydrocarbon concentrations in Unionidae mussels from northern Lake George. *Environ. Sci. Technol.*, **14**, 465–468.

Houba, C., and Remacle, J. (1982). Factors influencing toxicity of cadmium to *Tetrahymena pyriformis:* Particulate or soluble form and degree of complexation. *Environ. Pollut. Ser. A.*, **28**, 35–43.

Hutchinson, T. C., and Czyrska, H. (1975). Heavy metal toxicity and synergism to floating aquatic weeds. *Verh. Int. Ver. Theol. Angew. Limnol.*, **19**, 2102–2111.

Jenne, E. A., and Luoma, S. N. (1977). Forms of trace elements in soils, sediments and associated waters: An overview of their determination and biological availability. In *Biological implications of metals in the environment. Proceedings Fifteenth Annual Hanford Life Science Symposium.*, Drucker, H., and Wildung, R. E. (eds.). pp. 110–143.

Kägi, J. H. R., and Nordberg, M. (1979). *Metallothionein. Proceedings of the First International Meeting on Metallothionein and Other Low Molecular Weight Metal-Binding Proteins*. Birhauser Verlag, Basel.

Khalid, R. A. (1980). Chemical mobility of cadmium in sediment–water systems. In *Cadmium in the Environment, Part I, Ecological Cycling,* Nriagu, J. O. (ed.). Wiley, New York, pp. 257–304.

Khazaeli, M. B., and Mitra, R. S. (1981). Cadmium binding component in *Escherichia coli* during accommodation to low levels of this ion. *Appl. Environ. Microbiol.*, **41**, 46–50.

Khummongkol, D., Canterford, G. S., and Fryer, C. (1982). Accumulation of heavy metals in unicellular algae. *Biotechnol. Bioeng.*, **24**, 2643–2660.

Kinkade, M. L., and Erdman, H. E. (1975). The influence of hardness components (Ca^{2+} and Mg^{2+}) in water on the uptake and concentration of cadmium in a simulated freshwater ecosystem. *Environ. Res.*, **10**, 308–317.

Kito, H., Ose, Y., Mizuhira, V., Sato, T., Ishikawa, T., and Tazawa, T. (1982a). Separation and purification of (Cd, Cu, Zn)–metallothionein in carp hepato-pancreas. *Comp. Biochem. Physiol.* **73C**, 121–127.

Kito, H., Tazawa, T., Ose, Y., Sato, T., and Ishikawa, T. (1982b). Formation of metallothionein in fish. *Comp. Biochem. Physiol.,* **73C**, 129–134.

Kito, H., Tazawa, T., Ose, Y., Sato, T., and Ishikawa, T. (1982c). Protection by metallothionein against cadmium toxicity. *Comp. Biochem. Physiol.,* **73C**, 135–139.

Kumada, H., Kimura, S., Yokote, M., and Matida, Y. (1973). Acute and chronic toxicity, uptake and retention of cadmium in freshwater organisms. *Bull. Freshwater Fish. Res. Lab.*, **22**, 157–165.

Kumada, H., Kimura, S., and Yokote, M. (1980). Accumulation and biological effects of cadmium in rainbow trout. *Bull. Japan Soc. Sci. Fish.*, **46**, 97–103.

Macaskie, L. E., and Dean, A. C. R. (1982). Cadmium accumulation by microorganisms. *Environ. Technol. Lett.*, **3**, 49–56.

Manly, R., and George, W. O. (1977). The occurrence of some heavy metals in populations of the freshwater mussel *Anodonta anatina* (L.) from the River Thames. *Environ. Pollut.*, **14**, 139–154.

Mantoura, R. F. C., Dicksen, A., and Riley, J. P. (1978). The complexation of metals with humic materials in natural waters. *Est. Coast. Mar. Sci.*, **6**, 387–408.

Marafante, E. (1976). Binding of mercury and zinc to cadmium-binding protein in liver and kidney of goldfish (*Carassius auratus* L.). *Experientia*, **32**, 149–150.

Marshall, J. S. (1978). Population dynamics of *Daphnia galeata mendotae* as modified by chronic cadmium stress. *J. Fish. Res. Bd. Can.*, **35**, 461–469.

Mayes, R., and McIntosh, A. (1975). The use of aquatic macrophytes as indicators of trace metal contamination in fresh water lakes. In *Trace Substances in Environmental Health IX*, Hemphill, D. D. (ed.). pp. 157–167.

Mayes, R. A., McIntosh, A. W., and Anderson, V. L. (1977). Uptake of cadmium and lead by a rooted aquatic macrophyte (*Elodea canadensis*). *Ecology* **58**, 1176–1180.

McFarlane, G. A., and Franzin, W. G. (1980). An examination of Cd, Cu and Hg concentrations in livers of northern pike, *Esox lucius,* and the white sucker, *Catostomus commersoni* from five lakes near a base metal smelter at Flin Flon, Manitoba. *Can. J. Fish. Aquat. Sci.*, **37**, 1573–1578.

McIntosh, A., and Bishop, W. (1976). Distribution and effects of heavy metals in a contaminated lake. Purdue University Water Resources Research Center Technical Rep. 85.

McIntosh, A. W., Shephard, B. K., Mayes, R. A., Atchison, G. J., and Nelson, D. W. (1978). Some aspects of sediment distribution and macrophyte cycling of heavy metals in a contaminated lake. *J. Environ. Qual.*, **7**, 301–305.

Michibata, H. (1981). Effect of water hardness on the toxicity of cadmium to the egg of the teleost *Oryzias latipes*. *Bull. Environ. Contam. Toxicol.*, **27**, 187–192.

Miller, G. E., Wile, I., and Hitchin, G. G. (1983). Patterns of accumulation of selected metals in members of the soft-water macrophyte flora of central Ontario Lakes. *Aquat. Bot.*, **15**, 53–64.

Mount, D. I., and Stephan, C. E. (1967). A method of detecting cadmium poisoning in fish. *J. Wildl. Manag.*, **31**, 168–172.

Murphy, B. R., Atchison, G. J., and McIntosh, A. W. (1978a). Cadmium and zinc in the muscle of bluegill (*Lepomis macrochirus*) and largemouth bass (*Micropterus solmoides*) from an industrially contaminated lake. *Environ. Pollut.*, **17**, 253–257.

Murphy, B. R., Atchison, A. W., and Kolar, D. J. (1978b). Cadmium and zinc content of fish from an industrially contaminated lake. *J. Fish. Biol.*, **13**, 327–335.

Ornes, W. H., and Wildman, R. B. (1979). Effects of cadmium (II) on aquatic vascular plants. *Trace Subst. Environ. Health*, **13**, 304–312.

Parker, J. I., Stanlaw, K. A., Marshall, J. S., and Kennedy, C. W. (1982). Sorption and sedimentation of zinc and cadmium by seston in southern Lake Michigan U.S.A. *J. Great Lakes Res.*, **8**, 520–531.

Pärt, P., and Svanberg, O. (1981). Uptake of cadmium in perfused rainbow trout (*Salmo gairdneri*) gills. *Can. J. Fish. Aquat. Sci.*, **38**, 917–924.

Pascoe, D., and Beattie, J. H. (1979). Resistance to cadmium by pretreated rainbow trout alevins. *J. Fish Biol.*, **14**, 303–308.

Pascoe, D., and Mattey, D. L. (1977). Studies on the toxicity of cadmium to the three-spined stickleback *Gasterosteus aculeatus* L. *J. Fish. Biol.,* **11,** 207–215.

Peterson, R. H., Metcalfe, J. L., and Ray, S. (1983). Effects of cadmium on yolk utilization, growth and survival of Atlantic salmon alevins and newly feeding fry. *Arch. Environ. Contam. Toxicol.,* **12,** 37–44.

Pfister, R. M. (1982). Evaluation of bacterial binding and release of cadmium from aquatic sediments. U.S. Department of Interior Report 712437.

Poldoski, J. E. (1979). Cadmium bioaccumulation assays. Their relationship to various ionic equilibria in Lake Superior water. *Environ. Sci. Technol.,* **13,** 701–706.

Poldoski, J. E., and Glass, G. E. (1975). Considerations of trace element chemistry for streams in the Minnesota–Ontario border area. In *Proceedings of the Symposium of the International Conference on Heavy Metals in the Environment, Toronto,* pp. 901–921.

Rachlin, J. W., Jensen, T. E., Baxter, M., and Jani, V. (1982). Utilization of morphometric analysis in evaluating response of *Plectonema boryanum* (Cyanophyceae) to exposure to eight heavy metals. *Arch. Environ. Contam. Toxicol.,* **11,** 323–333.

Rai, L. C., Gaur, J. P., and Kumar, H. D. (1981). Phycology and heavy-metal pollution. *Biol. Rev.,* **56,** 99–151.

Rehwoldt, R., and Karimian-Teherani. (1976). Uptake and effect of cadmium on zebrafish. *Bull. Environ. Contam. Toxicol.,* **15,** 442–446.

Remacle, J. (1981). Cadmium uptake by freshwater bacterial communities. *Wat. Res.,* **15,** 67–71.

Remacle, J., Houba, C., and Ninane, J. (1982). Cadmium fate in bacterial microcosms. *Wat. Air Soil Pollut.,* **18,** 455–465.

Roberts, K. S., Cryer, A., Kay, J., Solbe, J. F. de L. G., Wharfe, J. R., and Simpson, W. R. (1979). The effects of exposure to sublethal concentrations of cadmium on enzyme activities and accumulation of the metal in tissues and organs of rainbow and brown trout (*Salmo gairdneri,* Richardson and *Salmo trutta fario* L.). *Comp. Biochem. Physiol.,* **62C,** 135–140.

Roch, M., and Maly, E. J. (1979). Relationship of cadmium-induced hypocalcemia with mortality in rainbow trout (*Salmo gairdneri*) and the influence of temperature on toxicity. *J. Fish. Res. Bd. Can.,* **36,** 1297–1303.

Rombough, P. J., and Garside, E. T. (1982). Cadmium toxicity and accumulation in eggs and alevins of Atlantic salmon, *Salmo salar. Can. J. Zool.,* **60,** 2006–2014.

Rowe, D. W., and Massaro, E. J. (1974). Cadmium uptake and time dependent alterations in tissue levels in the white catfish *Ictalurus catus* (Pisces: Ictaluridae). *Bull. Environ. Contam. Toxicol.,* **11,** 244–249.

Sakaguchi, T., Tsujii, T., Nakajima, A., and Horikoshi, T. (1979). Accumulation of cadmium by green microalgae. *Eur. J. Appl. Microbiol. Biotechnol.,* **8,** 207–215.

Sangalang, G. B., and Freeman, H. C. (1979). Tissue uptake of cadmium in brook trout during chronic sublethal exposure. *Arch. Environ. Contam. Toxicol.,* **8,** 77–84.

Shephard, B. K., McIntosh, A. W., Atchison, G. J., and Nelson, D. W. (1980). Aspects of the aquatic chemistry of cadmium and zinc in a heavy metal contaminated lake. *Wat. Res.,* **14,** 1061–1066.

Silverberg, B. A. (1976). Cadmium-induced ultrastructural changes in mitochondria of freshwater green algae. *Phycologia,* **15,** 155–159.

Simpson, W. R. (1981). A critical review of cadmium in the marine environment. *Prog. Oceanogr.,* **10,** 1–70.

Smith, B. P., Hejtmancik, E., and Camp, B. J. (1976). Acute effect of cadmium on *Ictalurus punctatus* (Catfish). *Bull. Environ. Contam. Toxicol.,* **15,** 271–277.

Soeder, C. J., Payer, H. D., Runkel, K. H., Beine, J., and Briele, E. (1978). Sorption and concentration of toxic minerals by mass cultures of Chlorococcales. *Mitt. Int. Vereim. Limnol.,* **21,** 575–584.

Spehar, R. L. (1976). Cadmium and zinc toxicity to flagfish, *Jordanella floridae*. *J. Fish. Res. Bd. Can.*, **33**, 1939–1945.

Spehar, R. L., Anderson, R. L., and Fiandt, J. T. (1978a). Toxicity and bioaccumulation of cadmium and lead in aquatic invertebrates. *Environ. Pollut.*, **15**, 195–208.

Spehar, R. L., Leonard, E. N., and Defoe, D. L. (1978b). Chronic effects of cadmium and zinc mixtures on flagfish (*Jordanella floridae*). *Trans. Am. Fish. Soc.*, **107**, 354–360.

Sullivan, J. F., Murphy, B. R., Atchison, G. J., and McIntosh, A. W. (1978). Time dependent cadmium uptake by fathead minnows (*Pimephales promelas*) during field and laboratory exposure. *Hydrobiologia*, **57**, 65–68.

Thomas, D. G., Solbe, J. F. de L. G., Kay, J., and Cryer, A. (1983a). Environmental cadmium is not sequestered by metallothionein in rainbow trout. *Biochem. Biophys. Res. Commun.*, **110**, 584–592.

Thomas, D. G., Cryer, A., Solbe, J. F. de L. G., and Kay, J. (1983b). A comparison of the accumulation and protein binding of environmental cadmium in the gills, kidney and liver of rainbow trout (*Salmo gairdneri* Richardson). *Comp. Biochem. Physiol.*, **76C**, 241–246.

Thorpe, J. H., Giesy, J. P., and Wineriter, S. A. (1979). Effects of chronic cadmium exposure on crayfish survival, growth and tolerance to elevated temperatures. *Arch. Environ. Contam. Toxicol.*, **8**, 449–456.

van der Werff, M., and Pruyt, M. J. (1982). Long-term effects of heavy metals on aquatic plants. *Chemosphere*, **11**, 727–739.

Vinikour, W. S., Goldstein, R. M., and Anderson, R. V. (1980). Bioconcentration patterns of zinc, copper, cadmium and lead in selected fish species from the Fox River, Illinois." *Bull. Environ. Contam. Toxicol.*, **24**, 727–734.

Williams, D. R., and Giesy, J. P. (1978). Relative importance of food and water sources to cadmium uptake by *Gambusia affinis* (Poeciliidae). *Environ. Res.*, **16**, 326–332.

Wilson, D., Finlayson, B., and Morgan, N. (1981). Copper, zinc, and cadmium concentrations of resident trout related to acid-mine wastes. *Calif. Fish Game*, **67**, 176–186.

Wolverton, B. C., and MacDonald, R. C. (1978). Bioaccumulation and detection of trace levels of cadmium in aquatic systems by *Eichomia crassipes*. *Environ. Health Perspect.*, **27**, 161–164.

Woodworth, J., and Pascoe, D. (1983). Induction of cadmium-binding protein in the three-spined stickleback. *Aquat. Toxicol.*, **3**, 141–148.

Woodworth, J., Evans, A. S. A., and Pascoe, D. (1983). The production of cadmium-binding protein in three species of freshwater fish. *Toxicol. Lett.*, **15**, 289–295.

Wright, D. A. (1980). Cadmium and calcium interactions in the freshwater amphipod, *Gammarus pulex* (L.) *Arch. Environ. Contam. Toxicol.*, **10**, 321–328.

Wright, D. A., and Frain, J. W. (1981). The effect of calcium on cadmium toxicity in the freshwater amphipod, *Gammarus pulex* (L.). *Arch. Environ. Contam. Toxicol.*, **10**, 321–328.

Zitko, V., and Carson, W. G. (1976). A mechanism of the effects of water hardness on the lethality of heavy metals to fish. *Chemosphere*, **5**, 299–303.

6

TOXICITY OF CADMIUM TO FRESHWATER MICROORGANISMS, PHYTOPLANKTON, AND INVERTEBRATES

P. T. S. Wong

Great Lakes Fisheries Research Branch
Department of Fisheries and Oceans
Canada Centre for Inland Waters
Burlington, Ontario, Canada

6.1. Cadmium Toxicity to Freshwater Microorganisms
    6.1.1. Effects on Growth
    6.1.2. Effects on Other Physiological Parameters
6.2. Effects on Phytoplankton and Aquatic Macrophytes
    6.2.1. Effects on Phytoplankton
    6.2.2. Effects on Aquatic Macrophytes
6.3. Effects on Invertebrates
    6.3.1. Acute Toxicity
    6.3.2. Chronic Toxicity
        6.3.2.1. Effects on Zooplankton
        6.3.2.2. Effects on Other Invertebrates
6.4. Environmental Factors Affecting Toxicity of Cadmium
    6.4.1. Complexation
    6.4.2. pH
    6.4.3. Hardness
    6.4.4. Temperature, Light, and Oxygen

6.5. Interactions with Other Metals
  6.5.1. Microorganisms
  6.5.2. Phytoplankton
  6.5.3. Invertebrates
Acknowledgments
References

6.1. CADMIUM TOXICITY TO FRESHWATER MICROORGANISMS

Cadmium is not very toxic to microorganisms relative to other aquatic biota (Table 1). The range of cadmium concentrations necessary to cause toxic effects on microorganisms varies from 1,000 to 100,000 µg/L. Because of this insensitivity to cadmium, most of the research on cadmium with microorganisms is focused on the accumulation of the metal (Doyle et al., 1975; Khazaeli and Mitra, 1981) and isolation of tolerant microorganisms (Houba and Remacle, 1980).

6.1.1. Effects on Growth

The range of cadmium concentrations necessary to produce growth inhibition in bacteria ranged from 100 µg/L in *Micromonospora chalcea* to 10,000 µg/L in *Bacillus cereus* and *Enterobacter aerogenes* (Babich and Stotzky, 1977a). In similar studies, Doyle et al. (1975) found that some bacteria such as *Escherichia coli* and *B. cereus* could tolerate 80,000 µg/L while other

Table 1. Effects of Cadmium on Microorganisms

Species	Cadmium (µg/L)	Effect	Reference
(A) Bacteria			
Bacillus megaterium	1,000	Growth inhibition	Babich and Stotzky, 1977a
Enterobacter aerogenes	10,000	Growth inhibition	Babich and Stotzky, 1977a
Escherichia coli	6,000	CO_2 production inhibition	Zwarum, 1973
Lactobacillus acidophilus	40,000	Growth inhibition	Doyle et al., 1975
Micrococcus agilis	1,000	Growth inhibition	Babich and Stotzky, 1977a
Proteus vulgaris	5,000	Growth inhibition	Babich and Stotzky, 1977a
Staphylococcus aureus	40,000	Growth inhibition	Doyle et al., 1975
Streptococcus faecalis	40,000	Growth inhibition	Doyle et al., 1975
(B) Yeasts and fungi			
Aspergillus fischeri	100,000	Growth inhibition	Babich and Stotzky, 1977a
Penicillium vermiculatum	100,000	Growth inhibition	Babich and Stotzky, 1977a
Pythium sp.	4,000	Growth rate reduction	Duddridge and Wainright, 1980
Saccharomycess ellipsoideus	73,000	48-h LC_{50}	Nakamura and Ashida, 1959

bacteria such as *Staphylococcus aureus* and *Streptococcus faecalis* were inhibited. It appeared that gram-negative bacteria were, in general, more resistant to cadmium than the gram-positive bacteria (Cavalli, 1946; Babich and Stotzky, 1977a).

Wide variation was also found in the response of yeasts and fungi to cadmium. In yeasts, *Schizosaccharomyces octosporus* was sensitive to cadmium at concentrations as low as 100 µg/L whereas *Rhodotorula* sp. could tolerate cadmium levels as high as 500,000 µg/L (Avakyan, 1967). Using mycelial extension as a criterion for growth, the fungi were separated into three broad categories based on their sensitivity to cadmium: fungi capable of growth in the range of 10,000–100,000 µg/L; the second group capable of growth at 100,000 µg/L but inhibited by 1×10^6 µg/L; and the third group capable of growth at 1×10^6 µg/L (Babich and Stotzky, 1977a). Some common soil fungi such as *Trichoderma viride* and *Rhizopus stolonifer* occurred in the third group. There was, however, no correlation between the class of the fungus and its sensitivity to cadmium. The sensitivity of the microorganisms to cadmium was quite species specific. For example, *Bacillus megaterium* was unable to grow at 50,000 µg/L Cd, whereas *B. cereus* grew at 100,000 µg/L Cd (Babich and Stotzky, 1977a).

Some bacteria have the ability to adjust to levels of cadmium that initially inhibited their growth. For example, Mitra et al. (1975) showed that *E. coli* could accommodate cadmium-binding protein and compartmentalized the metal intracellularly. The purified protein had a molecular weight of about 39,000 and could bind 0.7 µg Cd/mg protein (Khazaeli and Mitra, 1981).

The resistance or sensitivity of an organism to cadmium may be gene controlled. The penicillinase plasmid in *S. aureus* was shown to contain genes conferring resistance to cadmium ions. The resistant cells allowed calcium, but not cadmium, to enter (Kondo et al., 1978). On the other hand, the resistance or sensitivity could be a physiological adaptation rather than genetic alteration. For example, the yeast, *Saccharomyces elliopsoideus,* after successive culturing in 73,000 µg/L Cd yielded a strain resistant to 146,000 µg/L. However, this increased resistance was not stable, and the tolerance was lost if the cells were not constantly exposed to cadmium (Nakamura and Ashida, 1959).

6.1.2. Effects on Other Physiological Parameters

Apart from growth, other physiological parameters of the microorganisms are also affected by high concentrations of cadmium. The metabolism of glucose by *E. coli* and by water and sediment microorganisms from the Chesapeake Bay and Colgate Creek was inhibited in the presence of 10,000–12,000 µg/L (Zwarum, 1973; Mills and Colwell, 1977). Cadmium also reduced the rates of RNA and protein synthesis in *E. coli* (Blundell and Wild, 1969). Unusual responses to cadmium have been reported for the yeast *Candida utilis*. Cells grown in the presence of cadmium exhibited increased lipolytic activity, presumably due to the stimulation of lipases (Lobyreva, 1973). Other enzymes such as acid and alkaline phosphatases were inhibited by cadmium (Juma and Tabatabai, 1977). Cadmium also influenced fungal spore formation and germination, mycelial growth, and fruiting body formation (Babich and Stotzky, 1978). The spore formation of *Aspergillus niger* was reduced to 35% of the control when the fungus was exposed to 1,000 µg/L Cd. This concentration of cadmium did not significantly alter the mycelial growth of the fungus (Babich and Stotzky, 1977a).

From these results, it is obvious that concentrations of cadmium in the natural environment would have no effect on microorganisms.

6.2. EFFECTS ON PHYTOPLANKTON AND AQUATIC MACROPHYTES

6.2.1. Effects on Phytoplankton

Phytoplankton appear to vary considerably in their sensitivity to cadmium. Several factors have contributed to this variability, many of which are related to experimental design. For example, a number of different species have been examined under different exposure conditions, and a variety of parameters have been measured (e.g., growth rate, morphological change, change in chlorophyll content, and biochemical activity). This situation has been complicated further by the use of complexing agents (e.g., EDTA) in the growth media (Bartlett et al., 1974; Cain et al., 1980) that have complexed the cadmium present and masked the true toxicity. Very few studies have followed the experimental protocol of the standard algal toxicity test (APHA 1985) with regard to recommended test organisms, culture techniques, temperature, illumination, and parameter to measure (growth rate). As a result, it is difficult to evaluate these results with regard to determining a no-observable-effect concentration.

Many of the studies on the toxicity of cadmium to phytoplankton have concerned the effects on growth. The data in Table 2 show the concentration of cadmium that produces a 50% reduction in growth rate (EC_{50}) during 96 h or longer (several days). The 96-h EC_{50} range from 6 to 3000 µg/L Cd for four species of algae, illustrating the variability in response. For four different species tested for longer periods, the EC_{50} are more uniform, ranging from 55 to 120 µg/L Cd. Many studies do not determine the EC_{50} but instead measure the concentration that inhibits growth.

The difference in sensitivity among species is apparently related to the uptake of cadmium and whether the uptake is intra- or extracellular (see Chapter 9). For example, Conway and Williams (1979) found that a concentration of 9 µg/L Cd had no effect on the growth rate of *Fragilaria crotonensis*, whereas for another diatom, *Asterionella formosa*, the rate of growth was reduced by about 50%. It was determined that the accumulation of cadmium was three times lower in *F. crotonensis* than in *A. formosa* and

Table 2. Toxicity of Cadmium to Freshwater Phytoplankton[a]

Species	EC$_{50}$ (μg/L) 96 h	EC$_{50}$ (μg/L) Long term	Reference
Selanastrum capricornutum	6		Sedlacek et al., 1983
Nitzschia closterium	476		Rachlin et al., 1982
Navicula inserta	3000		Rachlin et al., 1982
Chlorella saccharophila	105		Rachlin et al., 1982
Scenedesmus acuminata		110 (13 days)	Hutchinson, 1973
Chlorella vulgaris		55 (14 days)	Hutchinson, 1973
Chlamydomonas eugametos		120 (12 days)	Hutchinson, 1973
C. vulgaris		60 (33 days)	Rosko and Rachlin, 1977

[a] Effect for all species was 50% reduction in growth rate.

that the passive sorption of *F. crotonensis* may have resulted in less accumulation of cadmium in the cell itself. The work of Sakaguchi et al. (1979) further emphasizes the importance of the site of accumulation. They found that concentrations up to 20,000 μg/L Cd had little effect on the growth of *Chlorella regularis*, apparently because the cadmium was accumulated passively, that is, extracellularly.

Other parameters besides growth have also been shown to be dependent on the intracellular concentration of cadmium. For example, the degree of inhibition of CO_2 fixation and O_2 production rates of *Chlorella pyrenoidosa* exposed to concentrations of 250, 500, and 1,000 μg/L Cd were correlated with the intracellular cadmium accumulation (Hart and Scaife, 1977).

Differences in sensitivity between groups of algae (i.e., blue-green and green algae) (Hart and Scaife, 1977) may be related to the structural makeup of the individual cells. For example, the green algae have chloroplasts to carry out photosynthesis whereas the blue-greens do not. They carry out photosynthesis in the cytoplasm. Hart (1975) has suggested that the greater tolerance of the green algae is due to the binding of cadmium in the cytoplasm, thereby preventing it from reaching the chloroplast.

Besides growth, other parameters such as photosynthesis (primary productivity), morphology, nutrient utilization, and enzyme activity have been used as indicators of cadmium's toxicity. As pointed out by Wong et al. (1980), the degree of toxicity determined is dependent on the algal species used and the parameter measured. For instance, Stratton and Corke (1979a) found that photosynthesis and nitrogenase activity were not as sensitive indicators of cadmium toxicity as was growth. On the other hand, Wong et al. (1979) reported that the primary productivity of another alga, *Ankistrodesmus falcatus*, was more sensitive to cadmium than was growth. Conway and Williams (1979) found that the utilization of carbon and phosphorus was reduced somewhat at concentrations of 0.1 μg/L, whereas growth was unaffected.

Silverberg (1976) exposed *A. falcatus*, *C. pyrenoidosa*, and *Scenedesmus quadricauda* to 30, 50, and 100 µg/L Cd to determine whether any ultrastructural changes in the cell organelles could be correlated with cadmium. Only the mitochondria exhibited morphological changes (vacuolization, inclusions, degeneration). In contrast, Burnison et al. (1975) began to observe changes in the endoplasmic reticulum and mitochondria, as well as granule accumulation and the loss of cell processes in the cells of three species of green algae exposed to levels of more than 30 µg/L Cd. In another study, Rachlin et al. (1982) exposed *Plectonema boryanum* to 100,000 µg/L Cd for 4 h and then looked for structural changes. This type of study is typical of the high-concentration, short-exposure period experiments, which are of no use in assessing possible effects of long-term exposure to low concentrations of cadmium.

The effect of cadmium on enzyme activity has also been investigated. Alkaline phosphatase was inhibited by 20 µg/L Cd in *A. falcatus* (Burnison et al. 1975). The permease system responsible for transporting phosphate or silicate into the cells of *A. formosa* was disrupted by cadmium (Conway, 1978). The nitrogenase activity of *Anabaena inaequalis* was reduced by cadmium (Stratton and Corke, 1979a).

Although the results show a wide degree of variability, it appears that concentrations of cadmium in the natural environment, even in polluted areas, would probably have little impact on phytoplankton populations. However, it should be emphasized that standard test procedures such as those recommended by the APHA (1985) should be used so that the variability between studies can be reduced.

6.2.2. Effects of Aquatic Macrophytes

Cadmium has been shown to be relatively nontoxic to aquatic plants (Hazen and Kneip, 1980). The growth and mortality of four aquatic plant species exposed to 100 µg/L Cd for a period of up to 73 days were unaffected (van der Werff and Pruyt, 1982). The authors speculated that the high oxalic acid content of the plants was responsible for chelating the cadmium as a nontoxic cadmium-oxalate. Because of the insensitivity of aquatic plants to cadmium, most of the research has focused on their accumulation potential as possible indicator species (see Chapter 9).

6.3. EFFECTS ON INVERTEBRATES

6.3.1. Acute Toxicity

The acute toxicity of cadmium to freshwater invertebrates has been evaluated on three main groups: crustaceans, aquatic insect larvae, and gastropod mollusks. The results listed in Table 3 should be viewed with caution since almost all of them were derived from studies using static systems and nominal concentrations of cadmium. Based on the geometric mean of the 96-h LC_{50}, the crustaceans are the most sensitive (96-h LC_{50} = 62 μg/L) and the insect larvae the most tolerant (96-h LC_{50} = 5,900 μg/L). The gastropod mollusks are intermediate at 1,500 μg/L Cd.

Within the crustaceans, the geometric mean of the 96-h LC_{50} indicates that the cladocerans are the most sensitive (19 μg/L Cd), the copepods the most tolerant (250 μg/L Cd), and the amphipods intermediate (62 μg/L Cd).

The aquatic insect larvae exhibit a large difference in sensitivity to cadmium, ranging from a 96-h LC_{50} of 840 μg/L (Ephemeroptera, *Atalophebia australis*) to one of 233,000 μg/L (Odonata, *Ishnura heterosticta*) (Table 3). This correlates well with the uptake data for insect larvae (see Chapter 9). The Ephemeroptera accumulated the most cadmium, whereas the Odonata was among those that tended to take up the least. Clubb et al. (1975) found some species that did not suffer any mortality after a 7-day exposure to 17,500 μg/L Cd (e.g., *Atherix variegata, Hexastoma* sp., and *Brachycentrus americanus*).

The variability in the 96-h LC_{50} of gastropods is much less than between aquatic insect larvae, ranging from 300 to 8,400 μg/L Cd. However, it appears that the lethal toxicity continued past 96 h. For example, the LC_{50} for the snail, *Physa integra,* decreased from (μg/L Cd) more than 238 at 96 h to 114 at 7 days and finally to 10.4 at 28 days (Spehar et al., 1978). Similarly for *Physa gyrina* the LC_{50} declined from 1370 μg/L Cd at 96 h to 830 μg/L Cd at 9.5 days (Wier and Walter, 1976). This effect has also been observed in insect larvae (e.g., *Ephemerella grandis grandis*) (Table 3). In addition, Spehar et al. (1978) in a 28-day study with *Ephemerella* sp. noted that significant mortality did not occur until the last week of exposure, when more than 50% of the organisms died. Thorp and Lake (1974) have suggested that the 96-h test period is not long enough to reflect the true acute toxicity of cadmium to freshwater invertebrates. For some species in Table 3, this appears to be the case. A contributing factor to this phenomenon could be that the exuvium (exoskeleton) or shell of the organism had accumulated a significant amount of the cadmium initially present (see Chapter 9), thereby reducing the amount available for internal uptake and elevating the lethal threshold concentration.

6.3.2. Chronic Toxicity

Published studies on the chronic toxicity of cadmium to freshwater invertebrates are rare. Short-term tests are of little use in assessing the sublethal effects of cadmium. For example, Griffiths (1980) suggested that morphological changes in the gut diverticula of *Daphnia magna* could be used as an indicator of cadmium exposure. However, subjecting daphnids to high concentrations of cadmium (12 and 52 μg/L) for a short period of time (42 h) provides little useful information on the long-term effects of low-level cad-

Table 3. Acute Toxicity of Cadmium to Freshwater Invertebrates[a]

Species	LC$_{50}$ (μ/L)						Reference
	24 h	48 h	72 h	96 h	7/10 day	Subacute	
	Crustaceans						
Cladocera							
Daphnia pulex	—	115	—	—	—	—	Ingersoll and Winner, 1982
Daphnia pulex	—	—	—	47	—	—	Bertram and Hart, 1979
Daphnia magna	—	65	—	—	—	5 (21 days)	Biesinger and Christensen, 1972
Daphnia magna	—	58	—	5	—	—	Attar and Maly, 1982
Daphnia sp.	—	1,000	—	—	—	—	Qureshi et al., 1980
Daphnia hyalina	—	55	—	—	—	—	Baudouin and Scoppa, 1974
Daphnia galeata mendotae	—	40[b]	—	30[b]	—	—	Marshall, 1978
Cypris sp.	—	800	—	—	—	—	Qureshi et al., 1980
Moina macrocopa	—	71	—	—	—	—	Hatakeyama and Yasuno, 1981
Copepoda							
Cyclops sp.	—	—	—	340	—	—	Fennikoh et al., 1978
Cyclops abyssorum	—	3,800	—	—	—	—	Baudouin and Scoppa, 1974
Cypridopsis sp.	—	—	—	190	—	—	Fennikoh et al., 1978
Eudiaptomus padanus	—	550	—	—	—	—	Bandouin and Scoppa, 1974
Amphipoda							
Austrochiltonio subtenius	—	90	—	40	—	—	Thorp and Lake, 1974
Ascellus aquaticus	—	—	210	—	—	—	Braginskii and Shcherban, 1978
Hyalella azteca	—	—	—	85	—	—	Fennikoh et al., 1978
Gammarus sp.	—	—	—	70	—	—	Rehwoldt et al., 1973
Decapoda							
Paratya tasmaniensus	—	—	—	60	—	—	Thorp and Lake, 1974

Species							Reference
Gastropoda							
Biomphalaria glabrata	4,800	1,060	—	—	—	—	Bellavere and Gorbi, 1981
Physa integra	—	—	300	—	114	10.4 (28 days)	Spehar et al., 1978
Physa gyrina	7,600	4,250	>238	1,370	—	830 (9.5 days)	Wier and Walter, 1976
Amnicola sp.	—	—	—	8,400	—	—	Rehwoldt et al., 1973
Insects							
Ephemeroptera							
Atalophlebia australis	—	—	840	—	—	—	Thorp and Lake, 1974
Cloeon dipterum	—	930	—	—	—	—	Braginsky and Scherban, 1978
Emphemeralla sp.	—	—	1,000	—	—	<3.0 (28 days)	Spehar et al., 1978
Hexagenia rigida	—	—	2,000	—	—	—	Leonhard et al., 1980
Ephemerella subvaria	—	—	28,000	—	—	—	Warnick and Bell, 1969
Ephemerella grandis grandis	—	—	—	17,500	—	—	Clubb et al., 1975
Odonata							
Ischnura heterosticta	—	—	233,000	—	—	—	Thorp and Lake, 1974
Trichoptera							
Hyudropsyche betteni	—	—	—	—	—	>238 (28 days)	Spehar et al., 1978
Hydropsyche betteni	—	—	—	32,000 (10 days)	—	—	Warnick and Bell, 1969
Plecoptera							
Pteronarcys dorsata	—	—	—	—	—	>238 (280 days)	Spehar et al., 1978
Acroneuria lycorias	—	—	—	—	—	32,000 (14 days)	Warnick and Bell, 1969
Pteronarcella badia	—	—	18,000	—	—	—	Clubb et al., 1975
Diptera							
Chironomus sp.	—	—	1,200	—	—	—	Rehwoldt et al., 1973
Holorusia sp.	—	—	—	42,500 (7 days)	—	—	Clubb et al., 1975

[a] Total hardness ranges from 10 to 130 mg/L CaCO$_3$.
[b] EC$_{50}$: relative number of *D. galeata mendotae* reduced 50%. Total hardness = 200 mg/L CaCO$_3$.

Table 4. Sublethal Responses of Cladocerans (Crustacea) to Cadmium Poisoning[a]

Species	Sublethal Effect Concentration (μg/L)	Effect	Reference
Cladocera			
Daphnia magna	0.7 (21 days)	Reproductive impairment EC_{50}; 50% reduction	Biesinger and Christensen, 1972
Daphnia pulex	1	Reduced average number of broods/producing daphnid	Bertram and Hart, 1979
Daphnia pulex	1	Reduced total young	Bertram and Hart, 1979
Daphnia pulex	1	Reproductive impairment (reduced young)	Bertram and Hart, 1979
Daphnia galeata mendotae	1	Mean prenatal deaths significantly higher than controls	Marshall, 1978
Moina macrocopa	0.78 (20 days)	50% reproductive impairment	Hatakeyama and Yasuno, 1981
Daphnia galeata mendotae	7.7 (157 days)	EC_{50}; carrying capacity of population reduced 50%	Marshall, 1978
Daphnia galeata mendotae	55 (157 days)	EC_{50}; mean life expectancy reduced by 50% from that of controls	Marshall and Mellinger, 1980
Daphnia pulex	5	Reduced longevity adult daphnid	Bertram and Hart, 1979
Mixed crustacean (zooplankton) population	3.5 (24–31 days)	EC_{50}; 50% reduction of relative crustacean abundance	Marshall and Mellinger, 1980
	5.2 (21 days)	EC_{50}; 50% reduction in percentage similarity of crustacean population	Marshall and Mellinger, 1980

[a] Total hardness ranges from 45 to 82 mg/L $CaCO_3$.

mium pollution and is not a very sensitive indicator of cadmium intoxication (see Table 4).

6.3.2.1. Effects on Zooplankton

Most of the available studies are concerned largely with the long-term effects of cadmium on cladocerans. A number of sublethal responses have been examined (Table 4). The average longevity of adult daphnids has been shown to have been reduced at 5 μg/L Cd (Marshall, 1978; Bertram and Hart, 1979). Marshall (1978) and Marshall and Mellinger (1980) investigated the effects of cadmium on various parameters of zooplankton populations. They determined that significant reductions in the carrying capacity of the population, the relative crustacean abundance, and the community similarity occurred at concentrations in the range of 3.5–7.7 µg/L Cd (Table 2).

The most sensitive indication of cadmium stress was that of impaired reproduction (Table 4). Bertram and Hart (1979) found that for *Daphnia pulex* the average number of young per brood and the total number of progeny per experiment were the best reproductive parameters to assess the effects of cadmium because they were unaffected by the longevity of the adult. Both parameters were significantly reduced at the lowest concentration tested, 1 µg/L Cd. Hatakeyama and Yasuno (1981) also found the average number per brood to be significantly lower than that of the controls in *Moina macrocopa* at a concentration of 1.2–1.6 µg/L Cd. The mean number of prenatal deaths (total number of aborted eggs and embryos) was more than four times higher in *Daphnia galeata mendotae* exposed to 1 µg/L Cd than in the controls (Marshall, 1978). Similarly, Biesinger and Christensen (1972) determined that *D. magna* suffered a 50% reduction in reproductive capability during a 3-week exposure to 0.7 µg/L Cd. Hence, it appears that the threshold of reproductive impairment in cladocerans is less than 1 µg/L Cd. However, it should be pointed out that in most of the studies listed in Table 4, the concentrations of cadmium in the water were not measured and may not have accurately reflected those present.

Marshall (1978) reported that sublethal concentrations of cadmium had reduced the average number of individuals in the population of *D. galeata mendotae* and also reduced the biomass. The population responded by increasing the birth rate to compensate for the elevated mortality rate. Hence, cadmium-stressed populations of cladocerans may be able to adjust to low levels of cadmium. However, in the field, the ability to adjust could be compromised by other stress factors such as predation and the presence of other toxic substances.

It should also be noted that zooplankton may be exposed to cadmium through the food chain. Bertram and Hart (1979) grew the alga *C. pyrenoidosa* in a 500-µg/L Cd solution for 40 h and then used it as a food source for *D. pulex*. They found that although the average life span was unaffected, the average number of broods per adult and the average number of young per brood were significantly reduced for the organisms fed the cadmium-contaminated algae. In addition, the total number of offspring from daphnids fed contaminated algae was only 38% of that of the controls. However, the question remains as to how the amount of cadmium accumulated by the algae exposed to a very high concentration in the lab relates to the amounts found in algae exposed to much lower levels in the field.

6.3.2.2. *Effects on Other Invertebrates*

There is only fragmentary information available concerning the effects of sublethal levels of cadmium on invertebrates, other than cladocerans. Thorp et al. (1979) exposed the crayfish, *Cambarus latimandus,* to concentrations of cadmium of 5 and 10 µg/L over a 5-month period and found that those organisms at the highest concentration suffered a significantly increased rate of mortality (approaching 20%) relative to the controls. Long-term studies of this nature point out the weaknesses of short-term studies at high concentrations. For instance, Gillespie et al. (1977) reported the crayfish *Orconectes propinquus propinquus* to be resistant to cadmium because after an 8-day exposure to a concentration of 100 µg/L Cd only 1 of 15 organisms had died.

Abnormal behavior in the form of a vigorous curling motion was observed for caddisflies exposed to sublethal concentrations of cadmium of 85.5 and 238 µg/L. This may be related to the convulsive hyperactivity induced by cadmium in fish (see Chapter 7).

6.4. ENVIRONMENTAL FACTORS AFFECTING TOXICITY OF CADMIUM

There are very few studies on the effects of environmental factors on cadmium toxicity to freshwater microorganisms, phytoplankton, and invertebrates. Such factors should include complexation, pH, hardness, temperature, oxygen, and light. These factors may influence the toxicity of cadmium by affecting its chemical speciation or its chemical/physical mobility and bioavailability (Babich and Stotzky, 1983).

6.4.1. Complexation

Cadmium in the natural environment is subject to complexation by natural organic compounds, chelators, and ligands excreted by some organisms and clay minerals. In general, natural organic compounds such as humic acid bind cadmium and reduce its toxicity. Giesy et al. (1977) observed that cadmium was less toxic to the daphnid *Simocephalus serrulatus* in highly organic pond water than in soft well water. Similarly, the accumulation and toxicity of cadmium to the green alga *Selanastrum capricornutum* grown in a medium devoid of chelating agents in the laboratory was diminished by aquatic humus (Sedlacek et al., 1983). However, this situation may be complicated by seasonal influences in the field. Laegreid et al. (1983) investigated the toxicity of cadmium to *S. capricornutum* in a eutrophic lake from May to September and found that the toxicity progressively increased until the lake turned over in the autumn. Because they found no relationship between the dissolved organic cabon and the observed toxicity, they suggested that some of the toxicity was due to the complexed metal species themselves, specifically those of low molecular weight. The smallest molecular weight fractions of humic substances have increased the metal uptake in phytoplanton and zooplankton (Sedlacek et al 1983; Giesy et al., 1977). In addition, the ligand diethyldithiocarbamate increased the uptake of cadmium by a daphnid to an extent greater than that of the free ionic form (Poldoski, 1979) (see Chapter 9). Thus, it appears that although most organic fractions reduce the toxicity of cadmium, some may actually increase it by facilitating uptake.

Chelating agents such as NTA (nitrilotriacetic acid) also tend to reduce the toxicity of cadmium. For example, Hongue et al. (1983) found that the addition of NTA ameliorated the toxic effect of cadmium on the photosynthesizing ability of a natural freshwater phytoplankton community. Similarly, EDTA (ethylenediaminetetraacetic acid) reduced the toxicity of cadmium to the bacterium *Klebsiella pneumoniae* (Pickett and Dean, 1979).

The ability to excrete complexing ligands to regulate the concentrations of heavy metals in the environment has also been noted in phytoplankton and bacteria. Van den Berg et al. (1979) found that different algal species excreted different amounts of ligands. On the one hand, it was suggested that the concentrations of excreted ligands were threshold concentrations necessary to prevent any toxic action by cadmium on the individual species concerned. On the other, it was thought that the complexing ligands might also protect other species in the water column. For example, in a study of 22 lakes, it was found that the toxic effects of cadmium on algae were either strongly reduced or completely arrested once the complexing capacity of the lake water attained a level greater than 2 µmol/L of copper equivalent (Chiaudani and Vighi, 1978).

The clay minerals kaolinite and montmorillonite were both able to reduce the toxicity of cadmium to fungi and bacteria (Babich and Stotzky, 1977b) because of their cation exchange capacity. The cations (e.g., H^+, K^+, Mg^{2+}, and Ca^{2+}) on the clays exchanged and removed cadmium from the medium. The greater the cation exchange capacity, the greater the protection of clay minerals.

6.4.2. pH

Cadmium is more toxic to phytoplankton at lower rather than at higher pH. Gipps and Coller (1980) observed that increasing the pH from 6.5 to 8.3 decreased the toxicity of cadmium to *C. pyrenoidosa*. Similarly, cadmium was more toxic to algae at pH 7 than at pH 8 because the algae accumulated twice as much cadmium at the lower pH (Hart and Sciafe, 1977). Moshe et al. (1972) found that cadmium at 1000 µg/L was very toxic to algae in laboratory experiments but not in the experimental oxidation ponds. The authors attributed the negative effects to the high pH (above 8) in the oxidation ponds, which caused metal ions to precipitate in the form of hydroxides.

The toxicity of cadmium to microorganisms seems to be related to the pH of the environment. Alkaline pH levels generally increased the toxicity of the metal to the bacteria and fungi (Babich and Stotzky, 1977a). It was unclear whether this increased toxicity reflected the complex ionic and molecular species of hydroxylated cadmium that would predominate at pH

8–9. Thus, the normal generalization that cadmium exerts a great toxicity at low pH may be unsound, at least in the case of soil microorganisms.

6.4.3. Hardness

Water hardness affects the toxicity of cadmium to aquatic organisms, especially fish. In the case of invertebrates, cadmium is more toxic to the animals in soft water than in hard water. The 48-h LC_{50} for an invertebrate, *Tubifex tubifex*, increased from 2.8 μg/L Cd in distilled water to 31 μg/L Cd at a hardness of 34 mg/L (as $CaCO_3$) to 720 μg/L Cd at a hardness of 261 mg/L (Brkovic-Popovic and Popovic, 1977). Cadmium toxicity to rotifers and *D. magna* was also lower in hard water than in soft water (Buikema et al., 1974; Bellavere and Gorbi, 1981).

In phytoplankton, cadmium was found to be more toxic to several species of algae in soft water (20 mg/L as $CaCO_3$) than in hard water (300 mg/L as $CaCO_3$) (Burnison et al., 1975). Since the solubility of cadmium in soft and hard waters was calculated to be about the same, the differences were thought to be due to an antagonistic effect of calcium (Davies, 1976) and magnesium (Hutchinson and Czyrska, 1972).

6.4.4. Temperature, Light, and Oxygen

Higher temperatures generally increase the toxicity of cadmium. For example, the estuarine crab *Paragrapsus gaimaridii* was more sensitive to cadmium at 19 °C than at 5 °C (Sullivan, 1977). The increased metabolic rate and accumulation of cadmium in the animal might explain the increased toxicity. Cadmium toxicity may also be enhanced by light. The toxicity of cadmium to microcrustaceans was shown to be greater in "light" or shallow epilimnetic incubations than in "dark" or deep epilimnetic incubations in Lake Michigan water (Marshall and Mellinger, 1980). The enhanced toxicity of cadmium with light was probably an indirect effect resulting from the reduction of phytoplankton production (food supply) by cadmium. That the phytoplankton were more sensitive to cadmium in light than in dark was demonstrated by Kogan et al. (1975) and Monahan (1975). Cadmium was relatively nontoxic to *Chlorella vulgaris* at low light intensity but became quite toxic when light intensity was increased. Similarly, the growth of *Scenedesmus obtusuisculus* was more reduced by 50 μg/L Cd in the presence of light intensity of 4000 lux than at 2000 lux. The enhanced toxicity of cadmium to algae in a higher light intensity could be explained by the increased metabolic activities of the algae and hence the higher accumulation of cadmium in the algae with light.

Not much is known about the effect of oxygen on cadmium toxicity. Clubb et al. (1975) observed increased mortality and uptake of cadmium with increasing oxygen concentration in several species of aquatic insects. The enhanced toxicity might be related to the increased metabolic rate of the animal with increasing oxygen. The data also suggested that cadmium might have less effect on aquatic insects in an oxygen deprived environment than in one with normal oxygen levels.

6.5. INTERACTIONS WITH OTHER METALS

A small number of studies have looked at the interactive effects of other metals on cadmium toxicity that would be ecologically significant, since cadmium rarely exists singly in the environment. There is more than one scheme for classifying and naming effects of chemicals that act simultaneously. A simple approach using the toxic unit concept is given by Sprague (1970). If an organism is exposed to half the concentration of toxicant A necessary to produce a given response and half the concentration of toxicant B necessary for the same response, and if this combination just elicits the response, the actions of A and B are exactly additive. On the other hand, if it causes more than the given response, the actions of A and B are more than additive. If it does not cause the response, the toxicants are less than additive, show no interaction, or are antagonistic, and further experimentation is necessary. Further details on the terminology in describing the combined effects of toxicants are provided in an excellent review by Sprague (1970).

6.5.1. Microorganisms

The literature on the interactive effects of cadmium with other metals on microorganisms is scarce. The growth of *E. coli* was inhibited in a medium containing 2,000 µg/L Mg and 2,000 µg/L Cd. An increase of magnesium to 20,000 µg/L resulted in much less growth inhibition by cadmium. However, the elevated magnesium concentration did not have much of a protective effect if cadmium was over 6,000 µg/L (Abelson and Aldous, 1950). Another study shows that zinc would also reduce the growth inhibition of *E. coli* by cadmium, especially when the bacterium was exposed to zinc prior to the addition of cadmium (Mitra et al., 1975).

From these results, it is almost impossible to predict the effects on aquatic organisms of cadmium interacting with other metals. Various factors such as different species, concentrations of the metals, and sequence of metal addition can change the interactive effects.

6.5.2. Phytoplankton

The 10 metals, including cadmium, individually present at levels equivalent to the objectives set by the Water Quality Subcommittee of the International Joint Commission (Cd 0.2; As 50; Cr 50; Cu 5; Pb 25; Fe 300; Hg 0.2; Ni 25; Se 10; and Zn 30 µg/L) were not toxic to natural phytoplankton from Lake Ontario water or to several species of pure algal cultures. However, a mixture of all the metals at the above-mentioned concentrations strongly inhibited growth and productivity when present together (Wong et al., 1978, 1982). These more-than-additive effects showed that water quality objectives based on studies with single metals are insufficient for protecting phytoplankton if several metals are present simultaneously. Gachter (1976) also observed that the simultaneous action of four metals (Hg 1; Cu 0.3; Cd 6; and Zn 4 µg/L) greatly reduced the primary productivity of a lake water. The toxicity appeared to be more than additive. The toxic effects of a mixture of mercury, cadmium, and nickel on a freshwater blue-green alga, *Anabaena inaequalis,* were investigated by Stratton and Corke (1979b). When cadmium and mercury were used at sublethal concentrations, the metal mixture interacted more than additively in reducing cell yield after 6 days of exposure but less than additively in cell yield when the incubation was extended to 12 days. A mixture of cadmium, mercury, and nickel caused an additive effect at day 6 but a less-than-additive effect in cell yield at day 12. The results suggest that incubation time is important in determining the interactive effect of the metal mixture.

Hutchinson and Czyrska (1972, 1975) studied the effects of cadmium and zinc, alone and together, on two floating aquatic plants, duckweed (*Lemna valdiviana*) and a fern (*Salvinia natans*). A concentration of 50 µg/L Zn had a stimulatory effect on both plants. The stimulatory effect on zinc was suppressed by the presence of 10 and 30 µg/L Cd. Hutchinson and Stokes (1975) reported that the growth of *Chlorella* sp. was reduced to 55% of the controls by 50 µg/L Cd and to 70% by 50 µg/L Se when the metals were alone. In combination, growth was reduced only by 19%, indicating the antagonistic effect of cadmium and selenium. At 50 µg/L Cd and 100 µ/L Se, growth was stimulated compared with the controls. In another study, zinc was reported to reduce the toxic effect of cadmium on the growth of *Euglena gracilis* (Nakano et al., 1981). At the toxic levels of cadmium (5–100 µg/L), increasing zinc concentrations from 100 to 20,000 µg/L reduced the cadmium toxicity. It is not clear from the data whether the effect is less than additive or antagonistic. The growth rate of *S. capricornutum* treated with copper, zinc, and cadmium individually and in combination was examined by Bartlett et al. (1974). The growth of the alga was completely inhibited by the addition of nominal concentrations of individual metals (Cu 90 µg/L; Zn 120 µg/L; or Cd 80 µg/L). The combination of copper (50 µg/L) and cadmium (20 µg/L) resulted in less growth inhibition than an equal concentration of copper (70 µg/L), suggesting a less-than-additive effect for the Cu–Cd mixture. Kneip (1978) reported that the exposures of an ecosystem consisting of fish, plankton, and benthic organisms to nickel and Ni–Cd combinations did not show an additive effect in Ni–Cd combination. Mortality or suppression of population growth rates could be accounted for essentially from the exposure to cadmium alone. However, the author did not provide enough data to substantiate the conclusion. Pietilainen (1975) tested the toxicity of lead and cadmium in the laboratory using natural phytoplankton communities from brackish water. Addition of 100 µg/L Cd reduced the primary production by 40%, and 1000 µg/L Pb reduced the primary production by 43%. The combination of 100 µg/L Cd and 1000 µg/L Pb reduced the primary production only by 69%, suggesting that the effect is less than additive.

6.5.3. Invertebrates

Borgmann (1980) examined the effects of cadmium, copper, mercury, lead, arsenic, and zinc on the biomass production rates of natural assemblages of freshwater copepods. Concentrations for the single-metal tests were 45 µg/L Cd and Cu, 30 µg/L Hg, 6,000 µg/L Pb, 7,500 µg/L As, and 900 µg/L Zn. The metal mixture contained one-fifth these concentrations with either zinc or arsenic omitted from each test. The observed toxicity of the five-metal mixture was higher than the calculated toxicity of a mixture of five metals with each metal at one-fifth the concentration of the single-metal tests, indicating a more-than-additive effect of the metal mixture. The sublethal study of D'Agostino and Finney (1974) on the interactive effecs of copper and cadmium on the copepod *Tigriopus japonicus* indicated that cadmium sulfate (43 µg/L) inhibited the development of ovigerous females and hence the production of the F_2 generation. Cuprous chloride at 64 µg/L inhibited the development of ovigerous females. When the copepod was reared in media containing cadmium and copper simultaneously, 10 times less than the concentrations were needed to cause an equivalent impairment when the metals had been tested individually. The results suggest that cadmium and copper exerted a more-than-additive effect on the copepod.

On the other hand, Attar and Maly (1982) have claimed that cadmium and zinc acted antagonistically in causing lethality to *D. magna*. Their data are insufficient for calculating whether the effects are less than additive or antagonistic. The results by Thorp and Lake (1974) on the effect of cadmium and zinc on freshwater shrimp, *Paratya tasmaniensis,* clearly demonstrate the less-than-additive interaction of these two metals. The 96-h LC_{50} of 60 µg/L Cd and 1100 µg/L Zn were each taken as one toxic unit. The combined effect on the Cd–Zn mixture was less toxic than the aggregated effect of cadmium and zinc separately at concentrations less than about one toxic unit, whereas above this level the combined effect was nearly additive. The two metals also acted less than additively in combination when evaluating their effects on total crustacean density and species diversity in a zooplankton community in Lake Michigan water (Marshall et al., 1981).

ACKNOWLEDGMENTS

The author wishes to express his appreciation to Dr. J. B. Sprague and I. R. McCracken for their critical review and constructive comments on the manuscript and to Mrs. Caryl Fawcett for typing the manuscript.

REFERENCES

Abelson, P. H., and Aldous, E. (1950). Ion antagonisms in microorganisms: Interference of normal magnesium metabolism by nickel, cobalt, cadmium, zinc and manganese. *J. Bacteriol.*, **60**, 401–413.

APHA (1985). American Public Health Association, American Water Works Association and Water Pollution Control Federation. Bioassay Methods for Aquatic Organisms. In Standard Method for the Examination of Water and Wastewater. Greenberg, A. E., Connors, J. J. Jenkins, D. (eds.). APHA, Washington, DC, pp. 689–823.

Attar, E. N., and Maly, E. J. (1982). Acute toxicity of cadmium, zinc, and cadmium-zinc mixtures to *Daphnia magna*. *Arch. Environ. Contam. Toxicol.*, **11**, 291–296.

Avakyan, Z. A. (1967). Comparative toxicity of heavy metals for certain microorganisms. *Mikrobiologiya*, **36**, 446–450.

Babich, H., and Stotzky, G. (1977a). Sensitivity of various bacteria including actinomycetes and fungi to cadmium and the influence of pH on sensitivity. *Appl. Environ. Microbiol.*, **33**, 681–685.

Babich, H., and Stotzky, G. (1977b). Reductions in the toxicity of cadmium to microorganisms by clay minerals. *Appl. Environ. Microbiol.*, **33**, 696–705.

Babich, H., and Stotzky, G. (1978). Effects of cadmium on the biota: Influence of environmental factors. *Adv. Appl. Microbiol.*, **23**, 55–117.

Babich, H., and Stotzky, G. (1983). Developing standards for environmental toxicants: The need to consider abiotic environmental factors and microbe-mediated ecologic processes. *Environ. Health Persp.*, **49**, 247–260.

Bartlett, L., Rabe, F. W., and Funk, W. H. (1974). Effect of copper, zinc and cadmium on *Selenastrum capricornutum*. *Wat. Res.*, **8**, 179–185.

Baudouin, M. F., and Scoppa, P. (1974). Acute toxicity of various metals to freshwater zooplankton. *Bull. Environ. Contam. Toxicol.*, **12**, 745–751.

Bellavere, C., and Gorbi, J. (1981). A comparative analysis of acute toxicity of chromium, copper and cadmium to *Daphnia magna*, *Biomphalaria glabrata* and *Brachydanio rerio*. *Environ. Technol. Lett.*, **2**, 119–128.

Bertram, P. E., and Hart, B. A. (1979). Longevity and reproduction of *Daphnia pulex* (de Geer) exposed to cadmium-contaminated food or water. *Environ. Pollut.*, **19**, 295–305.

Biesinger, K. E., and Christensen, G. M. (1972). Effects of various metals on survival, growth, reproduction and metabolism of *Daphnia magna*. *J. Fish. Res. Bd. Can.*, **29**, 1691–1700.

Blundell, M. R., and Wild, D. G. (1969). Inhibition of bacterial growth by metal salts. A survey of effects on the synthesis of ribonucleic acid and protein. *Biochem. J.*, **115**, 207–212.

Borgmann, U. (1980). Interactive effects of metals in mixtures on biomass production kinetics of freshwater copepods. *Can. J. Fish. Aquat. Sci.*, **37**, 1295–1302.

Braginskii, L. P., and Shcherban, E. P. (1978). Acute toxicity of heavy metals to aquatic invertebrates at different temperatures. *Gidrobiol. Z. H.*, **14**, 86–92.

Brkovic-Popovic, I., and Popovic, M. (1977). Effects of heavy metals on survival and respiration rate of tubificid worms: Part 1. Effects on survival. *Environ. Pollut.*, **13**, 65–72.

Buikema, A. L. Jr., Cairns, J. Jr., and Sullivan, G. W. (1974). Rotifers as monitors of heavy metal pollution in water. Virginia Polytechnic Institute and State University. Blacksburg, VA, pp. 1–73.

Burnison, G., Wong, P. T. S., Chau, Y. K., and Silverberg, B. A. (1975). Toxicity of cadmium to freshwater algae. *Proc. Can. Fed. Biol. Soc. Winnipeg*, **18**, 182.

Cain, J. R., Paschal, D. C., and Hayden, C. M. (1980). Toxicity and bioaccumulation of cadmium in the colonial green algae *Scenedesmus obliquus*. *Arch. Environ. Contam. Toxicol.*, **9**, 9–16.

Cavalli, V. (1946). Sullattivita antibatterica del cloruro di cadmio. *Atti Accad. Fisiocrit. Siena Sez. Med. Fis.*, **14**, 578–588.

Chiaudani, G., and Vighi, M. (1978). The use of *Selenastrum capricornutum* batch cultures in toxicity studies. *Mitt. Int. Ver. Limnol.*, **21**, 316–329.

Clubb, R. W., and Gaufin, A. R., and Lords, J. L. (1975). Acute cadmium toxicity studies upon nine species of aquatic insects. *Environ. Res.*, **9**, 332–341.

Conway, H. L. (1978). Sorption of arsenic and cadmium and their effects on growth, micronutrient utilization, and photosynthetic pigment composition of *Asterionella formosa*. *J. Fish. Res. Bd. Can.*, **35**, 286–294.

Conway, H. L., and Williams, S. C. (1979). Sorption of cadmium and its effect on growth and the utilization of inorganic carbon and phosphorus of two freshwater diatoms. *J. Fish. Res. Bd. Can.*, **36**, 579–586.

D'Agostino, A., and Finney, C. (1974). The effect of copper and cadmium on the development of *Tigriopus japonicus*. In *Pollution and Physiology of Marine Organisms*, Vernberg, F. J., and Vernberg, W. B. (eds.). Academic, New York, pp. 445–463.

Davies, P. H. (1976). The need to establish heavy metal standards on the basis of dissolved metals. Workshops on Toxicity to Biota of Metal Forms in Natural Water. International Joint Commission, Great Lakes Research Advisory Board, pp. 93–126.

Doyle, J. J., Marshall, R. T., and Pfander, W. H. (1975). Effects of cadmium on the growth and uptake of cadmium by microorganisms. *Appl. Microbiol.*, **29**, 562–564.

Duddridge, J. E., and Wainwright, M. (1980). Heavy metal accumulation by aquatic fungi and reduction in viability of *Gammarus pulex* fed Cd^{2+} contaminated mycelium. *Wat. Res.*, **14**, 1605–1611.

Fennikoh, K. B., Hirshfield, H. I., and Kneip, T. J. (1978). Cadmium toxicity in planktonic organisms of a freshwater food web. *Environ. Res.*, **15**, 357–367.

Gachter, R. (1976). Untersuchungen uber die Beeinflussung der planktischen photosynthese durch anorganische Metallsalze im eutrophen Alpnachersee und der mesotrophen Horwerbucht. *Schweiz. Z. Hydrol.*, **38**, 97–119.

Giesy, J. P. Jr., Leversee, G. J., and Williams, D. R. (1977). Effects of naturally occurring aquatic organic fractions on cadmium toxicity to *Simocephalus serrulatus* (Daphnidae) and *Gambusia affinis* (Poeciliidae). *Wat. Res.*, **11**, 1013–1021.

Gillespie, R., Reisine, T., and Massaro, E. J. (1977). Cadmium uptake by the crayfish, *Orconectes propinquus* (Girard). *Environ. Res.*, **13**, 364–368.

Gipps, J. F., and Coller, B. A. W. (1980). Effect of physical and culture conditions on uptake of cadmium by *Chlorella pyrenoidosa*. *Austral. J. Mar. Freshwater Res.*, **31**, 747–755.

Griffiths, P. R. E. (1980). Morphological and ultrastructural effects of sublethal cadmium poisoning on *Daphnia*. *Environ. Res.*, **22**, 277–284.

Hart, B. A. (1975). Bioconcentration and toxicity of cadmium in *Chlorella pyrenoidosa*. In *The Effect of Cadmium on Freshwater Phytoplankton*. PB 257–547, Office of Water Research and Technology, Washington, DC, pp. 1–31.

Hart, B. A., and Scaifes, B. D. (1977). Toxicity and bioaccumulation of cadmium in *Chlorella pyrenoidosa*. *Environ. Res.*, **14**, 401–413.

Hatakeyama, S., and Yasuno, M. (1981). Effects of cadmium on the periodicity of parturition and brood size of *Moina macrocopa* (Cladocera). *Environ. Pollut. (Ser. A)*, **26**, 111–120.

Hazen, R. E., and Kneip, T. J. (1980). Biogeochemical cycling of cadmium in a marsh ecosystem. In *Cadmium in the Environment*, Part I, *Ecological Cycling*, Nriagu, J. O. (ed.). Wiley, New York pp. 399–424.

Hongue, D., Skogheim, O. K., Hindar, A., and Abrahamsen, A. (1980). Effects of heavy metals in combination with NTA, humic acid, and suspended sediment on natural phytoplankton photosynthesis. *Bull. Environ. Contam. Toxicol.*, **25**, 594–600.

Houba, C., and Remacle, J. (1980). The composition of saprophytic bacterial communities in freshwater systems contaminated by heavy metals. *Microbiol. Ecol.*, **6**, 55–69.

Hutchinson, T. C. (1973). Comparative studies of the toxicity of heavy metals to phytoplankton and their synergistic interactions. *Wat. Pollut. Res. Can.*, **8**, 68–75.

Hutchinson, T. C., and Czyrska, H. (1972). Cadmium and zinc toxicity and synergism to floating aquatic plants. *Wat. Pollut. Res. Can.*, **7**, 59–65.

Hutchinson, T. C., and Czyrska, H. (1975). Heavy metal toxicity and synergism to floating aquatic weeds. *Verh. Int. Ver. Limnol.*, **19**, 2101–2111.

Hutchinson, T. C., and Stokes, P. M. (1975). Heavy metal toxicity and algae bioassays. In *Water Quality Parameters*. ASTM STP 573, American Society for Testing and Materials, Philadelphia, PA, pp. 320–343.

Ingersoll, C. G., and Winner, R. W. (1982). Effect on *Daphnia pulex* of daily pulse exposure to copper or cadmium. *Environ. Toxicol. Chem.*, **1**, 321–328.

Juma, N. G., and Tabatabai, M. A. (1977). Effects of trace elements on phosphatase activity in soils. *Soil Sci. Soc. Am. J.*, **41**, 343–346.

Khazaeli, M. B., and Mitra, R. S. (1981). Cadmium-binding component in *Escherichia coli* during accommodation to low levels of this ion. *Appl. Environ. Microbiol.*, **41**, 46–50.

Kneip, T. J. (1978). Effects of cadmium in an aquatic environment. In *Proceedings of the First International Cadmium Conference, San Francisco, January 1977*, Cadmium Assoc., New York, NY. pp. 120–124.

Kogan, I. G., Anikeeva, I. D., and Vanlina, E. N. (1975). Effect of cadmium on *Chlorella* II. Modification of the UV irradiation effect. *Genetika (Moscow)*, **11**, 84–87 [*Sov. Genet.* (English translation), **11**, 1550–1553].

Kondo, I., Ishikawa, T., and Nakahara, H. (1978). Mercury and cadmium resistances mediated by the penicillinase plasmid in *Staphylococcus auereus*. *J. Bacteriol.*, **117**, 1–7.

Laborey, F., and Lavollay, J. (1977). Sur l'antitoxicité du calcium et du magnesium a l'égard du cadmium, dans la croissance *d'Aspergillus niger*. *C. R. Acad. Sci.*, **284D**, 639–642.

Laegreid, M., Alstad, J., Klaveness, D., and Selp, H. M. (1983). Seasonal variation of cadmium toxicity toward the alga *Selenastrum capricornutum* Printz in two lakes with different humus content. *Environ. Sci. Technol.*, **17**, 357–361.

Leonhard, S. L., Lawrence, S. G., Friesen, M. K., and Flannagan, J. F. (1980). Evaluation of the acute toxicity of the heavy metal cadmium to nymphs of the burrowing mayfly *Hexagenia rigida*. In *Advances in Ephemeroptera Biology*, Flannagan, J. F., and Marshall, K. E. (eds.). Plenum, New York, pp. 457–465.

Lobyreva, L. B. (1973). Effect of metal ions on lipolytic activity of the yeast, *Candida utilis* 295t. *Izv. Akad. Nauk, SSR Otd. Biol. Nauk*, **4**, 581–584 [*Biol. Abstr.*, **57**, 51473 (1974)].

Marshall, J. S. (1978). Population dynamics of *Daphnia galeata mendotae* as modified by chronic cadmium stress. *J. Fish. Res. Bd. Can.*, **35**, 461–469.

Marshall, J. S., and Mellinger, D. L. (1980). Dynamics of cadmium-stressed plankton communities. *Can. J. Fish. Aquat. Sci.*, **37**, 403–414.

Marshall, J. S., Mellinger, D. L., and Parker, J. I. (1981). Combined effects of cadmium and zinc on a Lake Michigan zooplankton community. *J. Great Lakes Res.*, **7**, 215–223.

Mills, A. L., and Colwell, R. R. (1977). Microbiological effects of metal ions in Chesapeake Bay water and sediment. *Bull. Environ. Contam. Toxicol.*, **18**, 99–103.

Mitra, R. S., Gray, R. H., Chin, B., and Bernstein, I. A. (1975). Molecular mechanisms of accommodation in *Escherichia coli* to toxic levels of Cd. *J. Bacterioil.*, **121**, 1180–1188.

Monahan, T. J. (1975). Effects of cadmium on the growth and morphology of *Scenedesmus obtusuisculus*. *J. Phycol.*, **12** (Suppl.), 98.

Moshe, M., Betzer, N., and Kott, Y. (1972). Effect of industrial wastes on oxidation pond performance. *Wat. Res.*, **6**, 1165–1171.

Nakamura, H., and Ashida, J. (1959). Adaptation of yeast to cadmium. I. An introductory approach to the resistance mechanism. *Mem. Coll. Sci., Univ. Kyoto, Ser. B*, **26**, 323–336.

Nakano, Y., Abe, K., and Toda, S. (1981). Effect of cadmium on growth of *Euglena gracilis* grown in the zinc-optimum range. *J. Environ. Sci. Health*, **16A**, 175–187.

Pickett, A. W., and Dean, A. C. R. (1979). Cadmium and zinc sensitivity and tolerance in *Klebsiella (Aerobacter) aerogenes*. *Microbios*, **15**, 79–91.

Pietilainen, K. (1975). Synergistic and antagonistic effects of lead and cadmium on aquatic primary production. In *Proceedings of the International Conference on Heavy Metals in the Environment*, Hutchinson, T. C. (ed.). University of Toronto, Toronto, Canada, pp. 861–873.

Poldoski, J. E. (1979). Cadmium bioaccumulation assays. Their relationship to various ionic equilibria in Lake Superior Water. *Environ. Sci. Technol.*, **13**, 701–706.

Qureshi, S. A., Saksena, A. B., and Singh, V. P. (1980). Acute toxicity of some heavy metals to fish food organisms. *Int. J. Environ. Stud.*, **14**, 325–327.

Rachlin, J. W., Jensen, T. E., Baxter, M., and Jani, V. (1982). Utilization of morphometric analysis in evaluating response of *Plectonema boryanum* (Cyanophyceae) to exposure to eight heavy metals. *Arch. Environ. Contam. Toxicol.*, **11**, 323–333.

Rehwoldt, R., Lasko, R. L., Shaw, C. and Wirhowski, E. (1973). The acute toxicity of some heavy metal ions toward benthic organisms. *Bull. Environ. Contam. Toxicol.*, **10**, 291–294.

Rosko, J. J., and Rachlin, J. W. (1977). The effect of cadmium, copper, mercury, zinc and lead on cell division, growth, and chlorophyll a content of the chlorophyte *Chlorella vulgaris*. *Bull. Torrey Bot. Club*, **104**, 226–233.

Sakaguchi, T., Tsujii, A., Nakjima, H., and Horikoshi, A. (1979). Accumulation of cadmium by green microalgae. *Environ. J. Appl. Microbiol. Biotechnol.*, **8**, 207–215.

Sedlacek, J., Källqvist, T., and Gjessing, E. (1983). Effect of aquatic humus on uptake and toxicity of cadmium to *Selanastrum capricornutum* Printz. In *Aquatic and Terrestrial Humic Materials*, Christman, R. F., and Gjessing, E. T. (eds.). Ann Arbor Science, Ann Arbor, Michigan, pp. 495–516.

Silverberg, B. A. (1976). Cadmium-induced ultrastructural changes in mitochondria of freshwater green algae. *Phycologia*, **15**, 155–159.

Spehar, R. L., Anderson, R. L., and Fiandt, J. T. (1978). Toxicity and bioaccumulation of cadmium and lead in aquatic invertebrates. *Environ. Pollut.*, **15**, 195–208.

Sprague, J. B. (1970). Measurement of pollutant toxicity to fish. II. Utilizing and applying bioassay results. *Wat. Res.*, **4**, 3–32.

Stratton, G. W., and Corke, C. T. (1979a). The effect of cadmium ion on the growth, photosynthesis, and nitrogenase activity of *Anabaena inaequalis*. *Chemosphere*, **5**, 277–282.

Stratton, G. W., and Corke, C. T. (1979b). The effect of mercury, cadmium and nickel ion combinations on a blue-green alga. *Chemosphere*, **10**, 731–740.

Sullivan, J. K. (1977). Effects of salinity and temperature on the acute toxicity of cadmium to the estuarine crab, *Paragrapsus gaimaridii* (Milne Edwards). *Austral. J. Mar. Freshwater Res.*, **28**, 739–748.

Thorp, V. J., and Lake, P. S. (1974). Toxicity bioassays of cadmium on selected freshwater invertebrates and the interaction of cadmium and zinc on the freshwater shrimp. *Paratya tasmaniensis* Riek. *Austal. J. Mar. Freshwater Res.*, **25**, 97–104.

Thorp, J. H., Giesy, J. P., and Wineriter, S. A. (1979). Effects of chronic cadmium exposure on crayfish survival, growth and tolerance to elevated temperatures. *Arch. Environ. Contam. Toxicol.*, **8**, 449–456.

van den Berg, C. M. G., Wong, P. T. S., and Chau, Y. K. (1979). Measurement of complexing materials excreted from algae and their ability to ameliorate copper toxicity. *J. Fish. Res. Bd. Can.*, **36**, 901–905.

van der Werff, M., and Pruyt, M. J. (1982). Long-term effects of heavy metals on aquatic plants. *Chemosphere,* **11,** 727–739.

Warnick, S. L., and Bell, H. L. (1969). The acute toxicity of some heavy metals to different species of aquatic insects. *J. Wat. Pollut. Contr. Fed.,* **41,** 280–284.

Wier, C. G., and Walter, W. M. (1976). Toxicity of cadmium in the freshwater snail, *Physa gyrina* Say. *J. Environ. Qual.,* **5,** 359–362.

Wong, P. T. S., Chau, Y. K., and Luxon, P. L. (1978). Toxicity of a mixture of metals on freshwater algae. *J. Fish. Res. Bd. Can.,* **35,** 479–481.

Wong, P. T. S., Burnison, G., and Chau, Y. K. (1979). Cadmium toxicity to freshwater algae. *Bull. Environ. Contam. Toxicol.,* **23,** 487–490.

Wong, P. T. S., Mayfield, C. I., and Chau, Y. K. (1980). Cadmium toxicity to phytoplankton and microorganisms. In *Cadmium in the Environment,* Part I. *Ecological Cycling,* Nriagu, J. O. (ed.). Wiley, New York, pp. 571–585.

Wong, P. T. S., Chau, Y. K., and Patel, D. (1982). Physiological and biochemical responses of several freshwater algae to a mixture of metals. *Chemosphere,* **11,** 367–376.

Zwarum, A. A. (1973). Tolerance of *Escherichia coli* to cadmium. *J. Environ. Qual.,* **2,** 353–355.

7

EFFECTS OF CADMIUM ON FRESHWATER FISH

J. B. Sprague

Department of Zoology
College of Biological Science
University of Guelph
Guelph, Ontario, Canada

7.1. Introduction
7.2. Acute Lethality to Salmonid Fish
7.3. Acute Lethality to Other Groups of Fish
7.4. Modifying Factors and Toxic Forms of Cadmium
    7.4.1. Inorganic Substances in Water
    7.4.2. Organic Substances in Water
    7.4.3. Other Modifying Factors
    7.4.4. Physiological Responses of Fish as Modifying Factors
7.5. Mode of Lethal Action
7.6. Combined Toxicity in Mixtures
7.7. Sublethal Life-cycle Exposures
7.8. Other Sublethal Studies
7.9. Sublethal Effects and Residues in Tissues
7.10. Relevance of Toxicity Data to Cadmium Levels in Surface Waters
7.11. Summary and Conclusions
Appendix: Acute Lethality and Its Relationship to Water Hardness
References

7.1. INTRODUCTION

The findings for cadmium toxicity to freshwater fish show a great deal of variation as is expected for any toxicant. Simplifications have been made in this chapter in order to look for overall patterns, in particular by taking geometric averages of several values to produce one concentration. For example, if an author did three tests and reported three similar LC_{50} under the same test conditions, the geometric average has been used to yield one LC_{50}. In sublethal tests, authors usually report the lowest concentration causing an effect and the somewhat lower concentration that did not result in a significant effect (the concentration of no observed effect, or NOEC). The geometric mean of those two concentrations has been used as an estimate of the threshold of sublethal effect. The influence of such simplification is considered negligible compared with between-lab and between-test variation and with the variation usually found in chemical measurements of cadmium.

Almost all toxicity evaluations have been done in the laboratory in relatively clean water, presumably with suspended materials and organic complexing agents virtually absent. Hence, the toxic concentrations reported in this chapter represent cadmium that is easily available to organisms, probably almost entirely ionic or dissolved cadmium. This may be satisfactory for applying to field situations, since most cadmium in fresh waters is thought to be bioavailable (see Section 7.4). However, there would be situations in which some of the cadmium measured in surface waters was complexed or particulate, and this should be borne in mind when utilizing laboratory findings to interpret conditions in the field.

7.2. ACUTE LETHALITY TO SALMONID FISH

Salmonids seem to be very sensitive to cadmium. Although most results are for rainbow trout, the brook trout and the Pacific salmons do not appear to differ greatly in tolerance. Incipient or threshold lethal levels are reported in the range 1–30 μg/L Cd.

There appears to be a relationship between sensitivity and type of water as characterized by total hardness. Hardness has been taken as a measure of a group of water characteristics that are generally correlated with each other (notably alkalinity, calcium and magnesium content, and pH). Greater toxicity in soft water may be seen in the 4–10-day LC_{50} in Table 1, which are the values plotted in the lower part of Figure 1. A line may be fitted to this relationship (details in Appendix) with the equation

$$\log_{10}(LC_{50}) = 0.8334[\log_{10}(\text{hardness})] - 0.6938.$$

From this, we would expect a threshold LC_{50} for salmonids of 2.5 μg/L in water of hardness 20 and 27 μg/L in water of hardness 350 mg/L. This order-

Table 1. Synopsis of Tests of Lethality of Cadmium to Salmonid Fish in Fresh Water[a]

Species	Hardness	LC_{50} 2 days	4 days	7/10 day	Subacute	Notes	Reference
Salmo gairdneri, rainbow trout	14	—	—	—	6	42 days	U.K., 1976
	20	(100)	—	—	—	No details but reliable	Brown, 1968
	20?	—	—	~10^b	—	—	Ball, 1967
	20	91	1.0	0.9	—	—	Calamari et al., 1980
	24	—	1.8	—	—	—	Chapman, 1978
	31	—	—	—	—	—	Davies, 1976
	54	—	—	—	5.2	Adults, 17 days	Chapman and Stevens, 1978
	"Soft"	(120)	—	—	—	Abstract only	Rausina et al., 1975
	80	358	6.0	5.3	—	—	Calamari et al., 1980
	82	9.1	—	16.3	—	—	Kumada et al., 1980
	82	—	30	30	—	Adults, incipient LC_{50}	Majewski and Giles, 1981
	125	—	—	15	—	—	Roch and Maly, 1979
	250	—	—	—	—	42 days	U.K., 1976
	290	~7000^b	~2500^b	~9^b	—	—	Ball, 1967
	300	(5000)	—	—	—	No details but reliable	Brown, 1968
	320	3700	—	—	—	—	Calamari et al., 1980
Salvelinus fontinalis, brook trout	44	—	<12 → 405^b	—	—	Unsatisfactory procedures	Benoit et al., 1976
	(41)	(2.5)	(<1.5)	—	—	Calcium hardness only	Carroll et al., 1979
	(340)	(~64)	(~27)	—	—	—	
Oncorhynchus kisutchi, coho trout	22	—	—	3.7	—	Adults	Chapman and Stevens, 1978
	33	—	3.1	—	—	—	Buckley et al., 1985
	44	—	5.3	—	—	—	Buckley et al., 1985
	89	—	10.4	9	—	In river water	Lorz et al., 1978
Oncorhynchus tshawytscha, chinook trout	20	—	1.1	—	—	—	Finlayson & Verrue, 1982
	24	—	3.5	2.0	—	—	Chapman, 1978

[a] Most tests are with "juveniles" or fingerlings, and tests with alevins or newly hatched fish are not included here. Subacute tests are taken to be exposures longer than 10 days. Hardness refers to total hardness of test water as mg/L of $CaCO_3$. All LC_{50} are in micrograms per liter of Cd. LC_{50} in parentheses lack some detail or are questionable as noted.
[b] "Flat" section in toxicity curve, as explained in text.

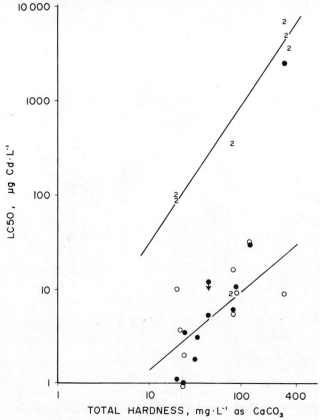

Figure 1. Acutely lethal concentrations of cadmium for salmonid fish. The symbol 2 represents a 2-day LC_{50}, closed points are 4-day LC_{50}, and open circles are 7/10-day LC_{50}. Results for recently hatched fry are not included here. The lower fitted line used 7/10-day values where available and 4-day values if the longer ones were not available. One 2-day value was not used to fit the upper line.

of-magnitude change is similar to the relationship for some other metals (Brown, 1968). The reason for the change in toxicity with type of water is discussed in Section 7.4.

For short exposures of 48 h, most authors have found that about 30–250 times more cadmium is required to kill salmonids than in the longer exposures (see Table 1, upper line of Fig. 1, and Appendix for derivation of the line). This difference is large, and results from a most unusual relationship between exposure time and acute lethality. This was well documented by Ball (1967), whose results are shown in Figure 2. On the right side of Figure 2, there is the usual relationship of increasing survival time at lower concentration (=decreasing LC_{50} with longer exposure times). This prevails for the first 1–6 days of exposure. Judging by most toxicants, one would expect this toxicity curve to turn upward at about 1000 μg/L Cd, but instead it runs

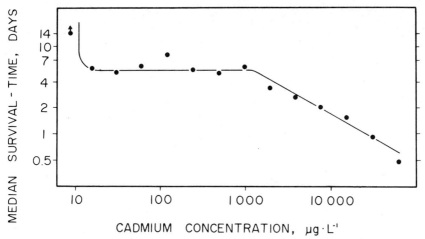

Figure 2. Toxicity curve for rainbow trout exposed to cadmium in hard water. The horizontal part of the curve shows the unusual feature that survival time is not related to concentrations spanning two orders of magnitude. The fitted line passes within the confidence limits of all median survival times. [Redrawn from *Water Research,* Volume 1, after Ball (1967) with the permission of Pergamon Journals.]

horizontally. There is little or no change in survival time over two orders of magnitude of concentration of cadmium. Many authors have found such "flat" parts to toxicity curves of cadmium for salmonid fish (Table 1). To generalize from their findings, it appears that the right-hand part of the toxicity curve results from the usual dose-related and rapid poisoning of some tissue or tissues that cease to function, causing death of the fish. Over the flat part of the curve, several authors have reported that death is associated with violent overreaction to external stimuli, uncoordinated swimming, and coma. This secondary mechanism of toxicity has been frequently noted in salmonids but rarely in other types of fish. It would seem to be associated with dysfunction of the nervous or muscular systems (see Section 7.5).

The LC_{50} for fingerlings or adults exposed for some weeks are generally fairly close to the lower line shown in Figure 1. Apparently, the secondary mechanism of poisoning is not a major factor at concentrations below the 10-d LC_{50}.

7.3. ACUTE LETHALITY TO OTHER GROUPS OF FISH

For nonsalmonid fishes, it is difficult to say much except that results for cadmium lethality are variable (Fig. 3). Any relationship to water hardness that may exist is obscured by this variation. Inspection of data in Table 2 does not give any clear differences in tolerance by families of fish.

Results vary from 100 to 23,000 µg/L, but many are in the range 700–7000

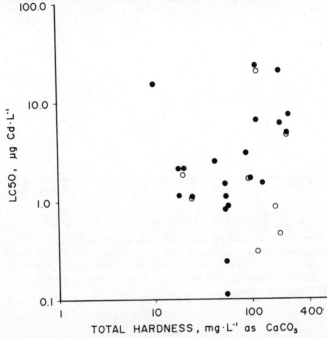

Figure 3. Acute lethality of cadmium to nonsalmonid fish in fresh water. Solid points represent 4-day LC_{50}, open circles are 7/10-day LC_{50}. If both were available from the same report, they are plotted in the figure.

µg/L, and the geometric mean is in the vicinity of 1660 µg/L. Obviously, these data are not satisfactorily described by a line relating LC_{50} to hardness. An attempt at fitting yielded a line that traversed the graph almost horizontally from an LC_{50} of 1500 µg/L in very soft water to one of 2000 µg/L in very hard water.

These 4–10-day LC_{50} for "other fish" are similar to the 2-day values for salmonids. The flat response seen with salmonids, apparently from a secondary mechanism of poisoning, was noted only for fathead minnows and sticklebacks, species that yielded some of the lower values in Figure 3. Sticklebacks in particular gave variable responses, including the highest one shown in figure 3, and the third lowest (Table 2).

7.4. MODIFYING FACTORS AND TOXIC FORMS OF CADMIUM

The chief modifying factor for toxicity of cadmium to freshwater fish is the general nature of the water and in particular those qualities associated with hardness. These govern the solubility of cadmium, the chemical species present, and the biological factors within the fish that determine the rates of cadmium uptake.

7.4.1. Inorganic Substances in Water

It might be thought that chemical changes of cadmium would explain its differing toxicity in various types of water. Earlier investigators pointed out that reduced toxicity in water of higher hardness/pH was accompanied by some precipitation of cadmium in the test tanks because of the lesser solubility in such water (Pickering and Gast, 1972; Eaton, 1974). Later research has shown, however, that speciation of cadmium does not explain the changes in toxicity. Davies (1976) investigated this by chemical and toxicity tests in soft water (Table 1) and in very hard water of pH 8.2 and total hardness 326 mg/L. The research deserves some consideration since it seems to be the only numerical comparison of toxicity with the chemical behavior of inorganic cadmium. Davies calculated that in the high-hardness water, the solubility of cadmium (in μg/L) would be 450 at pH 7, 140 at pH 7.5, 45 at pH 8, 14 at pH 8.5, and only 4.5 at pH 9. If more cadmium were added to the test tanks, it would form carbonate ($CdCO_3$), or hydroxide [$Cd(OH)_2$] at very high pH, both forms insoluble, although they may stay in the water as suspended precipitates. The soluble forms were calculated to be the free ion (Cd^{2+}) and another hydroxide [$Cd(OH)^+$]. At the pH of the experiment (8.2), the hydroxide was an order of magnitude lower in concentration than the ionic cadmium. (The two forms would be about equal in concentration at pH 9.) Cadmium was approximately 16.5 times less toxic to rainbow trout in the hard water than in the soft water. There was no way to explain this by the chemical speciation since the presumed toxic form, ionic cadmium, was present in the hard water at a concentration much higher than the LC_{50} in soft water.

Davies ended by implying a biological antagonism for cadmium toxicity in hard water, and other authors agree that the chemical form of cadmium is of lesser importance. Voyer et al. (1975) concluded that suspended carbonates were toxic. Calamari et al. (1980) demonstrated that ionic cadmium does not have the same toxicity in different types of water, wrote of the "relative independence of toxic effects from the chemical species," and concluded that "all the inorganic forms of [the] metal contribute to toxicity." To add to the argument that the chemical species of cadmium do not govern toxicity, Pickering and Gast (1972) suggested that *dissolved* cadmium might be more toxic at higher pH. This could well be the case since this has been shown for dissolved copper (Howarth and Sprague, 1978) and dissolved zinc (Bradley and Sprague, 1985).

We are left with the supposition that factors associated with water hardness act within the fish to govern the toxicity of cadmium. Two groups of researchers have shown this with elegance. Calamari et al. (1980) demonstrated for cadmium what has previously been known for zinc (Lloyd, 1965),

Table 2. Synopsis of Tests on Lethality of Cadmium to Species of Fish Other Than Salmonids in Fresh Water[a]

Species	Hardness	LC_{50}				Notes	Reference
		2 days	4 days	7/10 day	Subacute		
Pimephales promelas, fathead minnow	200	—	~6000[b]	450	—	—	Pickering and Gast, 1972
	130	—	1560	—	—	—	Benson and Birge, 1985
Carassius auratus, goldfish	220–270	—	7160	—	—	—	Benson and Birge, 1985
	20	2800	2100	1800	—	—	McCarty et al., 1978
Cyprinus carpio, carp	17	—	—	—	>100	30 days	Muramoto, 1981
	55	300	240	—	—	—	Rehwoldt et al., 1972
Ptychocheilus oregonesis, northern squawfish	25	—	1100	1050	—	—	Andros and Garton, 1980
Neomacheilus barbatulus, stoneloach	240	—	4800	4500	2000	54 days	Solbe and Flook, 1975
Ictalurus punctatus, channel catfish	"soft"	—	(3530)	—	—	Abstract only	Rausina et al., 1975
Catostomus commersoni, white sucker	18	1410	1110	—	—	—	Duncan and Klaverkamp, 1983
Gambusia affinis, mosquitofish	10	—	~15,500	—	—	—	Giesy et al., 1977
Jordanella floridae, flagfish	43	—	2500	—	—	—	Spehar, 1976

Species						Reference
Brachydenio rerio, zebrafish	100	1700	1700	—	—	Dave et al., 1981
Gasterosteus aculeatus, stickleback	115	—	6500^b	300–6000^b	—	Pascoe and Cram, 1977
	115	—	—	—	32	44-d, starved fish	Pascoe and Woodworth, 1980
	120	—	23,000^b	20,000^b	1	33 days	Pascoe and Mattey, 1977
Fundulus diaphanus, banded killifish	55	210	110	—	—	Rehwoldt et al., 1972
Fundulus heteroclitus, Mummichog	90?	>3100	~3000	—	—	Voyer et al., 1975
Anguilla rostrata, Eel	55	1100	820	—	—	Rehwoldt et al., 1972
Roccus americanus, white perch	55	1100	840?	—	—	Rehwoldt et al., 1972
Roccus saxatilis, striped bass	55	1500	1100	—	—	Rehwoldt et al., 1972
Lepomis gibbosus, pumpkinseed	55	2200	1500	—	—	Rehwoldt et al., 1972
Lepomis macrochirus, bluegill	18	2800	2250 (2530)	—	—	Bishop and McIntosh, 1981
	"soft"	—	—	—	—	Abstract only	Rausina et al., 1975
	180	—	—	850	850	138 days	Cearley and Coleman, 1974
	200	—	20,400	—	—	Eaton, 1974

^{a} All LC$_{50}$ are in micrograms per liter of Cd^{2+}. Tests with alevins or recently hatched fish are not included here. Hardness refers to total hardness of test water as milligrams per liter of CaCo$_3$.

^{b} "Flat" section to toxicity curve noted by authors.

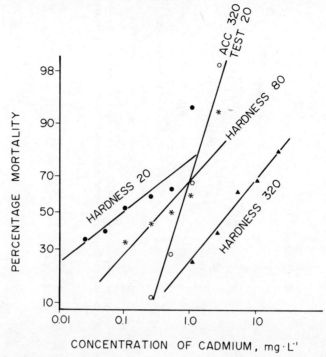

Figure 4. Acute lethality of water-borne cadmium to rainbow trout in waters of differing hardness. Three lines of similar slope are labeled with total hardness (mg/L $CaCO_3$) used in acclimation and testing of fish. The fourth line represents acclimation at hardness 320 and testing at hardness 20. [Redrawn from *Water Research,* Volume 14, after Calamari et al. (1980) with the permission of Pergamon Journals.]

that it takes several days for fish to acclimate to a new hardness of the water in which they are living. Before that, they react to a toxic metal as if they were in the "old" water, no matter what the hardness of the "new" water. The calcium content of fish tissue increases with that of the water surrounding them (Houston, 1959), and this apparently changes the intake rate of metal and its subsequent toxicity. Figure 4 shows that the mortality lines of Calamari et al. (1980) follow the expected pattern for cadmium lethality to fish acclimated and tested at hardness of 20, 80, or 320 mg/L. In contrast, fish acclimated at hardness 320 and then tested at hardness 20 showed a probit line of intermediate position and quite different slope. Calamari and colleagues did not find any change in chloride cells in the gills with various acclimations, so it was not an effect of increased excretion of cadmium by such cells in the harder water. The remaining explanation is decreased uptake in harder waters, and that has been documented by Part et al. (1985). In perfusion experiments with the dissected anterior portions of trout, they clearly showed that the flow of cadmium from the water into the gills was proportional to the calcium content of the water and, to a much lesser extent, to the magnesium content of the water. This classic work by Part and colleagues confirms the speculation by many authors, that calcium decreases the permeability of gills to metals, or at least to cadmium.

7.4.2. Organic Substances in Water

Organic matter in natural waters binds and detoxifies many metals, and apparently cadmium is no exception. Certainly, the standard chelating agents such as EDTA reduce both the uptake and toxicity of cadmium for fish (Muramoto, 1981). Buckley et al. (1985) tested coho salmon in river water and also in a 33% mixture of treated sewage in river water. Based on measurements of total cadmium, the LC_{50} in the sewage mixture averaged almost 1.6 times higher, a significant difference. Clearly, there was more binding and detoxification of cadmium in the mixture, presumably largely by organic matter. Buckley and associates also expressed the LC_{50} in terms of labile cadmium as measured by an ion exchange resin. Results indicated the presence of at least some free cadmium ion among the easily available forms. There was no significant difference between the "labile" LC_{50} for river water and sewage mixture, and the authors concluded that such measurement of the labile forms was a more accurate estimate of the toxicity than was measurement of total cadmium. The proportion of resin-labile cadmium was significantly correlated with alkalinity and total organic carbon of the waters, suggesting that both inorganic and organic binding may have taken place. Although Buckley et al. (1985) state that the nonlabile cadmium, according to the resin procedure, was 20% in river water and 30% in the sewage mixture, their toxicity data suggest that at least 28% of the cadmium in river water (and at least 39% in the mixture) may not have been bioavailable. Buckley et al. provide a good review of recent literature on chemical aspects of binding of cadmium and other metals.

Other results indicate a role for organic agents in binding cadmium. Humic acid prolonged the survival of salmon appreciably at the one concentration of cadmium tested (Gjessing, 1981). Suspended sediment was of overriding importance in reducing toxicity of cadmium for an aquatic invertebrate (*Daphnia*; Schuytema et al., 1984). Highly organic pond water reduced cadmium toxicity to another aquatic invertebrate but had little effect for mosquitofish because the binding capacity of the organic matter was small compared to the concentrations required to cause lethality for the fish (Giesy et al., 1977).

Unfortunately, prediction of the effects of organic binding on toxicity of cadmium by means of chemical tests appears to be somewhat uncertain at present. Buckley et al. (1985) point out that estimates of the cadmium-complexing ability of water vary with the chemical procedure used, and that the estimates are not related to the LC_{50} obtained.

It is apparent that organic matter binds only part of the cadmium in fresh water, and in some cases organic binding appears to be less important than might be thought. Even in a eutrophic lake polluted by cadmium and zinc, organic binding was effective for only 21% of total cadmium, and 50% of the metal persisted in the ionic form (Shephard et al., 1980). This finding substantiates the earlier work of Gardiner (1974). He found that Cd^{2+} made up 90% of the total cadmium in groundwater, an average of 54% in various river waters, and, surprisingly, 33% of the total in various sewage effluents. Organically bound cadmium was not present in groundwater, made up about 18% of the total in river water, and 38% in sewage effluents. $CdCO_3$ was the second most important inorganic form.

From all of the above findings, it appears that the greater proportion of any cadmium pollution of surface waters will remain in toxic forms; ionic cadmium itself is usually half of total cadmium or more.

7.4.3. Other Modifying Factors

The level of dissolved oxygen affects toxicity; cadmium's lethality was increased by a factor of 2.3 for one-quarter saturation with oxygen, compared to that at saturation (as judged by the LC_{50}; Voyer et al., 1975).

Added stress increases toxicity as might be expected. Mean survival time in 32 μg/L Cd was only 13 days in fish that were starved and parasite laden, compared to 44 days for starved fish without parasites (Pascoe and Woodworth, 1980).

7.4.4. Physiological Responses of Fish as Modifying Factors

Acclimation to cadmium can also modify the lethal level. Several investigators have found that preexposure to low levels of cadmium increased the survival times in a given lethal concentration, but such findings may or may not be indicative of a change in tolerance (i.e., a change in the lethal threshold concentration). For example, Pascoe and Beattie (1979) found that unacclimated rainbow trout alevins were killed by 100 μg/L Cd in about 2 days, but alevins acclimated to 10 μg Cd/L required a 7-day exposure to cause lethality. Duncan and Klaverkamp (1983) documented a true increase in tolerance of almost 2.5 times following acclimation to 0.66 of the normal lethal level. They found that preexposure to 0.19 of the LC_{50} did not result in a change in tolerance, a finding that agrees with the thresholds for stimulating an acclimation response to other metals (Chapman, 1985; Dixon and Sprague, 1981). Considerable variation was found in the acclimation response of fathead minnows to very low levels of cadmium by Benson and Birge (1985). A significant tolerance increase by a factor of 1.8 following acclimation to 10 μg Cd/L (only 0.0064 of the normal LC_{50}) seems unusual since preexposure to higher levels of up to 100 μg Cd/L stimulated smaller increases in tolerance that were not significantly different from the control response.

An apparent modifying factor within the fish that assists in increasing the tolerance is development of increased amounts of metal-binding proteins. A great deal of recent research on many organisms including fish makes it clear that these proteins are part of the system used to handle metals within the organisms, and cadmium is certainly one of the metals. In particular, metallothioneins are manufactured by fish, and their increased levels are closely associated with increased tolerance of lethal levels of metals (Chapman, 1985; Klaverkamp et al., 1984). Metallothioneins also increase during sublethal exposures and no doubt assist in the recovery of physiological homeostasis in a matter of days, as documented by Giles (1984).

A series of perceptive papers from the University of Victoria, British Columbia, culminated in one by Roch et al. (1986), which integrates their findings and provides entrance to the literature on metallothioneins (MT) in fish. Roch et al. show linear relationships between the level of MT in salmonid fish and the severity of sublethal exposure to zinc, copper, and cadmium. Although cadmium played a minor role in the mixture, this metal alone stimulates development of metal-binding protein in trout (Dixon and Sprague, 1981), and there is little doubt that there would be a relationship similar to that shown by Roch et al. (1986). Their relationship for metal mixtures prevailed in laboratory exposures (Roch and McCarter, 1984a) and also in field exposures for an area affected by mining pollution (Roch and McCarter, 1984b). Using the confidence limits of their regressions, Roch et al. (1986) were able to estimate the concentration of metal below which the MT concentrations of exposed fish could not be distinguished from those of nonexposed fish. They argued that this might be taken as an estimae of the "safe" or no-effect level of metal in the water. Their estimate for cadmium was <0.2 µg/L at a water hardness of 25 mg/L. That value is considerably lower than the effect threshold of 1.5 g/L estimated below for similar water but is relatively close to the recommendation of 0.4 µg Cd/L of soluble cadmium given by the American Fisheries Society for sensitive species (Davies et al., 1979).

7.5. MODE OF LETHAL ACTION

Destruction of gill tissue has been found to be the most rapidly acting mechanism of lethal effect for many metals, but this has not been clearly demonstrated for cadmium. Mummichogs removed from a lethal exposure did not show gill damage, except at extremely high concentrations (Voyer et al., 1975). No great effect on gill morphology was found in 8-month exposures of trout to 8 µg/L, which would be about half of the lethal threshold (Hughes et al., 1979). The lack of evidence for gill damage is somewhat surprising since

Mount and Stephan (1967) showed that there was a clear cutoff point for accumulation of cadmium in fish gill; those that died of acute cadmium poisoning had more than 130 μg Cd/g dry weight in the gill, while survivors had less than that.

There is some information on the "secondary" mechanism of acute lethality, that is, that which is associated with a flat section of the toxicity curve for salmonids. Benoit et al. (1976) made daily observations of brook trout in a lethality test. All test concentrations from 12 to 180 μg/L resulted in 90% mortality in 4 days. In another test kept isolated for 4 days and in which no observations were made, no mortality of fish occurred in a higher range of concentrations from 82 to 405 μg/L. When daily observations were made for a further 3 days on these same higher concentrations, they all produced 80% mortality. Hence the LC_{50} depended on the actions of the observer. Benoit and co-workers reviewed the numerous other findings of this apparent effect on the nervous system of salmonids and some other fish manifested by hyperirritability and convulsive uncoordinated swimming. Cearley and Coleman (1974) credit this secondary effect to inhibition of acetylcholinesterase and death by paralysis of the respiratory control system. In nature, salmonids would presumably be exposed to many stimuli, and the lower or secondary threshold of lethality would be relevant to survival of cadmium pollution.

A perceptive piece of research by Roch and Maly (1979) gives an alternative explanation of the secondary mechanism of acute lethality. These authors obtained a flat region of the toxicity curve for rainbow trout, extending over the range 30–3200 μg/L. In the middle of that range (300 μg/L), they demonstrated declines as great as 50% in the calcium content of the plasma in trout. The authors pointed out that the decline in calcium fits the symptoms of the "secondary" mortality very well since this can "severely impair nerve and muscle function, and hypersensitivity and . . . tetany are common symptoms of hypocalcemia in humans." Roch and Maly (1979, p. 1301) are almost certainly correct. With a decline of similar proportion in plasma calcium of humans, "the nerve fibres become more and more excitable and may discharge spontaneously, causing muscles to twitch and go into spasms or even tetany" (Solomon and Davis, 1978, p. 526).

Roch and Maly (1979) also suggested that the hypocalcemia might be caused by renal damage, a plausible hypothesis in view of the high cadmium levels and kidney damage known for other vertebrates, but this is apparently not the case. From an intensive study of electrolyte and water balance in cadmium-exposed fish, Giles (1984) concluded that kidney damage was not the cause of low plasma calcium, although the reduction might be caused by "impairment in the mechanisms responsible for calcium influx." Giles further concluded that his observations supported the suggestion of McCarty and Houston (1976), that sublethal cadmium concentrations resulted in a shift of water and electrolytes from the extracellular to the intracellular fluid spaces.

7.6. COMBINED TOXICITY IN MIXTURES

There has been so little research on effects of cadmium acting simultaneously with other toxicants that this section can deal with both lethal and sublethal findings on freshwater fish. The most deliberate study of lethal effects of cadmium together with other metals was by Finlayson and Verrue (1982). With copper, the results suggested a less-than-additive combined effect on salmon, or no interaction (the lethal concentration of cadmium was unchanged by the presence of more than one-third of the lethal level of copper). With zinc, no interaction was also suggested. Trimetal mixtures showed definite antagonism, that is, the presence of one metal reduced the toxicity of others. Low levels of copper appeared to be the operative factor. For example, median mortality of salmon was caused by a mixture containing 1.6 times the LC_{50} of cadmium alone, plus 2.6 times the LC_{50} of zinc and 0.56 times the LC_{50} of copper; if those numbers are simply added, they indicate a total presence of metals at 4.8 times their individual lethal levels. Finlayson and Verrue (1982) were somewhat more cautious in their interpretation based on statistical analysis and assigned one of their copper–cadmium tests to an "additive" category, whereas there was a suggestion of less-than-additive effects as indicated above. One of their main conclusions was that the exact kind of combined action depended very much on the proportions of the metals.

Those results contrast with the few previous findings with mixtures that included cadmium. Eaton (1973) reports that the lethal effects of cadmium on fathead minnows were approximately additive with the effects of zinc and copper present in the mixture, as had been found by Bandt (1946) for cadmium and zinc. Subsequently, Roch and McCarter (1984a,b) studied the lethal effects of a trimetal mixture with zinc and copper, but levels of cadmium were so low that they considered its influence to be negligible. They did point out that the metals were additive in the mixture, unlike the findings of Finlayson and Verrue (1982).

Sublethal effects are of more interest, however, and apparently cadmium does not contribute as fully as might be expected to the toxicity of metal mixtures. Cadmium and zinc mixtures had no greater effect on survival of larval flagfish than did zinc alone, and although there was some joint action on the reproductive performance of adults, it was considerably less than additive (Spehar et al., 1978). The chronic effects of cadmium in life-cycle tests with fathead minnows were apparently reduced when zinc and copper were also present (i.e., antagomism; Eaton, 1973). Low levels of cadmium and copper did not appear to have a more adverse effect on gill ATPase of coho salmon than copper alone. However, there was some joint action, perhaps less than additive, in causing mortality of the yearling coho when transferred from fresh to salt water (Lorz et al., 1978).

Since there do not seem to be many situations in which a severe level of cadmium is the primary freshwater pollutant (see Section 7.10), any concern would seem to be its contribution to sublethal effects as part of a mixture of pollutants. Indeed, situations with a single pollutant would be the exception, and the usual case would be the presence of several toxicants at once. From the small amount of information presented above, the presence of cadmium in mixtures may not be a great problem for freshwater fish. However, joint sublethal effect remains a poorly understood subject for any aquatic organism and deserves considerably more research for cadmium as well as other toxicants.

7.7. SUBLETHAL LIFE-CYCLE EXPOSURES

It has become clear that life-cycle tests provide the most sensitive assessment of meaningful sublethal effects of a toxicant on fish (McKim, 1977; Sprague, 1976). Groups of newly hatched or juvenile fish are reared to maturity at constant levels of the toxicant. Reproductive performance is assessed, and second-generation fish are reared for a month or more at the same concentration as their parents. Growth, survival, and abnormalities are assessed throughout the experiment. Any deleterious effect within the fish should be detected through one or other of the assessments of performance of the whole organism. Beyond this, embryo-larval and early juvenile responses have been shown to be the most sensitive, or among the most sensitive, of the entire life cycle and "will estimate the chronic life-cycle NOEC [no-observed-effect concentration] within a factor of two in most cases" (McKim, 1977). In all seven cases of cadmium exposure included in McKim's review, the response of young stages was exactly predictive of results for life-cycle exposure.

Accordingly, results for chronic and early-life-stage exposure of fish have been gathered in Table 3 and are plotted in Figure 5. The large number of results for water hardness of 45 mg/L reflects the preponderance of work from the Duluth lab of USEPA, which uses Lake Superior water. Despite the scantiness of data at other hardness values, lines fitted to the data for salmonids and other fishes indicate that sublethal cadmium toxicity is greater in soft water. Slopes of the lines for the two categories of fishes are similar and are not greatly different from the slope of the line for lethal thresholds of salmonids. The difference between salmonids and other fish appears to be real, but it is not a major difference.

The values used in calculation (details in Appendix) are geometric averages of the lowest effective concentration and the no-effect concentration, presumed to be a threshold of effect (EC) in micrograms per liter. The line fitted to salmonids is

$$\log_{10}(EC) = 0.9694[\log_{10}(hardness)] - 1.194,$$

and that for other fish is

$$\log_{10}(EC) = 0.8767[\log_{10}(\text{hardness})] - 0.5431.$$

Thus, in very soft water (hardness 20), we might expect a threshold of sublethal effect on salmonids to be in the vicinity of 1.2 μg/L Cd, and in very hard water (350), we might expect a threshold of chronic effect to be about 19 μg/L Cd. For other fish, the comparable values are 4.0 and 49 μg Cd/L.

It must be emphasized that these are estimates of the threshold between "effect" and "no effect." The values can be expected to be slightly higher than water quality criteria recommended as "safe" for aquatic organisms, and indeed they are. For example, no-effect levels recommended by the American Fisheries Society (Davies et al., 1979) for the same two hardnesses of water are, for sensitive organisms, 0.4 and 15 μg Cd/L (instead of 1.2 and 19) and, for less sensitive species, 0.8 and 30 μg Cd/L (instead of 4 and 49). The change with hardness of the recommended levels of Davies et al. (1979) is similar to that for salmon in Figure 5, but their change with hardness for less sensitive species is more extreme than estimated here for other fish.

If the sublethal thresholds yielded by the above formulas are taken as ratios of the acutely lethal concentrations derived previously (using the poorly fitting lethality line for other fish), the following values are obtained: for salmonids: soft water 0.48, hard water 0.70; for other fish: soft water 0.0027, hard water 0.024. The ratios are anything but uniform. It appears that for salmonids, the threshold concentration causing sublethal effects is not much lower than the 4/10-day LC_{50} for juvenile fish. Apparently, the "secondary" mechanism of acute poisoning that was discussed above, operational over exposures of about 7 days, elicits a very sensitive response. Indeed, in life-cycle tests, the "hyperactive symptom" tended to develop at spawning time, especially in males, at the lowest concentrations causing developmental effects in young fish for bluegills (Eaton, 1974), brook trout (Benoit et al., 1976), and flagfish (Spehar, 1976).

The sublethal–lethal ratios for other fish are more similar to those expected for toxic pollutants, although the soft-water ratio of 0.0027 is usually low. The difference between ratios for hard and soft waters may be an artifact since lethality data resulted in a very uncertain slope with respect to hardness.

7.8. OTHER SUBLETHAL STUDIES

The sublethal effects covered above were whole-organism ones such as reproductive success and growth. Other sublethal studies have been done with freshwater fish, but they do not change the general picture of concentration versus effect.

Table 3. Sublethal Effects of Cadmium on Freshwater Fish as Demonstrated by Life-Cycle Exposures and Egg-Larval-Fry Exposures[a]

Species	Hardness	Chronic (Life Cycle)			Young Stages			Response	Reference
		NOEC	($\bar{x}$)	Effect	NOEC	($\bar{x}$)	Effect		
Salmo salar, Atlantic salmon	13	—	—	—	0.18	(0.57)	1.8	60-days growth, survival	Peterson et al., 1983
Salmo gairdneri, rainbow trout	100	0.7	(1.0)	1.5	3.6	(4.5)	5.0	≈60-days egg larval	Dave et al., 1981
	31	13.5	(16.8)	21	—	—	—	Some mortality of	Davies, 1976
	326	—	—	—	—	—	—	≈19-months (?) fingerlings	Davies, 1976
Salvelinus fontinalis, brook trout	37	—	—	—	1.0	(1.7)	3.0	Growth	Sauter et al., 1976
	44	1.7	(2.4)	3.4	—	—	—	Adult death	
								Juvenile growth	
	45	—	—	—	0.55	(0.62)	0.7	Growth and biochemistry of alevins; parents exposed	Benoit et al., 1976 Christensen, 1975
	45	—	—	—	3.8	(6.7)	11.7	31-days growth	Eaton et al., 1978
	188	—	—	—	7	(9.2)	12	Growth, survival	Sauter et al., 1976
Salmo trutta, brown trout	45	—	—	—	3.8	(6.7)	11.7	33-days growth	Eaton et al., 1978
Salvelinus namaycush, lake trout	45	—	—	—	4.4	(7.3)	12.3	31-days growth	Eaton et al., 1978
Oncorhynchus tsawytscha, chinook trout	45?	—	—	—	1.0	(1.1)	1.3	19-weeks growth, survival	USEPA, 1975

Species						Endpoint	Reference
Oncorhynchus kisutchi, coho trout	45	—	—	1.3	(2.1)	3.4	27-days growth	Eaton et al., 1978
	45	—	—	4.1	(7.2)	12.5	27-days growth	Eaton et al., 1978
Oncorhynchus nerka, sockeye salmon	83	—	—	3.3	(4.3)	5.6	Egg fry	Servizi and Martens, 1978
Oncorhynchus gorbuscha, pink salmon	83	—	—	—	[2,700]	—	fry 7-days LC$_{50}$	Servizi and Martens, 1978
Catastomus commersoni, white sucker	45	—	—	4.2	(7.1)	12	30-days growth	Eaton et al., 1978
Ictalurus punctatus, catfish	37	—	—	11	(13.6)	17	60-days survival	Sauter et al., 1976
	185	—	—	12	(14.2)	17	60-days survival	Sauter et al., 1976
Pimephales promelas, fathead minnow	50	—	—	—	—	2.6	30-days LC$_{50}$	USEPA, 1975
	200	37	(46)	—	—	—	Embryo development	Pickering and Gast, 1972
Jordanella floridae, flagfish	44	4.1	(5.8)	57	—	—	Spawning	Spehar, 1976
	45	—	—	8.1	—	—	Larval survival	Spehar et al., 1978
	?	7.4	(11.2)	8.5	—	—	Embryo larval	Carlson and Tucker, reported in McKim, 1977
	?	3.0	(4.4)	16.9	—	—	Embryo larval	
				6.5				
Micropterus dolomieu, smallmouth bass	45	—	—	4.3	(7.4)	12.7	30-days growth	Eaton et al., 1978
Esox lucius, northern pike	45	—	—	4.2	(7.1)	12	28-days growth	Eaton et al., 1978
Stizostedion vitreum, walleye	35	—	—	9.0	(15.)	25	30-days survival	Sauter et al., 1976
Lepomis macrochirus, bluegill	200	31	(50)	80	—	—	Larval and adult mortality	Eaton, 1974

[a]The values are in microgram per liter Cd for no-observed-effect concentration (NOEC), effect, and the geometric mean of those values ($\bar{x}$).

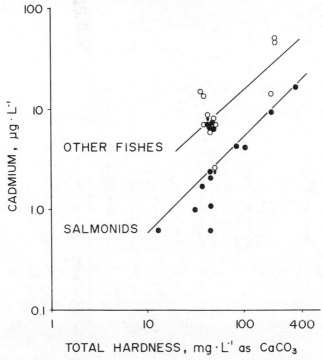

Figure 5. Thresholds of sublethal effect of cadmium on freshwater fishes, as determined in life-cycle and embryo-larval exposures. Solid points are for salmonids; open circles are for other kinds of fish. Geometric means of "effect" and "no-effect" levels are plotted.

Behavioral reactions to a pollutant are important because they may occur at concentrations quite unrelated to physiological effects. Fathead minnows proved more vulnerable to predation after exposure to 25 µg/L Cd for 21 days but not after exposure to 18 µg/L (water hardness 41 mg/L, Sullivan et al., 1978). This effect level is well above the line fitted to chronic effects for "other fish" in Figure 5. In other words, this measure of behavioral and functional performances showed a less sensitive response than did the chronic exposures discussed above. Short exposures of 24 h required much higher concentrations to increase vulnerability of fathead minnows, the effect level being between 249 and 370 µg/L. An assortment of behavioral changes were studied in bluegills by Henry and Atchinson (1979a,b), but the exposure was to a mixture of cadmium and zinc. It appeared that strong behavioral effects were evident at 66 µg/L Cd and higher at a hardness of 340 mg/L. Since this value neglects the contribution of zinc, it would appear that the behavioral response is again much less sensitive than that fitted by the line in Figure 5. Some changes were noted by Henry and Atchinson at 21 µg/L, but their statistical analysis did not document whether the changes were significant at that concentration.

Behavioral changes at the time of spawning were noted in laboratory exposures of several species of fish, as mentioned above. Again, these occurred at about the same concentrations that caused chronic effects on development or survival. Avoidance reactions of fish to cadmium have apparently not been studied.

Turning to other measures of sublethal effect, Lorz et al. (1978) did a series of tests on yearling coho salmon in water of hardness 89 mg/L. Fish exposed to 6 µg/L Cd for 6 days showed mortality when transferred to seawater, but 4 µg/L did not have this effect. This finding falls almost exactly on the previously fitted line for sublethal effects on salmonid fish (Fig. 5). Six-day exposures in the range 0.75–4.5 µg/L Cd appeared to cause slight reductions in subsequent success of downstream migration, but results were variable, and after 1- or 2-month exposures, there did not seem to be harmful effects since migratory performance after exposure to 4.5 µg/L was better than for control fish.

Respiratory responses of fish change in water containing cadmium relative to those in clean water. Bluegills in hard water showed increased coughing at 50 µg/L Cd, a concentration that fits the line for "other fish" in Figure 5 (Bishop and McIntosh, 1981). Respiratory rate increased in adult rainbow trout exposed to 6.4 µg/L Cd but not to 3.6 µg/L (Majewski and Giles, 1981). The tests being in water of hardness 80 mg/L, this effect level also fits the salmonid line in Figure 5. Thus, it appears that respiratory/coughing rates of fish may provide a "short-cut" procedure for predicting levels of significant sublethal effect.

Trout gonad cells cultured *in vitro* showed slower growth at 5000 µg/L but were stimulated at 1000 µg/L Cd (Sanford, 1974). The significance for tissues in a living fish is not clear, although these concentrations would not be likely in a real environment.

A number of investigations of biochemical processess in freshwater fish have documented changes with cadmium exposure, but not at lower concentrations than those discussed above for whole-organism responses. Succinic oxidase was inhibited in the mitochondria of bluegill liver at 590 µg/L Cd (Hiltibran, 1971). Goldfish in soft water (21 mg/L) showed distinct reduction in ability to form hemoglobin and red blood cells after a 2-week exposure to 18 µg CD/L (Houston and Keen, 1984). Smith et al. (1976) exposed catfish to initial cadmium concentration of up to 800 µg/L and assessed hematocrit, an enzyme, and certain other parameters in the blood, finding no significant differences from the control. The biomass–water loading of the fish tanks was at least 10 times, and perhaps 200 times, too high so the absence of effect cannot be related to ambient cadmium concentrations. Adult rainbow trout showed increases in hematocrit, impairment of oxygen transfer, and decline in erythrocyte ATP at 6.4 µg/L Cd, the same level that changed respiratory rate (Majewski and Giles, 1981). However, coho salmon did not show changes in (Na, K)-stimulated activity of ATPase at 4.5 µg/L Cd, which caused whole-organism effects (Lorz et al., 1978). The oxidation of lactate by gills of rainbow trout was decreased by 48-h exposures to 64 µg/L

Cd in very soft water (Bilinski and Jonas, 1973), but this concentration is extremely high compared to those causing other sublethal effects.

A high concentration (45 μg/L) for 25 days resulted in changes in plasma and muscle electrolytes of goldfish, but they compensated, and after 50 days, only sodium was low (McCarty and Houston, 1976). Giles (1984) exposed rainbow trout to 3.6 and 6.4 μg Cd/L at a total hardness of 82 mg/L for 6 months and screened many variables to assess electrolyte and water balance. At the lower concentration, there were transient effects during the first few days, but at the higher concentration, there were significantly lowered levels of plasma calcium, sodium, and potassium and increased magnesium. Giles gives an excellent discussion of the physiological changes and their meanings but apparently did not relate them to a measure of whole-organism performance. Haux and Larsson (1984) exposed trout to 10 and 100 μg Cd/L during periods of 4–9 months and also found decreased plasma calcium and increased magnesium, as well as anemia, hyperglycemia, and disturbed carbohydrate metabolism. Since the fish were held in brackish water, their exact findings and cadmium concentrations may not apply to a freshwater exposure, but in general, they support the previously described findings. Embryos and/or alevins of brook trout chronically exposed to 0.7 μg/L Cd showed decreases in weight, increases in protein content, and increased activity of acetylcholine esterase (ACH) (Christensen, 1975). This effect value is, in fact, the lowest value (mean 0.62 μg/L) plotted for hardness of 45 mg/L in Figure 5 and was used in fitting the line for salmonids since it was a concentration causing a whole-organism effect. Cadmium did not elicit a stress response of plasma cortisol in coho salmon, even at concentrations that would kill the fish (Schreck and Lorz, 1978). Three-month exposures of brook trout to 1 μg Cd/L delayed maturation of male brook trout by about 2 weeks and also impaired the clearance of hormones (Sangalang and Freeman, 1974).

As a general comment, many of the within-organism studies are not too informative. Some of them seem to have an approach of screening a number of biochemical parameters to see if anything turned up rather than selecting certain ones to test a developed hypothesis on mechanism of effect. In any case, it appears that the life cycle and larval results provide a reasonable estimate of the most sensitive and meaningful dose–effect relationships available at present.

Pickering and Gast (1972) observed circulatory problems among newly hatched fathead minnows in a chronic exposure. In many larvae, the heart was beating but red blood cells were not circulating, and many blood clots appeared throughout the vascular system. This observation would seem to provide a lead if it were desired to study mechanisms of sublethal effect.

A synoptic listing of physiological and biochemical effects of cadmium on aquatic organisms is presented in Tables 3.8 and 3.9 of Coombs (1979). Simpson (1981) reviews the same topic for marine organisms, but the mechanisms are relevant to freshwater organisms, and indeed he includes some such work.

7.9. SUBLETHAL EFFECTS AND RESIDUES IN TISSUES

Cadmium must, obviously, enter the tissues of a fish in order to affect it sublethally. Conversely, it should be possible to identify localities where fish are receiving potentially harmful exposures by measuring cadmium residues in resident fish. From the discussion of tissue levels in Chapter 5, it seems that fish would reflect concentrations in the water since they accumulate little cadmium from their food. The same chapter makes it clear that cadmium levels in fish muscle show little change with different levels of exposure, so measurements of that tissue would give a poor prediction of waterborne concentrations. Similarly, whole-body levels of cadmium would not be useful since they are largely governed by the muscle levels. Kidney, liver, or gills would be the best tissues to use as environmental monitors since they show the greatest changes with exposure level. Indeed, it was shown two decades ago that gill concentration of cadmium was a sensitive autopsy method to determine if a fish had been killed by cadmium (Mount and Stephan, 1967).

Unfortunately, using tissue residues to monitor sublethal levels of cadmium appears to have complications. For example, liver of brook trout showed about 10 μg Cd/g dry weight after long exposure to the lowest concentration causing sublethal effects in soft water (3.4 μg/L) (Benoit et al., 1976). This is considerably higher than the fractions of μg/g found in livers of fish from uncontaminated waters. For bluegills, liver contained 325 μg Cd/g (about 33 times the above-mentioned level for trout) at the lowest concentration causing sublethal effect in hard water (80 μg Cd/L; Eaton, 1974). Other native species would presumably have their own critical levels, and good information is not available for them. It was described in Section 7.4 that uptake of cadmium (as well as its toxicity) is governed by the calcium content of the water.

In view of this complexity, there can be no universal scale for monitoring sublethal cadmium pollution by tissue residues in fish. Any such scales would have to be local ones. A more promising approach would be measurement of metallothioneins in fish as suggested by Roch et al. (1986), although they point out that this assessment should also be done on a local basis. Although metallothionein measurements are not easy to carry out, the same can be said for chemical measurements of low levels of cadmium in water.

7.10. RELEVANCE OF TOXICITY DATA TO CADMIUM LEVELS IN SURFACE WATERS

It is a little difficult to compare the cadmium levels causing biological effects with the natural levels in fresh water. The instrumentation in many laboratories has detection limits above the natural levels, and it is common to see measurements reported, for example, as "<0.5 µg/L." Recent careful measurements suggest 0.04–0.06 µg Cd/L as mean values for total cadmium in surface waters (Laxen, 1984; Nriagu et al., 1981). Of this, the particulate cadmium averaged only 0.007–0.008 µg/L, according to the same authors. Each paper emphasizes the extreme care required to avoid adding cadmium to the sample by contamination from apparatus and reagents. Laxen (1984) measured the range of total cadmium in 20 streams and lakes of northwestern England as 0.015–0.12 µg/L, with particulate cadmium in 15 of those locations ranging from <0.002 to 0.059 µg/L. Nriagu et al. (1981) measured similar levels of labile cadmium (acidified to pH 2) in 16 samples of Lake Ontario water, ranging from 0.033–0.069 µg/L. The particulate cadmium in nine open-water samples ranged from 0.004 to 0.014 µg/L, and in six samples near harbors, the range was 0.014–0.058 µg/L.

If we generalize from the above that natural waters may average about 0.04 µg/L of free or labile cadmium, the threshold of lethal effect for salmonids in soft water is two orders of magnitude higher. The threshold for sublethal effect on salmonids in soft water (hardness 20 mg/L) is about 30 times higher than the natural level.

Even polluted situations do not seem to have exceptionally high levels of cadmium. One lake in Saskatchewan, Canada, has been reported to receive fallout from a smelter, with an average cadmium concentration of 0.5 µg/L (filtered samples; McFarlane and Franzin, 1978). This in itself might not be harmful, since at the prevailing water hardness of 50 mg/L, it represents only about 17% of the predicted sublethal threshold for salmonids, or 6% of the threshold for other fish. By contrast, zinc concentration in the filtered lakewater was about double the threshold for sublethal effect, and copper concentration was about equal to the sublethal threshold. It is understandable that most species of fish were rare or absent and that white sucker populations showed signs of stress. However, cadmium appears to have played only a minor role. Roch and McCarter (1984b) reported 0.8 µg Cd/L in a mine-polluted lake, about half of the threshold of sublethal effect for salmonids considering the hardness of the water. Shephard et al. (1980) measured an average of 3.2 µg/L of total cadmium after dilution of an electroplating effluent into a small eutrophic lake in Indiana, compared to 1.2 µg/L in a supposedly unpolluted part of the lake. About 20% of the cadmium was associated with suspended matter, and of the remaining dissolved cadmium, about one-third was bound as carbonate or by organic matter. The remaining 1.7 µg/L was presumably in a toxic form (mostly Cd^{2+}) (representing about 13% of the sublethal effect level at the hardness of the lake) for salmonid fish if they were able to exist in a warm eutrophic lake.

Thus, it appears difficult to find situations in which there is direct toxicity to freshwater fish of cadmium acting alone. The remaining possibilities for damage by cadmium would be its toxicity to other kinds of aquatic organisms, the possibility of accumulation in food chains, and the contribution made by cadmium to the toxicity of a mixture of pollutants (Section 7.6).

7.11. SUMMARY AND CONCLUSIONS

1. The available body of data on toxicity of cadmium to freshwater fish seems adequate. In particular, the pattern for sublethal effects is consistent enough to allow the prediction of harmful effects of water-borne cadmium.

2. Thresholds of acute lethality to salmonid fish may be expected at about 2.5 μg Cd/L in very soft water, ranging up to 27 μg/L in very hard water. Thresholds of acute lethality are much higher for nonsalmonid fish, with a geometric average of about 1660 μg/L. Lethal concentrations are widely scattered for these other fish, with no apparent relation to hardness of the water or family of fish.

3. The sensitivity of salmonids to acute lethality is apparently caused by a secondary mechanism of poisoning, characterized by violent and uncoordinated activity on external stimulus. This is probably an effect on the nervous system caused by reduced levels of calcium in the plasma. The decrease in acute lethality in hard water is because of reduced cadmium intake at the gills when salmonids are acclimated to water of high calcium content.

4. Life-cycle exposures of salmonid fish indicate that thresholds of sublethal effect range from about 1.2 μg/L Cd in very soft water to 19 μg/L in very hard water. These values are little lower than the thresholds for lethality, an unusual situation resulting from the secondary mechanism of acute poisoning. For other fish, thresholds of sublethal effect would be expected at about 4 μg/L Cd in soft water and 50 μg/L in very hard water, two or three orders of magnitude lower than lethal concentrations. "Safe" levels would be somewhat lower than these thresholds of effect.

5. In most surface waters, a large proportion of total cadmium is present as the free ion or other toxic forms. Binding as nontoxic inorganic or organic forms is of lesser importance compared to a metal such as copper. Therefore, initial toxicological assessments can be based on chemical measurements of total cadmium or dissolved cadmium.

6. The concept of using tissue levels of cadmium in fish to assess pollution that is sublethally dangerous to the fish is complicated by differences between species, between types of water, and by binding of cadmium to proteins with a reduction in toxicity. Measuring the elevation of metallothioneins in the fish appears to provide an integrated assessment of exposure to metals. Direct chemical measurement of cadmium in the water seems an expeditious approach at present.

7. There do not seem to be published records of pollution situations in which cadmium acting alone is responsible for toxicity to freshwater fish. Other concerns would be more important, for example, contribution of cadmium to the toxicity of mixtures of pollutants, toxicity to other groups of organisms, or accumulation in food chains with effects on the consumers.

8. A point arising from item 7 is that there is very little information on joint toxic action of cadmium with other toxicants, and in general, there is

poor understanding of the topic of combined sublethal action of toxicants on fish and other aquatic organisms. Because of this lack of scientific knowledge, the topic is virtually ignored in compendia of objectives for water quality (e.g., Thurston et al., 1979), although most cases of water pollution involve several substances acting simultaneously. Sublethal effects of combinations of toxicants should have a high priority in research on aquatic toxicology.

APPENDIX: ACUTE LETHALITY AND ITS RELATIONSHIP TO WATER HARDNESS

The 4- and 7–10-day LC_{50} for salmonids (from Table 1) were used to fit a line relating $\log(LC_{50})$ to $\log(\text{hardness})$ (Fig. 1, lower). If values for both time periods were available in the same report, only the longer one was used in calculations. Some values were not used: the value of Rausina et al. (1975) because this was published only as an abstract and water hardness was not given and two values from Carroll et al. (1979) because the very small volumes of test water were considered to be unsatisfactory procedure. For brook trout, a value of 12 µg/L was used for the range presented by Benoit et al. (1976). This gave 14 LC_{50} at a variety of hardnesses.

The relationship for salmonid 4/10-day LC_{50} was

$$\log_{10}(LC_{50}) = 0.8334[\log_{10}(\text{hardness})] - 0.6938.$$

Similar procedures were followed for the 2-day LC_{50} of Table 1. Six values were used. Those of Carroll et al. (1979) were omitted as before. The single value of Kumada et al. (1980) was excluded because it seemed unusual, inasmuch as it was similar to the longer exposure values of other authors. Values of Brown (1968) were not supported by experimental details but were used because of the quality of work from Brown's laboratory at that time. The equation was

$$\log(LC_{50}) = 0.00467 + 1.4[\log(\text{hardness})]$$

A relationship for all other species of fish was developed from the values shown in Table 2. The 4- or 7–10-day LC_{50} were used in the same way as for salmonids. The two values of Rausina et al. (1975) were omitted for the reasons given previously. A value of 300 was used for sticklebacks from the range given by Pascoe and Cram (1977).

This left 22 values and a line of the form

$$\log_{10}(LC_{50}) = 3.033 + 0.1021[\log_{10}(\text{hardness})].$$

In calculating the line of best fit for sublethal effects on salmonids shown in Fig. 5, 14 values from Table 3 were used, leaving out the relatively short LC_{50} for fry of pink salmon since it was not exactly equivalent to the other sublethal effects. In the line for other fish, 11 values were used, leaving out the 2 for which hardness was not available. The "effect" levels of 2.6 and 8.5 µg/L for fatheads and flagfish were used, lacking a mean value between effect and no-effect concentrations.

REFERENCES

Andros, J. D., and Garton, R. R. (1980). Acute lethality of copper, cadmium, and zinc to northern squawfish. *Trans. Am. Fish. Soc.*, **109**, 235–238.

Ball, I. R. (1967). The toxicity of cadmium to rainbow trout (*Salmo gairdneri* Richardson). *Wat. Res.*, **1**, 805–806.

Bandt, H. J. (1946). Ueber verstarkte Schadwirkungen auf Fische, insbesondere uber erhohte Giftwirking durch Kombination von Abwassergiften. *Beitr. Wasser. Abwasser-u. Fischereichem.*, **1**, 15–23.

Benoit, D. A., Leonard, E. N. Christensen, G. M., and Fiandt, J. T. (1976). Toxic effects of cadmium on three generations of brook trout (*Salvelinus fontinalis*). *Trans. Am. Fish. Soc.*, **105**, 550–560.

Benson, W. H., and Birge, W. J. (1985). Heavy metal tolerance and metallothionein induction in fathead minnows: Results from field and laboratory investigations. *Environ. Toxicol. Chem.*, **4**, 209–217.

Bilinski, E., and Jonas, R. E. E. (1973). Effects of cadmium and copper on the oxidation of lactate by rainbow trout (*Salmo gairdneri*) gills. *J. Fish. Res. Bd. Can.*, **30**, 1553–1558.

Bishop, W. E., and McIntosh, A. W. (1981). Acute lethality and effects of sublethal cadmium exposure on ventilation frequency and cough rate of bluegill (*Lepomis macrochirus*). *Arch. Environ. Contam. Toxicol.*, **10**, 519–530.

Bradley, R. W., and Sprague, J. B. (1985). The influence of pH, water hardness, and alkalinity on the acute lethality of zinc to rainbow trout (*Salmo gairdneri*). *Can. J. Fish. Aquat. Sci.*, **42**, 731–736.

Brown, V. M. (1968). The calculation of the acute toxicity of mixtures of poisons to rainbow trout. *Wat. Res.*, **2**, 723–733.

Buckley, J. A., Yoshida, G. A., Wells, N. R., and Aquino, R. T. (1985). Toxicities of total and chelex-labile cadmium to salmon in solutions of natural water and diluted sewage with potentially different cadmium complexing capacities. *Wat. Res.*, **19**, 1549–1554.

Calamari, D., Marchetti, R., and Vailati, G. (1980). Influence of water hardness on cadmium toxicity to *Salmo gairdneri* Rich. *Wat. Res.*, **14**, 1421–1426.

Carroll, J. J., Ellis, S. J., and Oliver, W. S. (1979). Influences of hardness constituents on the acute toxicity of cadmium to brook trout (*Salvelinus fontinalis*). *Bull. Environ. Contam. Toxicol.*, **22**, 575–581.

Cearley, J. E., and Coleman, R. L. (1974). Cadmium toxicity and bioconcentration in largemouth bass and bluegill. *Bull. Environ. Contam. Toxicol.*, **11**, 146–151.

Chapman, G. A. (1978). Toxicities of cadmium, copper and zinc to four juvenile stages of chinook salmon and steelhead. *Trans. Am. Fish. Soc.*, **107**, 841–847.

Chapman, G. A. (1985). Acclimation as a factor influencing metal criteria. In *Aquatic Toxicology and Hazard Assessment: Eighth Symposium,* Bahner, R. C., and Hanson, D. J. (eds.).

American Society for Testing and Materials, Philadelphia, PA, ASTM STP 891, pp. 119–136.

Chapman, G. A., and Stevens, D. G. (1978). Acutely lethal levels of cadmium, copper and zinc to adult male coho salmon and steelhead. *Trans. Am. Fish. Soc.*, **107**, 837–840.

Christensen, G. M. (1975). Biochemical effects of methylmercuric chloride, cadmium chloride, and lead nitrate on embryos and alevins of the brook trout, *Salvelinus fontinalis*. *Toxicol. Appl. Pharmacol.*, **32**, 191–197.

Coombs, T. L. (1979). Cadmium in aquatic organisms. In *The Chemistry, Biochemistry, and Biology of Cadmium*, Webb, M. (ed.). Elsevier/North-Holland, New York, pp. 93–139.

Dave, G., Anderson, K., Berglind, R., and Hasselrot, B. (1981). Toxicity of eight solvent extraction chemicals and of cadmium to water fleas, *Daphnia magna*, rainbow trout, *Salmo gairdneri*, and zebrafish, *Brachydanio rerio*. *Comp. Biochem. Physiol.*, **69C**, 83–98.

Davies, P. H. (1976). The need to establish heavy metal standards on the basis of dissolved metals. In *Toxicity to biota of metal forms in natural waters. Proceedings of Workshop Held in Duluth, Minnesota, October 7–8, 1975*, Andrew, R. W., Hodson, P. V., and Konasewich, D. E. (eds.). International Joint Commission, Windsor, Ontario, pp. 93–126.

Davies, P. H., Calabrese, A., Goettl, J. P. Jr., Hodson, P. V., Kent, J. C., McKim, J. M., Pearce, J. B., Reish, D. J., Sprague, J. B., and Sunda, W. G. (1979). Cadmium. In *A Review of the EPA Red Book: Quality Criteria for Water*, Thurston, R. V., Russo, R. C., Fetterolf, C. M. Jr., Edsall, T. A., and Barber, Y. M. (eds.). American Fisheries Society, Bethesda, MD, pp. 51–66.

Dixon, D. G., and Sprague, J. B. (1981). Copper bioaccumulation and hepato-protein synthesis during acclimation to copper by juvenile rainbow trout. *Aquat. Toxicol.*, **1**, 69–82.

Duncan, D. A., and Klaverkamp, J. F. (1983). Tolerance and resistance to cadmium in white suckers (*Catostomus commersoni*) previously exposed to cadmium, mercury, zinc, or selenium. *Can. J. Fish. Aquat. Sci.*, **40**, 128–138.

Eaton, J. G. (1973). Chronic toxicity of a copper, cadmium and zinc mixture to the fathead minnow (Pimephales promelas Rafinesque). *Wat. Res.*, **7**, 1723–1736.

Eaton, J. G. (1974). Chronic cadmium toxicity to the bluegill (*Lepomis macrochirus* Rafinesque). *Trans. Am. Fish. Soc.*, **103**, 729–735.

Eaton, J. G., McKim, J. M., and Holcombe, G. W. (1978). Metal toxicity to embryos and larvae of seven freshwater fish species. I. Cadmium. *Bull. Environ. Contam. Toxicol.*, **19**, 95–103.

Finlayson, B. J., and Verrue, K. M. (1982). Toxicities of copper, zinc and cadmium mixtures to juvenile chinook salmon. *Trans. Am. Fish. Soc.*, **111**, 645–650.

Gardiner, J. (1974). The chemistry of cadmium in natural water—I. A study of cadmium complex formation using the cadmium specific-ion electrode. *Wat. Res.*, **8**, 23–30.

Giesy, J. P. Jr., Leversee, G. J., and Williams, D. R. (1977). Effects of naturally occurring aquatic organic fractions on cadmium toxicity to *Simocephalus serrulatus* (Daphnidae) and *Gambusia affinis* (Poeciliidae). *Wat. Res.*, **11**, 1013–1020.

Giles, M. A. (1984). Electrolyte and water balance in plasma and urine in rainbow trout (*Salmo gairdneri*) during chronic exposure to cadmium. *Can. J. Fish. Aquat. Sci.*, **41**, 1678–1685.

Gjessing, E. T. (1981). The effect of aquatic humus on the biological availability of cadmium. *Arch. Hydrobiol.*, **91**, 144–149.

Haux, C., and Larsson, A. (1984). Long-term sublethal physiological effects on rainbow trout, *Salmo gairdneri*, during exposure to cadmium and after subsequent recovery. *Aquat. Toxicol.*, **5**, 129–142.

Henry, M. G., and Atchison, G. J. (1979a). Behavioral changes in bluegill (*Lepomis macrochirus*) as indicators of sublethal effects of metals. *Environ. Biol. Fish.*, **4**, 37–42.

Henry, M. G., and Atchison, G. J. (1979b). Influence of social rank on the behavior of bluegill, *Lepomis macrochirus* Rafinesque exposed to sublethal concentrations of cadmium and zinc. *J. Fish. Biol.*, **15**, 309–315.

Hiltibran, R. C. (1971). Effects of cadmium, zinc, manganese, and calcium on oxygen and phosphate metabolism of bluegill liver mitochondria. *J. Wat. Pollut. Contr. Fed.*, **43**, 818–823.

Houston, A. H. (1959). Osmoregulatory adaptation of steelhead trout (*Salmo gairdneri* Richardson) to sea water. *Can. J. Zool.*, **37**, 729–743.

Houston, A. H., and Keen, J. E. (1984). Cadmium inhibition of erythropoiesis in goldfish, *Carassius auratus*. *Can. J. Fish. Aquat. Sci.*, **41**, 1829–1834.

Howarth, R. S., and Sprague, J. B. (1978). Copper lethality to rainbow trout in waters of various hardness and pH. *Wat. Res.*, **12**, 455–462.

Hughes, G. M., Perry, S. F., and Brown, V. M. (1979). A morphometric study of effects of nickel, chromium, and cadmium on the secondary lamellae of rainbow trout gills. *Wat. Res.*, **13**, 665–679.

Klaverkamp, J. F., Macdonald, W. A., Duncan, D. A., and Wagemann, R. (1984). Metallothionein and acclimation to heavy metals in fish: A review. In *Contaminant Effects on Fisheries*, Cairns, V. W., Hodson, P. V., and Nriagu, J. O. (eds.). Advances in Environmental Science and Technology, Vol. 16. Wiley, New York, pp. 99–113.

Kumada, H., Kimura, S., and Yokote, M. (1980). Accumulation and biological effects of cadmium in rainbow trout. *Bull. Jap. Soc. Sci. Fish.*, **46**, 97–103.

Laxen, D. P. H. (1984). Cadmium in freshwaters: Concentrations and chemistry. *Freshwater Biol.*, **14**, 587–595.

Lloyd, R. (1965). Factors that affect the tolerance of fish to heavy metal poisoning. In *Biological Problems in Water Pollution, Third Seminar, 1962*. U.S. Public Health Service, Washington, DC, Publ. 999-WP-25, pp. 181–187.

Lorz, H. W., Williams, R. H., and Fustich, C. A. (1978). Effects of several metals on smolting of coho salmon. U.S. Environmental Protection Agency, Ecological Research Series, EPA-600/3-78-090.

Majewski, H. S., and Giles, M. A. (1981). Cardiovascular-respiratory responses of rainbow trout (*Salmo gairdneri*) during chronic exposure to sublethal concentrations of cadmium. *Wat. Res.*, **15**, 1211–1217.

McCarty, L. S., and Houston, A. H. (1976). Effects of exposure to sublethal levels of cadmium upon water-electrolyte status in the goldfish (*Carassius auratus*). *J. Fish. Biol.*, **9**, 11–19.

McCarty, L. S., Henry, J. A. C., and Houston, A. H. (1978). Toxicity of cadmium to goldfish, *Carassius auratus*, in hard and soft water. *J. Fish. Res. Bd. Can.*, **35**, 35–42.

McFarlane, G. A., and Franzin, W. G. (1978). Elevated heavy metals: A stress on a population of white suckers, *Catastomus commersoni*, in Hamell Lake, Saskatchewan. *J. Fish. Res. Bd. Can.*, **35**, 963–970.

McKim. J. M. (1977). Evaluation of tests with early life stages of fish for predicting long-term toxicity. *J. Fish. Res. Bd. Can.*, **34**, 1148–1154.

Mount, D. I., and Stephan, C. E. (1967). A method for detecting cadmium poisoning in fish. *J. Wildlife Manag.*, **31**, 168–172.

Muramoto, S. (1981). Influence of complexans (EDTA, DTPA) on the toxicity of cadmium to fish at chronic levels. *Bull. Environ. Contam. Toxicol.*, **26**, 641–646.

Nriagu, J. O., Wong, H. K. T., and Coker, R. D. (1981). Particulate and dissolved trace metals in Lake Ontario. *Wat. Res.*, **15**, 91–96.

Part, P., Svanberg, O., and Kiessling, A. (1985). The availability of cadmium to perfused rainbow trout gills in different water qualities. *Wat. Res.*, **4**, 427–434.

Pascoe, D., and Beattie, J. H. (1979). Resistance to cadmium by pretreated rainbow trout alevins. *J. Fish. Biol.*, **14**, 303–308.

Pascoe, D., and Cram, P. (1977). The effect of parasitism on the toxicity of cadmium to the three-spined stickleback, *Gasterosteus aculeatus* L. *J. Fish. Biol.*, **10**, 467–472.

Pascoe, D., and Mattey, D. L. (1977). Studies on the toxicity of cadmium to the threespined stickleback *Gasterosteus aculeatus* L. *J. Fish. Biol.*, **11**, 207–215.

Pascoe, D., and Woodworth, J. (1980). The effects of joint stress on sticklebacks. *Z. Parasitenkd.*, **62**, 159–163.

Peterson, R. H., Metcalfe, J. L., and Ray, S. (1983). Effects of cadmium on yolk utilization, growth, and survival of Atlantic salmon alevins and newly feeding fry. *Arch. Environ. Contam. Toxicol.*, **12**, 37–44.

Pickering, Q. H., and, Gast, M. H. (1972). Acute and chronic toxicity of cadmium to the fathead minnow (*Pimephales promelas*). *J. Fish. Res. Bd. Can.*, **29**, 1099–1106.

Rausina, G., Goode, J. W., Keplinger, M. L., and Calandra, J. C. (1975). Four-day static fish toxicity studies conducted with thirteen cadmium pigments in three species of freshwater fish. *Toxicol. Appl. Pharmacol.*, **33**, 188.

Rehwoldt, R., Menapace, L. W., Nerrie, B., and Alessandrello, D. (1972). The effect of increased temperature upon the acute toxicity of some heavy metal ions. *Bull. Environ. Contam. Toxicol.*, **8**, 91–96.

Roch, M., and Maly, E. J. (1979). Relationship of cadmium-induced hypocalcemia with mortality in rainbow trout (*Salmo gairdneri*) and the influence of temperature on toxicity. *J. Fish. Res. Bd. Can.*, **36**, 1297–1303.

Roch, M., and McCarter, J. A. (1984a). Hepatic metallothionein protein and resistance to heavy metals by rainbow trout (*Salmo gairdneri*)—I. Exposed to an artificial mixture of zinc, copper and cadmium. *Comp. Biochem. Physiol.*, **77C**, 71–75.

Roch, M., and McCarter, J. A. (1984b). Hepatic metallothionein production and resistance to heavy metals by rainbow trout (*Salmo gairdneri*)—II. Held in a series of contaminated lakes. *Comp. Biochem. Physiol.*, **77C**, 77–82.

Roch, M., Noonan, P., and McCarter, J. A. (1986). Determination of no-effect levels of heavy metals for rainbow trout using hepatic metallothionein. *Wat. Res.*, **20**, 771–774.

Sanford, W. C. (1974). Effects of cadmium salts on the reproductive potential of male rainbow trout as determined by in vivo and in vitro techniques. Office of Water Resources Research, Washington, DC, Project Completion Report, Project No. A-042-Oklahoma.

Sangalang, G. B., and Freeman, H. C. (1974). Effects of sublethal cadmium on maturation and testosterone and 11-ketotestosterone production *in vivo* in brook trout. *Biol. Reprod.*, **11**, 429–435.

Sauter, S., Buxton, K. S., Macek, K. J., and Petrocelli, S. R. (1976). Effects of exposure to heavy metals on selected freshwater fish. U.S. Environmental Protection Agency. EPA-600/3-76-105. Duluth, MN.

Schreck, C. B., and Lorz, H. W. (1978). Stress response of coho salmon (*Oncorhynchus kisutch*) elicited by cadmium and copper and potential use of cortisol as an indicator of stress. *J. Fish. Res. Bd. Can.*, **35**, 1124–1129.

Schuytema, G. S., Nelson, P. O., Malveg, K. W., Nebeker, A. V., Krawczyk, D. F., Ratcliff, A. K., and Gakstatter, J. H. (1984). Toxicity of cadmium in water and sediment slurries to *Daphnia magna*. *Environ. Toxicol. Chem.*, **3**, 293–308.

Servizi, J. A., and Martens, D. W. (1978). Effects of selected heavy metals on early life of sockeye and pink salmon. International Pacific Salmon Fisheries Commission, New Westminster, British Columbia, Progress Report 39.

Shephard, B. K., McIntosh, A. W., Atchison, G. J., and Nelson, D. W. (1980). Aspects of the aquatic chemistry of cadmium and zinc in a heavy metal contaminated lake. *Wat. Res.*, **14**, 1061–1066.

Simpson, W. R. (1981). A critical review of cadmium in the marine environment. *Progr. Oceanogr.*, **10**, 1–70.

Smith, B. P., Hejtmancik, E., and Camp. B. J. (1976). Acute effect of cadmium on *Ictalurus punctatus* (catfish). *Bull. Environ. Contam. Toxicol.*, **15**, 271–277.

Solbe, J. F. de L. G., and Flook, V. A. (1975). Studies on the toxicity of zinc sulphate and of cadmium sulfate to stone loach *Noemacheilus barbatulus* (L.) in hard water. *J. Fish. Biol.*, **7**, 631–637.

Solomon, E. P., and Davis, P. W. (1978). *Understanding Human Anatomy and Physiology*. McGraw-Hill, New York.

Spehar, R. L. (1976). Cadmium and zinc toxicity to flagfish, *Jordanella floridae*. *J. Fish. Res. Bd. Can.*, **33**, 1939–1945.

Spehar, R. L., Leonard, E. N., and Defoe, D. L. (1978). Chronic effects of cadmium and zinc mixtures on flagfish (*Jordanella floridae*). *Trans. Am. Fish. Soc.*, **107**, 354–360.

Sprague, J. B. (1976). Current status of sublethal tests of pollutants on aquatic organisms. *J. Fish. Res. Bd. Can.*, **33**, 1988–1992.

Sullivan, J. F., Atchison, G. J., Kolar, D. J., and McIntosh, A. W. (1978). Changes in the predator–prey behavior of fathead minnows (*Pimephales promelas*) and largemouth bass (*Micropterus salmoides*) caused by cadmium. *J. Fish. Res. Bd. Can.*, **35**, 446–451.

Thurston, R. V., Russo, R. C., Fetterolf, C. M. Jr., Edsall, T. A., and Barber, Y. M. Jr. (eds.). (1979). *A Review of the EPA Red Book: Quality Criteria for Water*. American Fisheries Society, Bethesda, MD.

U.K. (1976). *Water Pollution Research, 1973*. United Kingdom Ministry of the Environment, Her Majesty's Stationery Office, London.

USEPA (1975). *Quarterly Research Report, March 31, 1975*. U.S. Environmental Protection Agency, National Water Quality Laboratory, Duluth, MN.

Voyer, R. A., Yevich, P. P., and Barszcz, C. A. (1975). Histological and toxicological responses of the mummichog. *Fundulus heteroclitus* (L.) to combinations of levels of cadmium and dissolved oxygen in freshwater. *Wat. Res.*, **9**, 1069–1074.

8

EFFECTS OF CADMIUM ON MARINE BIOTA

D. W. McLeese

Department of Fisheries and Oceans
Biological Station, St. Andrews
New Brunswick, Canada

J. B. Sprague

Department of Zoology
College of Biological Science
University of Guelph
Guelph, Ontario, Canada

S. Ray

Department of Fisheries and Oceans
Biological Station, St. Andrews
New Brunswick, Canada

8.1. Fish
    8.1.1. Acute Toxicity to Different Life Stages
    8.1.2. Sublethal Effects
        8.1.2.1. Enzyme Activity
        8.1.2.2. Ion Balance in Blood Plasma
        8.1.2.3. Metallothioneins
8.2. Invertebrates
    8.2.1. Mollusks
    8.2.2. Crustaceans
    8.2.3. Annelids
    8.2.4. Echinoderms

8.3. Phytoplankton
8.4. Factors Modifying Toxicity
    8.4.1. Salinity
    8.4.2. Temperature
    8.4.3. Salinity and Temperature
    8.4.4. Other Identities
    8.4.5. Multiple Toxicity
8.5. Summary
Acknowledgments
References

8.1. FISH

8.1.1. Acute Toxicity to Different Life Stages

The acute toxicity of cadmium has been determined for the egg, larval, and juvenile stages of a few fish species. Eggs of Atlantic silverside (*Menidia menidia*), mummichog (*Fundulus heteroclitus*), winter flounder (*Pseudopleuronectes americanus*), Baltic herring (*Clupea harengus*), Baltic flounder (*Pleuronectes flesus*), and garpike (*Belone belone*) appear to survive until hatching equally as well as the controls at concentrations of cadmium ranging up to 1000 µg/L (Middaugh and Dean, 1977; Voyer et al., 1977, 1982; von Westernhagen et al., 1974, 1975; von Westernhagen and Dethlefsen, 1975). At successively higher concentrations, eggs of these species suffer progressively greater mortalities. However, a measurement of the number of viable larvae (those that are free of the chorion, lacking morphological deformities and able to swim) is a more accurate assessment of the toxicity of cadmium, since mortality only becomes apparent once the larvae have emerged. For example, Voyer et al. (1982) found that eggs of the winter flounder exhibited an embryonic survival of 89% upon exposure to a nominal concentration of 1000 µg/L Cd, while the viable hatch was only 36%. Similar responses were noted for the garpike and Baltic herring (von Westernhagen et al., 1974, 1975). The resistance of eggs is thought to be due to the cadmium-binding effect of the chorion, as evidenced by the lack of accumulation of cadmium in the emerging larvae (von Westernhagen and Dethlefsen, 1975; von Westernhagen et al., 1974, 1975).

The acute toxicity of cadmium to the larval stage has been investigated only in a study by Middaugh and Dean (1977). Mummichogs and Atlantic silversides were exposed to cadmium 1, 7, and 14 days after emergence from the egg. The mummichog was very resistant to cadmium at all times; the 48-h LC_{50} varied from 9000 (7-day-old) to 32,000 µg/L Cd (14-day-old). The Atlantic silverside, although not as resistant as the mummichog, was still quite tolerant; the 48-h LC_{50} declined from 3800 μg/L for 1-day-old larvae to 2200 μg/L Cd for 14-day-old larvae, suggesting that with increasing age at the time of exposure there was an increased susceptibility to cadmium. In both species, the adults were more tolerant (Table 1).

The juvenile stage is more resistant than the larval stage. For example, the 48-h LC_{50} ranged from 13,000 μg/L for the Atlantic silverside to 60,000 μg/L for the mummichog. Tests of longer duration (i.e., 96 and 168 h) with other species gave similar results (Table 1). Juveniles of marine fish are much more resistant than juveniles of freshwater species (see Chapter 7), especially salmonids, perhaps because of the different speciation of cadmium in the respective waters.

It appears then, that at concentrations of dissolved cadmium found in the open ocean (see Chapters 1 and 4) or even in grossly polluted estuarine areas (Hazen, 1981) that none of the life stages tested would be exposed to acutely lethal conditions.

However, demersal eggs could be exposed to high levels of cadmium in the sediments. For example, the sediment concentrations in Foundry Cove at the mouth of the Hudson River in New York often exceed 1000 mg/kg dry weight, although the concentrations of dissolved cadmium typically range from 5 to 15 μg/L. Thus, the total amount present is of little concern, but rather the fraction that is bioavailable. The latter is dependent on the quantity leached from the sediment and the concentration in the aqueous phase, which in turn is governed by the cation exchange capacity and the organic carbon content (Ray and McLeese, 1983).

8.1.2. Sublethal Effects

8.1.2.1. *Enzyme Activity*

Studies on the effects of cadmium on various enzymes in marine fish are not very informative as each one is concerned largely with an isolated enzyme rather than with a mechanism of action or with ecological significance. In addition, it is difficult to arrive at a consensus on the effect of cadmium on enzymatic activity because of the variability between studies. Tests using excessively high concentrations ranging from 1000 to 27,000 μg/L (Jackim et al., 1970; Pruell and Engelhardt, 1980) are of little use in predicting the effects in the natural environment and hence are omitted from this discussion.

It appears that in many cases where cadmium exists in the microgram-per-liter range, effects on enzymatic activity are minimal. For example, Dawson et al. (1977) exposed juvenile striped bass, *Morone saxatilus,* to nominal concentrations of 0.5, 2.5, and 5.0 μg/L Cd over a 90-day period and examined the activities of three enzymes, aspartate aminotransferase (AAT), glucose-6-phosphate dehydrogenase (G6PdH), and malic enzyme (ME) in the muscle and liver tissues. Although a significant drop in activity was observed after 30 days for both AAT and G6PdH in the liver of fish

Table 1. Acute Lethality of Cadmium to Marine Fish Arranged in Apparent Order of Tolerance

Species	Life Stage	Temperature (°C)	Salinity (‰)	LC$_{50}$ (μg/L)			Reference
				48 h	96 h	168 h	
Menidia menidia, Atlantic silverside	Adult	20	20	13,000	—	—	Middaugh and Dean, 1977
Aldrichetta forsteri, yellow-eye mullet	Juvenile	19.5	34	—	—	16,000	Negilski, 1976
Atherinasoma microstoma, small-mouthed hardyhead	Juvenile	19.5	34	—	—	21,000	Negilski, 1976
Fundulus majalis, striped killifish	Juvenile	20	20	—	21,000	—	Eisler, 1971
Agonus cataphractus		—	—	—	33,000	—	Portman and Wilson, 1971
Fundulus heteroclitus, mummichog	Juvenile	20	20	—	55,000	—	Eisler, 1971
	Adult	20	20	60,000	—	—	Middaugh and Dean, 1977

exposed to 5.0 μg/L Cd, no significant changes in activity were noted for either enzyme after 90 days at any of the concentrations tested. Similarly, no effects were noted for either AAT or ME in muscle tissue.

The activities of three enzymes, leucine aminopeptidase (LAP), carbonic anhydrase (CA), and G6PdH were investigated in the kidney and hematopoietic tissue of the winter flounder, *P. americanus,* exposed to 5.0 and 10.0 μg/L Cd over a 60-day period (Gould, 1977). The only significant alteration was elevated activity of carbonic anhydrase in fish exposed to 10 μg/L Cd. At much higher concentrations (i.e., 50 and 100 μg/L Cd), MacInnes et al. (1977) found that in the liver of the cunner, *Tautogolabrus adspersus,* the activity of G6PdH was significantly elevated at both concentrations while that of AAT was significantly reduced at 100 μg/L. The significance of these results remains unclear since concentrations of dissolved cadmium of this magnitude are unlikely to be encountered, even in contaminated areas.

Johansson-Sjöbeck and Larsson (1978) examined the hematological response and activity of δ-aminolevulinic dehydratase (ALA-D) in the blood and blood-forming tissues in the flounder, *P. flesus,* exposed for 63 days to measured concentrations of 6.2, 48, and 517 μg/L Cd. The ALA-D activity was significantly increased in renal tissue at all cadmium concentrations, apparently in response to the reduced values of hematocrit, hemoglobin, and RBC count that occurred. This was thought to be a compensatory mechanism for the observed anemic response since it is indicative of an increased biosynthesis of hemoglobin. However, as pointed out by Johansson-Sjöbeck and Larsson (1978), the functional significance of these disturbances is difficult to determine.

8.1.2.2. Ion Balance in Blood Plasma

Larsson et al. (1981) exposed the flounder, *P. flesus,* to measured concentrations of cadmium of 6.2, 48, and 517 μg/L for 63 days and examined the effects on ions in the blood plasma. The major electrolytes Na^+ and Cl^- were unaffected, and the water content of the liver and muscle tissues was unchanged, indicating that osmoregulation was not affected. Other levels of blood plasma ions exhibited changes. The concentrations of inorganic phosphate and magnesium were elevated significantly at the two highest concentrations tested, whereas those of potassium and calcium were depressed significantly. These results suggest that renal regulation of ions was impaired by cadmium.

These findings are similar to those of Roch and Maly (1979) for the freshwater rainbow trout. At a concentration of 300 μg/L Cd, the calcium concentration in the plasma declined by 40% at temperatures ranging from 6–18°C, while magnesium concentrations increased in the range 12–18 °C. The trout also showed symptoms of hypersensitivity, erratic swimming, and tetanic contractions, as was noted for flounders exposed to 500 μg/L Cd (Larsson et al., 1981). Given the role of calcium in regulating neuromuscular function, it is possible that the observed hypocalcemia was a contributing factor to the above-noted hyperexcitability. What caused the hypocalcemia to occur is unknown, but the authors of both studies suggested that impaired renal resorptive ability of the tubules was responsible. Roch and Maly (1979) also suggested that the hypocalcemia was the cause of death of these fish (see also Section 7.5.).

8.1.2.3. Metallothioneins

As do many other organisms, marine fish develop increased levels of metal-binding proteins in response to cadmium exposure. An appreciable amount of recent research has investigated such a group of proteins, the metallothioneins. Since increase in metallo-proteins would seem to represent an adaptive response rather than a deleterious effect, the topic is not reviewed here but is discussed in Sections 5.7.2. and 9.7.1.1.

8.2. INVERTEBRATES

8.2.1. Mollusks

For acute lethality to adult or juvenile gastropods, the geometric average LC_{50} is 5900 µg/L Cd ($\pm$ s.d. 1900–18,000) and for bivalves it is 1700 µg/L Cd ($\pm$ s.d. 490–5800). The distributions of the LC_{50} for the two groups are not significantly different, so the overall short-term mean LC_{50} for mollusks is 2500 µg/L ($\pm$ s.d. 680–9200). The analysis of variance was based on logarithms of the LC_{50}, since variance increased with LC_{50}. Use of averages to characterize the data on lethality is a simplification, but it is warranted in view of the usual within-lab and between-lab variation. For example, Table 2 shows that two LC_{50} of cadmium for the mussel *Mytilus edulis* obtained in two laboratories were different by a factor of 16.

Long exposures in lethal tests of adults or subadults should yield toxicity numbers between acute and chronic values, and this proves to be the case. The 1-month LC_{50} of cadmium for *Mytilus galloprovincialis* was about 350 µg/L, one-fifth the value of the 6-d LC_{50} of 1700 µg/L (Pavicic and Jarvenpää, 1974). No apparent cessation of long-term mortality had occurred by the end of the month. For the bay scallop (*Argopecten irradians*), the 1.3-month LC_{50} was 530 µg/L Cd, but a much lower concentration of 78 µg/L affected growth (Pesch and Stewart, 1980). The long-term LC_{50} for the two species are quite close, differing by a factor of only 1.5.

Growth and development of the highly sensitive embryo-larval stages is considered a very meaningful assessment of sublethal damage, as established in freshwater toxicology (Chapter 7). Test results on survival, growth, and development of embryos and larvae can be expected to predict within a factor of 2 the threshold concentration for life-cycle exposures among freshwater fish (McKim, 1977), and there is no reason to expect that the situation

Table 2. Acutely Lethal Toxicity of Cadmium to Subadult and Adult Marine Invertebrates[a]

Group and Species	Exposure Time (days)	LC_{50} (μg/L)	Reference
Mollusca, Gastropoda			
Urosalpinx cinerea, oyster drill	4	6600	Eisler, 1971
Bullia digitalis	4	900	Cuthbert et al., 1976
Nassarius obsoletus, mudsnail	7	16000	Eisler and Hennekey, 1977
	4	10500	Eisler, 1971
	3	7000	MacInnes and Thurberg, 1973
Mollusca, Bivalvia			
Mytilus edulis, blue mussel	4	25000	Eisler, 1971
	4	1600	Ahsanullah, 1976
	7	~3000	Flatau and Aubert, 1979
M. galloprovincialis	6	1700	Pavicic and Jarvenpää, 1974
Argopecten irradians, bay scallop	4	1500	Nelson et al., 1976
Neotrigonia margaritacea	4	2200	Ahsanullah, 1976
Crassostrea rhizophorae, mangrove oyster	7	800	Chung, 1980
Cardium edule, common cockle	2	2600	Portman and Wilson, 1971
Mya arenaria, soft-shell clam	7	150	Eisler, 1977b
	7	700	Eisler and Hennekey, 1977
	4	2200	Eisler, 1971
Crustacea, Copepoda			
Acartia clausi	4	390	Moraitou-Apostolopoulou et al., 1979
A. simplex	1	1370	Arnott and Ahsanullah, 1979
A. spinicauda	2	50	Madhupratap et al., 1981
A. tonsa	4	169	Sosnowski and Gentile, 1978
Nitocra spinipes	4	1800	Bengtsson, 1978
Paracalanus parvus	1	2710	Arnott and Ahsanullah, 1979
Scutellidium sp.	1	660	Arnott and Ahsanullah, 1979
Tartanus forcipatus	2	130	Madhupratap et al., 1981
Tisbe holothuriae	2	~1500	Verriopoulos and Moraitou-Apostolopoulou, 1981
Crustacea, Malacostraca ("shrimps," etc.)			
Amphipoda			
Allorchestes compressa	4	300	Ahsanullah, 1976
Corophium acherusicum	4	1400	Bellan-Santini and Reish, 1976
Marinogammarus obtusatus	4	13300	Wright and Frain, 1981
Podocerus fulanus	4	320	Bellan-Santini and Reish, 1976
Isopoda			
Paracerceis sculpta	4	1250	Bellan-Santini and Reish, 1976
Decapoda, shrimps			
Penaeus duorarum	4	3500	Nimmo et al., 1977b
Crangon crangon, brown shrimp	2	5500	Portman, 1968, 1970
	5–6	~350	Price and Uglow, 1979

Table 2. (Continued)

Group and Species	Exposure Time (days)	LC$_{50}$ (μg/L)	Reference
C. septemspinosa	4	320	Eisler, 1971
Palaemon sp.	4	6600	Ahsanullah, 1976
Palaemonetes pugio	7	890	Sunda et al., 1978
P. vulgaris,	4	420	Eisler, 1971
grass shrimp	4	760	Nimmo et al., 1977b
Decapoda, lobsters, and crabs			
Homarus americanus, lobster	6.5	5000	McLeese, 1981
Pagurus longicarpus, hermit crab	4	320	Eisler, 1971
	17	700	Eisler and Hennekey, 1977
Callinectes sapidus, blue crab	4	11600	Frank and Robertson, 1979
Carcinus maenas,	4	4100	Eisler, 1971
green crab	7	~3000	Flatau and Aubert, 1979
Eurypanopeus depressus, mud crab	3	4900	Collier et al., 1973
Paragrapsus gaimardii	4	34300	Sullivan, 1977
Paragrapsus quadridentatus	5/7	16700	Ahsanullah and Arnott, 1978
Uca pugilator,	6	29600	Vernberg et al., 1974
fiddler crab	4	46600	O'Hara, 1973
Annelids			
Ophryotrocha labronica	4	8000	Brown and Ahsanullah, 1971
Ophryotrocha diadema	—	3500	Reisch and Carr, 1978
Nereis virens	4	11000	Eisler, 1971
Ctenodrilas serratus	4	3500	Reisch and Carr, 1978
Neanthes arenacoedentala, juvenile	4	12500	Reisch and Martin, 1976
Neanthes arenacoedentala, adult	4	12000	Reisch and Martin, 1976
Capitella capitella, adult	4	7500	Reisch and Martin, 1976
Echinoderms			
Asterias forbesi	4	800	Eisler, 1971
Patiriella exigua	4	≥10000	Ahsanullah, 1976

[a] These data were selected to represent full-strength seawater, or at least relatively high salinity, at a summer temperature. Results of lethal tests on newly hatched or very young stages are not included here. General experimental procedures were judged adequate, although many results are for nominal or "added" concentrations, which could give a somewhat higher LC$_{50}$ than concentrations measured in the test tank. Some results have been adjusted to reflect statements on measured concentrations or interpolated from results presented by authors.

Table 3. Effects of Cadmium on Larvae or Other Young Stages of Shellfish[a]

Species	Lowest Effect Concentration (μg/L)	Effect	Reference
Mollusca, Bivalvia, Mytilidae (mussels)			
Mytilus edulis	~75	Restricted salinities for survival, growth, development	Lehnberg and Theede, 1979
M. galloprovincialis	2000	Embryo, 6-days survival development	Pavicic and Jarvenpää, 1974
Mollusca, Bivalvia, Ostreiidae (oysters)			
Crassostrea cucullata	(≥100)	No effect, embryo development	Watling, 1981
C. gigas	(≥100)	No effect, embryo development	Watling, 1981
	920	EC_{50}, embryo development	Cardwell et al., 1979
	20	Survival, growth, 5-day-old larvae	Watling, 1978
	50	LC_{50}, 5-day-old larvae	Watling, 1978
C. margaritacea	(≥100)	No effect, embryo development	Watling, 1981
C. virginica	15	Abnormal development larvae	Zaroogian and Morrison, 1981
	3800	LC_{50}, embryos	Calabrese et al., 1973
Mollusca, Bivalvia, Veneroida	60	Embryo development, EC_{50}	Cardwell et al., 1979
Tresus capax, horse clam	370	Embryo LC_{50}	Cardwell et al., 1979
T. nuttalli, horse clam	590	Embryo development, EC_{50}	Cardwell et al., 1979
	1700	Embryo LC_{50}	Cardwell et al., 1979
Protothaca staminea, littleneck clam	1290	Embryo development, EC_{50}	Cardwell et al., 1979
	13900	Embryo LC_{50}	Cardwell et al. 1979
Crustacea, Copepoda			
Pseudodiaptomus coronatus	5	Four generations exposure, reproduction half that of controls; no effect, growth, feeding	Paffenhöfer and Knowles, 1978
Crustacea, Cirripedia (barnacles)	100	Larval development, swimming	Lang et al., 1981
Balanus improvisus	140	Larval LC_{50}, unfavorable salinity Reproduction	Lang et al., 1981

Table 3. (Continued)

Species	Lowest Effect Concentration (μg/L)	Effect	Reference
Crustacea, Mysidacea (mysids)	11	Survival, 23-day life cycle	Nimmo et al., 1978
Mysidopsis bahia	16	4-days LC_{50}, age 1 day	Nimmo et al., 1977a
Crustacea, Amphipoda			
Marinogammarus obtusatus	3500	4-days LC_{50}, age 2 days	Wright and Frain, 1981
Crustacea, Decapoda, Caridae (shrimps)			
Palaemonetes pugio	300	Larval survival in unfavorable salinity	Middaugh and Floyd, 1978
Pandalus platyceros, spot shrimp	4970	EC_{50}, embryo development	Cardwell et al., 1979
Crustacea, Decapoda, Astacidae (large-clawed lobsters)	30	Larvae, long-term LC_{50}	McLeese, 1981
Homarus americanus	78	4-days LC_{50}, age 1 day	Johnson and Gentile, 1979
	180	4-days LC_{50}, newly hatched	Dorband et al., 1976
Crustacea, Decapoda, Brachyura (crabs)			
Cancer magister, Dungeness crab	1040	EC_{50}, embryo development	Cardwell et al., 1979
Eurypanopeus depressus	10	Survival, development, swimming	Mirkes et al., 1978
Callinectes sapidus, blue crab	50	Slow development, normal salinity, survival low	Rosenberg and Costlow, 1976
Paragrapsus quadridentatus	490	4-days LC_{50}, age 1 day	Ahsanullah and Arnott, 1978
Rhithropanopeus harrisii, mud crab	50	Development	Rosenberg and Costlow, 1976
	150	Survival	Rosenberg and Costlow, 1976

[a] Includes two life-cyle tests. Unfavorable signifies less than optimal.

is much different for marine invertebrates. Sixteen embryo-larval exposures of mollusks (Table 3) did not demonstrate significant differences in effect levels for mussels, oysters, and clams. The geometric average for the nine specific effect levels for all mollusks is 230 μg/L (±s.d. 34–1600). The average is an order of magnitude lower than the average acute LC_{50} for adults. The data did not lend themselves to accurate estimation of no-effect levels, which would be somewhat lower than the above average. The lowest effective concentrations that appeared to cause meaningful effects on the organism were tabulated in Table 3 and were used in calculating the average of 230 µg/L. One adjustment was made. There were five studies in which it was possible to compare the sublethal effect concentration with the LC_{50} for larval mollusks, and the ratio averaged 0.3. That factor was applied to the LC_{50} in Table 3 to estimate the average of 230 µg/L for sublethal effects. The correction procedure seems more appropriate than mixing the LC_{50} with sublethal concentration or neglecting the data on larval LC_{50}.

Despite the adjustment, it is clear that there is a wide range of effect concentrations listed for larvae in Table 3. Some values may be anomalous because of problems of chemical analysis, presence of other toxicants in the test water, or some other unfavorable condition in the experiment. Still the average of 230 µg/L obviously would not be appropriate for all organisms and responses studied. A reasonable compromise might be to adopt the value of the mean − 1.0 s.d. as an approximation of a low sublethal effect concentration of cadmium for mollusks. That value is 34 µg/L, or two orders of magnitude lower than the LC_{50} for adults.

There are many other studies of sublethal changes in response to cadmium. Table 4 summarizes a range of responses ranging from whole-organism activity to within-organism biochemistry and cell structure. Many of the effects listed for mollusks are associated with cadmium concentrations higher than the levels just described for effects on larvae, and thus they do not contribute to an estimate of the threshold for sublethal effect. The lowest values for mollusks, 78 and 100 µg/L, are somewhat higher than the level of 34 µg/L suggested above for the "effect" threshold for molluskan larvae.

The within-organism changes listed in Table 4 (tissue damage at a rather high concentration and decrease in amino acids) are general ones and do not give much information on the mechanisms of effect. The studies of whole-organism activity (burrowing, distress symptoms, etc.) are useful since they indicate that behavioral effects would apparently occur only at concentrations higher than those affecting larvae.

8.2.2. Crustaceans

Acute LC_{50} for subadult and adult crustaceans range from 50 to 47,000 µg/L Cd (Table 2). Crabs were most tolerant, with a geometric average LC_{50} of 6300 µg/L (±s.d. 1300–31,000). That is significantly higher than the average of 520 µg/L (±s.d. 130–1900) for copepods. The copepod LC_{50} are undoubtedly biased toward lower concentrations since the 2–4-day exposures are more subacute than acute for these short-lived organisms. However, these data will be accepted since such exposures are more or less standard. The LC_{50} for shrimp, amphipods, and isopods were similar and averaged 1100

Table 4. Lowest Concentrations of Cadmium Associated with Various Behavioral, Physiological, and Structural Responses of Mollusks and Crustaceans

Organism	Lowest Effect Level (µg/L)	Effect	Reference
Behavior and activity			
Nassarius obsoletus	500	Distress symptoms at these or higher concentrations	MacInnes and Thurberg, 1973
Bullia digitalis	500	Stress symptoms, burrowing	Cuthbert et al., 1976
Mytilus edulis	1050	Failure to produce byssal threads, 2 weeks	Briggs, 1979
	500	EC_{50}, byssal thread production	Martin et al., 1975
Macoma balthica	1000	Reduced burrowing activity	Eldon et al., 1980
	500	Increased burrowing activity	Eldon et al., 1980
Acarti clausi	200–400	Feeding rate reduced	Moraitou-Apostolopoulou et al., 1979
Growth and regeneration			
Argopecten irradians, bay scallop	78	40 days EC_{50}, growth	Pesch and Stewart, 1980
Palaemonetes pugio, grass shrimp	60	Growth reduced 42%	Pesch and Stewart, 1980
Uca pugilator, fiddler crab	100	Reduced leg regeneration	Weis, 1976, 1980
	1	Swimming rate of stage I larvae (not stage III or IV)	Vernberg et al., 1974
Respiration			
Crassostrea virginica, eastern oyster	300	Reduced respiration, excised gill tissue after 14 days exposure of adults	Engel and Fowler, 1979a
Acartia clausi, copepod	800	Increased respiration	Moraitou-Apostolopoulou et al., 1979
Palaemonetes pugio	(50)	No effect, respiration	Nimmo et al., 1977b
Crangon crangon, brown shrimp	5	Increased ventilation after 13 days exposure	Price and Uglow, 1979
Homarus americanus, American lobster	3000	Increased respiration, excised gill tissue, 30 days exposure of adults	Thurberg et al., 1977

Table 4. (*Continued*)

Organism	Lowest Effect Level (µg/L)	Effect	Reference
Carcinus maenas, green crab	500	Reduced respiration, excised gill tissue, 2 days exposure of adults	Thurberg et al., 1973
Cancer irroratus, rock crab	120	Reduced respiration, excised gill tissue, 2 days exposure of adults	Thurberg et al., 1973
Eurypanopeus depressus	7000	No effect, adult respiration, but decrease in excised gill tissue	Collier et al., 1973
Uca pugilator, fiddler crab	1	Zoeal stages I and III, increased respiration, stage IV depressed during rearing	Vernberg et al., 1974
Osmoregulation			
Idotea baltica	(10000)	No effect on blood osmotic concentration	Jones, 1975
I. emarginata	(10000)	No effect on blood osmotic concentration	Jones, 1975
I. neglecta	(10000)	No effect on blood osmotic concentration	Jones, 1975
Jaera albifrons, isopods	(20000)	No effect on blood osmotic concentration	Jones, 1975
	(20000)	No effect on blood osmotic concentration	Jones, 1975
Homarus americanus, American lobster	(6)	No effect, osmolality after 30 days	Thurberg et al., 1977
Carcinus maenas, green crab	500	Elevation of blood serum osmolality	Thurberg et al., 1973
Tissue and cell structure			
Crassostrea virginica, eastern oyster	100	No effect, fine-structure gill tissue	Engel and Fowler, 1979b
Macoma balthica	1000	Cell damage in broken-off parts of siphons, 1 days exposure	Eldon et al., 1980
Jaera nordmanni, isopod	10000	Cell structure, gills, 8 h, in 10% seawater	Bubel, 1976
Palaemonetes vulgaris	75	Blackened necrotic gills after 10 days	Nimmo et al., 1977b
Palaemon serratus	5000	Slight changes in gill cells after 2 days	Papathanassiou and King, 1983
Penaeus duorarum, pink shrimp	760	Blackened necrotic patches of cells in gills after 15 days	Couch, 1977

Table 4. (*Continued*)

Organism	Lowest Effect Level (μg/L)	Effect	Reference
Paratya tasmaniensis, shrimp	30	Mitochondrial and other cell changes in gills after 4 days	Lake and Thorp, 1974
Biochemical physiology			
Mytilus edulis, blue mussel	150	Decrease in muscle amino acid after 9 days	Briggs, 1979
Homarus americanus, American lobster	6	Increase in Mg–ATPase after 30 days, not in NaK–ATPase	Tucker, 1979
	—	Decrease in NaK–ATPase after 4 days, then 7 days in clean water; no effect, adenylate energy charge	Haya et al., 1983
	6	Slight increase in activity, several enzymes, after 30 days, then 7 days in clean water	Gould, 1980
Cancer irroratus, rock crab	1000	Increase of aminotransferase after 4 days, as Cd chloride, not as Cd nitrate	Gould et al., 1976
	10000	*In vitro* addition to gill tissue inhibited	Tucker and Matte, 1980
	(1000)	ATPase (and no effect) after 30 min	

μg/L (±s.d. 650–9600), intermediate between copepod and crab averages but significantly different from both.

Long-term tests with cadmium and crustaceans yielded the following LC_{50}s:

Copepod, *Acartia clausi*: longevity affected at 100 μg/L (Moraitou-Apostolopoulou et al., 1979).

Shrimp, *Penaeus duorarum*: 1-m LC_{50} = 720 μg/L (Nimmo et al., 1976).

Grass shrimp, *Palaemonetes pugio*: 1.3-m LC_{50} about 250 μg/L (Pesch and Stewart, 1980).

Hermit crab, *Pagurus longicarpus*: 2-m LC_{50} = 70 μg/L (Pesch and Stewart, 1980).

These results for crustaceans vary by an order of magnitude. The crab appears to be the most sensitive, slightly more so than the copepod, although crabs were most resistant overall in acute tests with cadmium.

There were 14 studies of sublethal effects on crustacean larvae (Table 3). The average effect concentration for the three tests with amphipods and shrimp is 1500 µg/L, which is considered anomalous and rejected since it is higher than the acute LC_{50} for this group. The values for lobsters and crabs did not prove different from those for copepods so all 11 values were lumped to yield an average of 52 µg/L ($\pm$s.d. 11–250). The value of 11 µg/L, one standard deviation below the mean, may perhaps be taken as an approximation of the lowest sublethal effect concentration for most crustaceans.

Two sets of results in Table 3 are for complete life-cycle tests. Nimmo et al. (1978) reared mysids (*Mysidopsis bahia*) from 1 day old to maturity at 23 days and recorded mortality and reproductive success. As it happened, the concentration of 11 µg/L Cd that severely affected survival was also the concentration that caused appreciable delay in reproduction as well as fewer young. Perhaps fortuitously, the value corresponds exactly with the overall value suggested above to characterize the larval tests. A more recent chronic test with *M. bahia* showed 10 µg/L as the lowest effect level, agreeing with the value above, but survival of young stages was the most sensitive effect, not reproduction (Gentile et al., 1982). Paffenhöfer and Knowles (1978) found that when the estuarine copepod *Pseudodiaptomus coronatus* was subjected to 5 µg/L Cd, the cadmium-exposed females produced only 50% as many nauplii as those in the control group, although the survival from hatching to adult was unaffected, as was growth rate. Apparently, reproductive ability or survival of young may be the most sensitive parameter of crustacean response to cadmium.

There are some population and "ecosystem" studies, that should be most useful for predicting harmful sublethal levels of cadmium in real environments. Rather than looking at a single species, Kuiper (1981) exposed coastal plankton communities together, in large plastic bags, to cadmium concentrations of 1–53 µg/L. The dominant species were calanoid copepods, *A. clausi, Centropages hamatus, Temora longicornis,* and *Enterpina acutifrons*. Concentrations of 1 and 5 µg/L Cd had no apparent effect on the development or biomass of the copepods. However, at 53 µg/L, the nauplii and copepodites of *C. hamatus* as well as the copepodites of *A. clausi* were significantly reduced, whereas no effect was seen on *T. longicornis* and *E. acutifrons*. Consequently, the species composition shifted and the latter two species became dominant. It appeared that either the nauplii and copepodites were more sensitive to cadmium than adults of the affected species or the reproductive ability of adults was impaired, agreeing with the inference drawn above from laboratory tests. The effect concentration of 53 µg/L agrees almost exactly with the average of 52 µg/L calculated from laboratory tests with larvae, again perhaps fortuitously.

In the only other population experiment found, 500 µg/L Cd caused eventual extinction of pure cultures of a copepod, while 333 µg/L caused extinction in 3 out of 15 populations (Hoppenheit and Sperling, 1977). The experiment continued for almost 7 months, during which time the cultures were "harvested" to various degrees, simulating predatory consumption of the copepods. It was not clear whether 222 µg/L Cd caused permanent changes in populations, such as ratios of adults to immature individuals, but if so, the species appeared to adapt to that level of cadmium over 20 generations (Hoppenheit, 1977). The copepod *Tisbe holothuriae* was considered to be relatively tolerant of environmental change, perhaps because the culture had undergone selection during several years in the laboratory, and hence the results may not be completely predictive to real-world environments.

Among the other sublethal responses listed for crustaceans in Table 4, most are again well above the whole-organism sublethal levels of 11 and 52 µg/L described above. Indeed, many results for physiological response are associated with cadmium concentrations in the lethal range, which makes the experiments seem a pointless exercise.

Table 4 also lists some very low concentrations of cadmium. Two authors noted effects on enzymes at 6 µg/L, below the expected lower limit. In both cases, however, there was an absence of effect on other enzymes, and the overall significance to the organism was not clear. Finally, two organisms showed ventilatory and respiratory changes at concentrations of 5 and 1 µg/L. The 5 µg/L is for a respiratory increase and perhaps can be regarded as a change that may or may not be harmful. The other value, 1 µg/L, is for respiration of larval fiddler crabs, and again, we have the contradictory situation that respiration increased in two of the larval stages but decreased in another stage. However, other changes were reported, notably decreased swimming rate in one of three stages. The cadmium did not seem to cause any change in overall mortality of fiddler crab larvae, but there were changes in optimal temperature–salinity combinations. The lower limit of favorable temperature appeared to shift downward by 3 or 4 °C, and the lower limit of favorable salinity may have shifted upward by a few parts per thousand. On the other hand, a small percentage of the cadmium-exposed larvae had a wider temperature tolerance than the controls but a narrower tolerance of salinity. It is difficult to assign a degree of harmfulness of 1 µg/L Cd to fiddler crab larvae, but the value is the lowest one associated with a change. This interesting research of Vernberg et al. (1974) was not included in Table 3 because data on actual changes in survival, growth, or development were not apparent.

8.2.3. Annelids

Annelids appear to be similar to mollusks in terms of sensitivity to cadmium. For five species of annelids, the 96-h LC_{50} ranges from 3500 to 12,000 µg/L (Table 2), and the geometric average LC_{50} is 6400 µg/L.

Juvenile and adult *Neanthes* have similar 96-h LC_{50} for cadmium, that is,

12,500 and 12,000 µg/L. In contrast, larvae (trochophore) of *Capitella* are more sensitive than adults (96-h LC_{50} of 220 and 7500 µg/L, respectively).

8.2.4. Echinoderms

It is impossible to generalize about the resistance of echinoderms to cadmium since the two examples (Table 2) indicate (1) low resistance (96-h LC_{50} of 800 µg/L for *Asterias*) and (2) high resistance (96-h $LC_{50} \geq 10,000$ µg/L for *Patiriella*).

8.3. PHYTOPLANKTON

Information regarding the toxicity of cadmium to marine algae is scant. Acute toxicity data are not available. However, two studies have attempted to simulate natural conditions by exposing whole communities of plankton to cadmium in large polyethylene bags maintained in the sea. Kremling and Petersen (1978) and Kuiper (1981) found that concentrations of 1.3 and 1.0 µg/L Cd, respectively, had no effect on the species composition, dominance, or biomass of the populations tested. At higher concentrations of 5 and 53 µg/L, Kuiper (1981) again found no effect, except that in one experiment the chlorophyll content was higher than in the controls.

Soyer and Prevot (1981) examined the dinoflagellate *Prorocentrum micans* for histopathological changes using electron microscopy after exposure to cadmium at concentrations ranging from 10 to 100 µg/L. They found as a nonspecific response to a toxicant that the number of peripheral vacuoles and lysosomes increased. As a specific response to cadmium, up to 50% of the mitochondria showed evidence of change (e.g., an electron density less than the controls, visible vacuolation, and fewer cristae). Chloroplasts and nuclei exhibited no change.

The toxicity of cadmium varies depending on the species of phytoplankton tested (Braek et al., 1980; Fisher and Fabris, 1982) and is apparently related to the amount accumulated intracellularly. The accumulation can be altered by algae through the excretion of organic compounds, which are capable of complexing cadmium (Fisher and Fabris, 1982). For example, cadmium in the exudates from *Cricosphaera elongata* significantly reduced the cadmium concentration within the cells (Härdstedt-Roméo and Gnassia-Barelli, 1980). Similarly, waters rich in dissolved organic carbon can protect phytoplankton by complexing the cadmium present (Fisher and Frood, 1980).

Another possible mechanism involved in reducing toxicity concerns the competition between ions for a limited number of uptake sites. For example, when the concentration of Mg^{2+} in the medium was high, the sorption of cadmium by *Phaeodactylum tricornutum* was reduced (Braek et al., 1980). Conversely, when Mg^{2+} was low, cadmium uptake was enhanced. Hence in the marine environment where high levels of nontoxic divalent cations (Ca^{2+}, Mg^{2+}) are present, some protection may be provided through competition because the uptake sites may be saturated initially with Ca^{2+} or Mg^{2+}.

8.4. FACTORS MODIFYING TOXICITY

8.4.1. Salinity

Salinity is the most studied modifying agent. For lethal effects of cadmium, the effect of salinity is clear-cut: it becomes less toxic as the salinity of the water increases. A typical finding is for a crab (*Paragrapsus*). At a salinity of 9‰, cadmium was 2.2 times as toxic as at a salinity of 35‰ (Sullivan, 1977). That result was for high temperature; at low temperature the ratio was 4.5. For other species of crustaceans there are similar ratios: 2.5 times (salinities of 15 and 33‰; Frank and Robertson, 1979) and 1.4–3.0 times depending on temperature (salinities of 10 and 30‰; Vernberg et al., 1974). These ratios are based on 4–10-day LC_{50}, and other researchers have found increased mortality at low salinity for a given level of metal (Engel and Shelton, 1979; Jones, 1975).

The same conclusion applies to lethal and sublethal effects on young stages of fish. For example, Voyer et al. (1982) determined that the percentage of viable hatches of winter flounder eggs exposed to nominal concentrations of 550 and 1000 μg/L Cd increased from 7 and 0% at 10‰ to 100 and 88%, respectively, at 32‰. Similar results were recorded earlier for the same species (Voyer et al., 1977) and for Atlantic silversides (Voyer et al., 1979). Atlantic silversides were also exposed to cadmium under a regime in which salinity fluctuated uniformly between 10 and 30‰ in a sinusoidal pattern over 12-h periods (Voyer et al., 1979). Embryo viability was related to the average salinity and corresponded with that of embryos exposed to a fixed salinity of 20‰.

Part of the explanation for the increased toxicity at low salinity levels may be the result of increased stress on animals that were not adapted to low-salinity regimes. However, there is clearly another factor that is purely chemical, at least for cadmium. Sunda et al. (1978) and Engel and Fowler (1979a) presented findings that cadmium is partially bound by the chloride ion and is nontoxic or less toxic than the free cadmium ion. More free cadmium was present at low salinity, varying in logarithmic fashion from about 23% of total cadmium at 5‰ salinity to only about 4% at salinity of 32‰. Toxicity appeared to increase on the basis of total cadmium, and these authors found the usual three- to five-fold change in lethality of total cadmium from high to low salinity (29 to 5‰). However, when their numerous results were plotted on the basis of free cadmium ions, all the lethality data converged in a single relationship between percentage mortality and free cadmium ion concentration, with an LC_{50} of about 0.05 μg/L.

For seawater of normal salinity and pH, it is not necessary to be concerned about the proportion of free cadmium ion that is present. All the toxicity results reviewed so far appear to be based on total cadmium, and predictions can be based on such a measurement. If, however, one is evaluating pollutional effects in a low-salinity estuary, it would be highly desirable to take the relationship into account. Roughly speaking, if one had a toxicity value obtained in 32‰ seawater, one should divide that toxicity number by 1.7 if the salinity were 20‰ and by about 3.2 if the salinity were 10‰.

8.4.2. Temperature

The lethal effects of cadmium are more severe at high temperature, according to the results available. The metal was 2.0 times as toxic in warm water for a copepod (22 vs. 14 °C; Moraitou-Apostolopoulou et al., 1979), 2.4 times for a crab, with a trend for more extreme temperature effect at high salinity (19 vs. 5 °C; Sullivan, 1977), and 2.7 times for fiddler crabs (20 vs. 10 °C; no overall trend with salinity; Vernberg et al., 1974). Similarly, the 96-h LC_{50} for the crab *Uca* was 10,100 µg/L at 30 °C and 46,600 µg/L at 20 °C, a 4.6 factor (O'Hara, 1973). Fish also appear to be more susceptible to acute levels of cadmium at higher temperatures. Mummichogs displayed a lower resistance at 20 °C than at 5 °C, when exposed in a 192-h bioassay at salinities of 5, 15, 25, and 30‰ (Eisler, 1971). The evidence is all compatible, with greater toxicity at high temperature, but the effect is fairly modest within the usual biokinetic range of the organisms.

8.4.3. Salinity and Temperature

Most of the sublethal work with invertebrates on modifying effects of salinity also involves temperature as a variable. There are excellent pieces of research displaying isopleth diagrams on temperature–salinity axes to demonstrate the interactive effects of these variables and cadmium on the development of shellfish larvae. Comprehensive as such research is, it becomes difficult to sort out a specific effect of any one of salinity, temperature, or metal. The situation is further complicated because some research uses estuarine organisms that are adapted to salinities lower than that of seawater and show a "worsening" of biological responses if salinity is raised.

As a general statement on salinity effect, most of this sublethal research showed a deleterious effect on some aspect of performance of the organism, at salinities below optimal ones, and the effect of the metal was more pronounced. In other words, the toxicity of the metal could be considered greater at low salinities compared with optimal ones. Sometimes effects were slight, sometimes appreciable, but the same general finding emerged in research by Middaugh and Floyd (1978), Rosenberg and Costlow (1976), and Thurberg et al. (1973). In one case, development of larvae fitted the above generalization, but for survival and growth, the effect of cadmium appeared to be worse at higher, usually optimal salinities (Lehnberg and Theede, 1979).

On the other hand, it must be added that performance of the organisms fell off at salinities higher than the optimum, and cadmium could be considered to have a more severe effect. This was certainly the case in three of the papers mentioned above and in the work of Vernberg et al. (1974), in which cadmium appeared to cause a narrowing of the favorable range of salinities. The organisms were all estuarine crustaceans (*Callinectes sapidus, Palaemonetes pugio, Uca pugilator,* and *Rhithropanopeus harrisii*).

In the sublethal experiments with both cadmium and wide-ranging salinity as variables, the primary factor governing response of the organisms seemed to be salinity, not the change in proportion of free-to-total cadmium at the various salinities. Results from one experiment (Rosenberg and Costlow, 1976) were replotted on the basis of free cadmium instead of total cadmium as used by the authors. While the variation of organism response at different salinities was somewhat reduced, it was clear that the salinity optimum of the species was more important than the low experimental levels of cadmium. This was true even if results were adjusted for the response of control organisms. What this means is that sublethal toxicity data for cadmium cannot be extrapolated to low-salinity situations simply on the basis of the calculated toxic form of the metal. The interaction of the cadmium effect with the organism's response to low salinity is complex, and toxicity tests at the salinity of interest are required to evaluate any estuarial pollution. This would scarcely be news to the authors cited above, most of whom discussed in some detail the complexities of their two- or three-factor experiments.

A general statement cannot be made on the modifying effect of temperature on cadmium toxicity from these three-factor experiments. Most of the research, excellent as it is, was not designed to show the effect of temperature on metal toxicity but rather the effect of metal on temperature–salinity tolerance. If one attempts to isolate a temperature effect, its influence on cadmium tolerance runs either way depending on the region of the response surface that is examined.

8.4.4. Other Identities

These other modifying conditions have received little attention, probably because they do not have major effects. One paper showed that very low levels of oxygen (1–4 mg/L) did not greatly change the degree of mortality of copepods exposed to cadmium (Verriopoulos and Moraitou-Apostolopoulou, 1981). No work was found on pH as a modifier of toxicity, perhaps because it is relatively constant in seawater. Nor have size and sex of test organisms received much study, and this suggests that no great influence has been noticed with regard to toxicity of metals. There was no apparent effect of size or sex on survival times of brown shrimp exposed to cadmium (Price and Uglow, 1979), and larger mussels lived only slightly longer in given concentrations of cadmium (Pavicic and Jarvenpää, 1974).

The calcium–magnesium content of fresh water can change the toxicity of metals by an order of magnitude. There appears to be a small effect of this kind when calcium is added to seawater, although the topic is largely of academic interest since the calcium level is not likely to change greatly in marine habitats. Calcium at 200 mg/L apparently reduced the sublethal toxicity of cadmium somewhat, since crabs showed improved regeneration of limbs during exposure in half-strength seawater (Weis, 1978). Similarly, the addition of 800 mg/L Ca prolonged the resistance time of *Marinogammarus obtusatus* exposed to cadmium by factors of 2–3 (Wright and Frain, 1981). The effect on the lethal threshold of cadmium would probably be small.

8.4.5. Multiple Toxicity

There is a small amount of information on the toxicity of cadmium acting jointly with other metals or less similar toxicants. The research does not lead to any general-purpose model but instead to all possible patterns of joint action, which contradict each other. For example, two studies have investigated the toxicity of cadmium in mixtures to fish embryos. The effects on percentage of viable hatch apparently depends on the metals involved. For example, Voyer et al. (1982) found that silver reduced the toxicity of cadmium to winter flounder eggs, and since silver was nontoxic to eggs up to levels of 180 μg/L, they suggested that there was competitive uptake, with a preferential accumulation of silver. On the other hand, von Westernhagen et al. (1979) determined that the toxicity of cadmium, copper, and lead to Baltic herring embryos was strictly additive.

The same lack of a single pattern is found in research with invertebrates. The acute toxicity of cadmium in combination with other metals was investigated in 48-h bioassays with the copepod *Tisbe holothuriae* (Moraitou-Apostolopoulou and Verripoulos, 1982). Combinations of copper and cadmium as well as chromium and cadmium produced mortality greater than expected on an additive basis, whereas a mixture of all three metals resulted in a greater toxicity than would have occurred if each was acting alone but less than in two-metal combinations. Cadmium was simply additive with the insecticide methoxychlor and PCB in 30-day lethal tests with shrimp (Nimmo et al., 1977a). Lead, zinc, and cadmium in a mixture with three other metals apparently resulted in less than additive effects on long-term mortality of clams, since either copper or zinc would have accounted for much of the mortality (Eisler, 1977a). There are two examples of antagonism. Cadmium prolonged the survival of mussels exposed to mercury in an acute test (Breittmayer et al., 1980). Cadmium plus zinc resulted in faster leg regeneration by crabs than when either was present singly, although zinc and mercury were additive in slowing down regeneration (Weis, 1980).

There is no way of summarizing these findings about multiple toxicity in a consistent way. The confusion is not unique to marine organisms and metals. There is not now in existence any satisfactory model for predicting joint action of toxicants in an aquatic habitat. The subject is much understudied, perhaps because it is a difficult one, and yet it is of utmost importance. Considering that any case of cadmium pollution would almost certainly involve several other toxicants, it is curious that studies of the toxicity of cadmium alone continue to emerge in the literature, with a mere sprinkling of studies for multiple toxicity.

8.5. SUMMARY

Lethal concentrations of cadmium for adult or subadult fish are greater than 10,000 µg/L Cd, while those for invertebrates are lower, in the range of a few hundred to a few thousand micrograms per liter. The early life stages of fish do not appear to be affected sublethally at 1000 µg/L, but mollusk larvae are affected at about 30 and crustacean larvae at about 11 µg/L. Ecosystem experiments indicate slight effects on crustaceans but no effects on algae at 50 µg/L Cd. Most concentrations causing meaningful biochemical and physiological effects are in the same general range as the values listed above for sublethal effects.

Toxicity increases 2–4.5 times as salinity drops from 35 to 9‰. This appears to be largely a function of the response of the organisms, although the proportion of the toxic free cadmium ion also increases in low-salinity waters. With increased temperature, toxicity increases by 2–5 times.

When cadmium is present with other toxicants, all possible results have been obtained, from strongly more-than-additive toxicity to antagonism; that is, one metal reduces the toxicity of another. Since any case of cadmium pollution would almost certainly involve other contaminants as well, the confusing topic of multiple toxicity deserves considerably more research.

ACKNOWLEDGMENTS

A large proportion of the references and information on mollusks and crustaceans was collected by the second author while supported by the International Lead Zinc Research Organization, Inc. Typing of the manuscript was done by the now-disbanded Environmental Secretariat of the National Research Council of Canada.

REFERENCES

Ahsanullah, M. (1976). Acute toxicity of cadmium and zinc to seven invertebrate species from Western Port, Victoria. *Aust. J. Mar. Freshwater Res.*, **27**, 187–196.

References

Ahsanullah, M., and Arnott, G. H. (1978). Acute toxicity of copper, cadmium, and zinc to larvae of the crab *Paragrapsus quadridentatus* (H. Milne Edwards), and implications for water quality criteria. *Aust. J. Mar. Freshwater Res.*, **29**, 1–8.

Arnott, G. H., and Ahsanullah, M. (1979). Acute toxicity of copper, cadmium and zinc to three species of marine copepod. *Aust. J. Mar. Freshwater Res.*, **30**, 63–71.

Bellan-Santini, D., and Reish, D. M. (1976). Utilisation de trois espèces de crustacés comme animaux tests de la toxicité de deux sels de métaux lourds. *C. R. Acad. Sci. Série D, Paris*, **282**, 1325–1327.

Braek, G. S., Malnes, D., and Jensen, A. (1980). Heavy metal tolerance of marine phytoplankton. IV. Combined effect of zinc and cadmium on growth and uptake in some marine diatoms. *J. Exp. Mar. Biol. Ecol.*, **42**, 39–54.

Breittmayer, J. P., Guido, R., and Tuncer, S. (1980). Effet du cadmium sur la toxicité du mercure vis-a-vis de la moule *Mytilus edulis* (L.). *Chemosphere*, **9**, 725–728.

Briggs, L. R. (1979). Effects of cadmium on the intracellular pool of free amino acids in *Mytilus edulis*. *Bull. Environ. Contam. Toxicol.*, **22**, 838–845.

Brown, B., and Ahsanullah, M. (1971). Effect of heavy metals on mortality and growth. *Mar. Pollut. Bull.*, **2**(12), 182–187.

Bubel, A. (1976). Histological and electron microscopical observations on the effects of different salinities and heavy metal ions, on the gills of *Jaera nordmanni* (Rathke) (Crustacea, Isopoda). *Cell. Tiss. Res.*, **167**, 65–95.

Calabrese, A., Collier, R. S., Nelson, D. A., and MacInnes, J. R. (1973). The toxicity of heavy metals to embryos of the American oyster *Crassostrea virginica*. *Mar. Biol.*, **18**, 162–166.

Cardwell, R. D., Woelke, C. E., Carr, M. I., and Sanborn, E. W. (1979). Toxic substance and water quality effects on larval marine organisms. State of Washington Department of Fisheries, Technical report number 45.

Chung, K. S. (1980). Acute toxicity of selected heavy metals to mangrove oyster *Crassostrea rhizophorae*. *Bull. Jap. Soc. Sci. Fish.*, **46**, 777–780.

Collier, R. S., Miller, J. E., Dawson, M. A., and Thurberg, F. P. (1973). Physiological response of the mud crab *Eurypanopeus depressus* to cadmium. *Bull. Environ. Contam. Toxicol.*, **10**, 378–382.

Couch, J. A. (1977). Ultrastructural study of lesions in gills of a marine shrimp exposed to cadmium. *J. Invest. Pathol.*, **29**, 267–288.

Cuthbert, K. C., Brown, A. C., and Orren, M. J. (1976). Toxicity of cadmium to *Bullia digitalis* (Prosobranchiata: Nassaridae). *Trans. Roy. Soc. S. Afr.*, **42**, 203–208.

Dawson, M. A., Gould, E., Thurberg, F. P., and Calabrese, A. (1977). Physiological response of juvenile striped bass, *Morone saxatilus*, to low levels of cadmium and mercury. *Chesapeake Sci.*, **18**, 353–359.

Dorband, W. R., Van Olst, J. C., Carlberg, J. M., and Ford, R. F. (1976). Effects of chemicals in thermal effluent on Homarus americanus maintained in aquaculture systems. In *Proceedings of the Seventh Annual Workshop World Mariculture Society, San Diego, California, 1976*, Louisiana State University, Division of continuing education, Baton Rouge, La. p. 20.

Eisler, R. (1971) Cadmium poisoning in *Fundulus heteroclitus* (Pices: Cyprinodontidae) and other marine organisms. *J. Fish. Res. Bd. Can.*, **28**, 1225–1234.

Eisler, R. (1977a). Toxicity evaluation of a complex metal mixture to the softshell clam *Mya arenaria*. *Mar. Biol.*, **43**, 265–276.

Eisler, R. (1977b). Acute toxicities of selected heavy metals to the softshell clam, *Mya arenaria*. *Bull. Environ. Contam. Toxicol.*, **7**, 137–145.

Eisler, R., and Hennekey, R. J. (1977). Acute toxicities of Cd^{2+}, Cr^{+6}, Hg^{2+}, Ni^{2+} and Zn^{2+} to estuarine macrofauna. *Arch. Environ. Contam. Toxicol.*, **6**, 315–323.

Eldon, J., Pekkarinen, M., and Kristoffersson, R. (1980). Effects of low concentrations of heavy metals on the bivalve *Macoma balthica*. *Ann. Zool. Fenn.*, **17**, 233–242.

Engel, D. W., and Fowler, B. A. (1979a). Factors influencing cadmium accumulation and its toxicity to marine organisms. *Environ. Health Perspect.*, **28**, 81–88.

Engel, D. W., and Fowler, B. A. (1979b). Copper and cadmium induced changes in the metabolism and structure of molluscan gill tissue. In *Marine Pollution: Functional Responses*, Vernberg, W. B., Thurberg, F. P., Calabrese, A., and Vernberg, F. J. (eds.). Academic, New York, pp. 239–256.

Engel, D. W., and Shelton, M. G. (1979). The effects of the interactions of radiation, salinity, and metals on grass shrimp, *Palaemonetes pugio*. In *Radiation Effects on Aquatic Organisms*, Egami, N. (ed.). University Park Press, Baltimore, MD, pp. 59–72.

Fisher, N. S., and Fabris, J. G. (1982). Complexation of Cu, Zn, and Cd by metabolites excreted from marine diatoms. *Mar. Chem.*, **11**, 245–255.

Fisher, N. S., and Frood, D. (1980). Heavy metals and marine diatoms: Influence of dissolved organic compounds on toxicity and selection for metal tolerance among four species. *Mar. Biol.*, **59**, 85–93.

Flatau, G., and Aubert, M. (1979). Etude de la toxicité directe et induite du cadmium en milieu marin. *Rev. Intern. Océanogr. Méd.*, **53–54**, 51–59.

Frank, P. M., and Robertson, P. B. (1979). The influence of salinity on toxicity of cadmium and chromium to the blue crab, *Callinectes sapidus*. *Bull. Environ. Contam. Toxicol.*, **21**, 74–78.

Gentile, S. M., Gentile, J. H., Walker, J., and Heltshe, J. F. (1982). Chronic effects of cadmium on two species of mysid shrimp: *Mysidopsis bahia* and *Mysidopsis bigelowi*. *Hydrobiologia*, **93**, 195–204.

Gould, E. (1977). Alteration of enzymes in winter flounder, *Pseudopleuronectes americanus*, exposed to sublethal amount of cadmium chloride. In *Physiological Responses of Marine Biota to Pollutants*, Vernberg, F. J., Calabrese, A., Thurberg, F. P., and Vernberg, W. B. (eds.). Academic, New York, pp. 209–224.

Gould, E. (1980). Low-salinity stress in the American lobster, *Homarus americanus*, after chronic sublethal exposure to cadmium: Biochemical effects. *Helgol. Wiss. Meeresunters.*, **33**, 36–46.

Gould, E., Collier, R. S., Carolus, J. J., and Givens, S. (1980). Heart transaminase in the rock crab, *Cancer irroratus*, exposed to cadmium salts. *Bull. Environ. Contam. and Toxicol.*, **15**, 635–643.

Härdstedt-Roméo, M., and Gnassia-Barelli, M. (1980). Effect of complexation by natural phytoplankton exudates on the accumulation of cadmium and copper by the Haptophyceae *Cricosphaera elongata*. *Mar. Biol.*, **59**, 79–84.

Haya, K., Waiwood, B. A., and Johnston, D. W. (1983). Adenylate energy charge and ATPase activity of lobster (*Homarus americanus*) during sublethal exposure to a zinc. *Aquat. Toxicol.*, **3**, 115–126.

Hazen, R. E. (1981). *Cadmium in an aquatic ecosystem*. Ph.D. Thesis. New York University.

Hoppenheit, M. (1977). On the dynamics of exploited populations of *Tisbe holothuriae* (Copepoda, Harpacticoidae) V. The toxicity of cadmium: Response to sublethal exposure. *Helgol. Wiss. Meeresunters.*, **29**, 503–523.

Hoppenheit, M., and Sperling, K.-R. (1977). On the dynamics of exploited populations of *Tisbe holothuriae* (Copepoda, Harpacticoidae) IV. The toxicity of cadmium: Response to lethal exposure. *Helgol. Wiss. Meeresunters.*, **29**, 328–336.

Jackim, E., Hamlin, S. M., and Sonis, S. (1970). Effects of metal poisoning on five liver enzymes in the killifish (*Fundulus heteroclitus*). *J. Fish. Res. Bd. Can.*, **27**, 328–336.

Johansson-Sjöbeck, M. L., and Larsson, A. (1978). The effect of cadmium on the hematology and on the activity of δ-aminolevulinic acid dehydratase (ALA-D) in blood and hematopoietic tissues of the flounder, *Pleuronectes flesus* L. *Environ. Res.*, **17**, 191–204.

Johnson, M. W., and Gentile, J. H. (1979). Acute toxicity of cadmium, copper, and mercury to larval American lobster *Homarus americanus*. *Bull. Environ. Contam. Toxicol.*, **22**, 258–264.

Jones, M. B. (1975). Synergistic effects of salinity, temperature and heavy metals on mortality and osmoregulation in marine and estuarine isopods (Crustacea). *Mar. Biol.*, **30**, 13–20.

Kremling, K., and Petersen, H. (1978). The distribution of Mn, Fe, Zn, Cd, and Cu in Baltic seawater: A study on the basis of one anchor station. *Mar. Chem.*, **6**, 155–170.

Kuiper, J. (1981). Fate and effects of cadmium in marine plankton communities in experimental enclosures. *Mar. Ecol. Prog. Ser.*, **6**, 161–174.

Lake, P. S., and Thorp, V. J. (1974). The gill lamellae of the shrimp *Paratya tasmaniensis* (Atyidae:Crustacea). Normal ultrastructure and changes with low levels of cadmium. *Eighth Int. Congr. Electr. Microsc.*, Canberra, Australia, **III**, 448–449.

Lang, W. H., Miller, D. C., Ritacco, P. J., and Marcy, M. (1981). The effects of copper and cadmium on the behavior and development of barnacle larvae. In *Biological monitoring of marine pollutants*, Vernberg, F. J., Calabrese, A., Thurberg, F. P., and Vernberg, W. B. (eds.). Academic, New York, pp. 165–204.

Larsson, A., Bengtsson, B.-E., and Haux, C. (1981). Disturbed ion balance in flounder, *Platichthys flesus* L. exposed to sublethal levels of cadmium. *Aquat. Toxicol.*, **1**, 19–35.

Lehnberg, W., and Theede, H. (1979). Kombinierte wirkungen von temperatur, salzgehalt und cadmium auf entwicklung, wachstum und mortalität der larven von *Mytilus edulis* aus der westlichen Ostsee. *Helgol. Wiss. Meeresunters.*, **32**, 179–199.

MacInnes, J. R., and Thurberg, F. P. (1973). Effects of metals on the behavior and oxygen consumption of the mud snail. *Mar. Bullut. Bull.*, **4**, 185–186.

MacInnes, J. R., Thurberg, F. P., Greig, R. A., and Gould, E. (1977). Long-term cadmium stress in the cunner, *Tautogolabrus adspersus*. *U.S. Dept. Comm., Nat. Ocean. Atmospher. Admin. Fish. Bull.*, **75**, 199–203.

Madhupratap, M., Achuthankutty, C. T., and Nair, S. R. S. (1981). Toxicity of some heavy metals to copepods *Acartia spinicauda* and *Tortanus forcipatus*. *Ind. J. Mar. Sci.*, **10**, 382–383.

Martin, J. M., Piltz, F. M., and Reisch, D. J. (1975). Studies on the *Mytilus edulis* community in Alamitos Bay, California. V. The effects of heavy metals on byssal thread production. *Veliger*, **18**, 183–188.

McKim, J. M. (1977). Evaluation of tests with early life stages of fish for predicting long-term toxicity. *J. Fish. Res. Bd. Can.*, **34**, 1148–1154.

McLeese, D. W. (1981). Cadmium and marine invertebrates. *Wat. Sci. Tech.*, **13**, 1085–1086.

Middaugh, D. P., and Dean, J. M. (1977). Comparative sensitivity of eggs, larvae and adults of the estuarine teleosts *Fundulus heteroclitus* and *Menidia menidia*, to cadmium. *Bull. Environ.*, **17**, 645–652.

Middaugh, D. P., and Floyd, G. (1978). The effect of prehatch and posthatch exposure to cadmium on salinity tolerance of larval grass shrimp, *Palaemonetes pugio*. *Estuaries*, **1**, 123–125.

Mirkes, D. Z., Vernberg, W. B., and DeCoursey, P. J. (1978). Effects of cadmium and mercury on the behavioral responses and development of *Eurypanopeus depressus* larvae. *Mar. Biol.*, **47**, 143–147.

Moraitou-Apostolopoulou, M., and Verripoulos, G. (1982). Individual and combined toxicity of three heavy metals Cu, Cd and Cr for the marine copepod *Tisbe holothuriae*. *Hydrobiologia*, **87**, 83–87.

Moraitou-Apostolopoulou, M., Verripoulos, G., and Lentzou, P. (1979). Effects of sublethal concentrations of cadmium as possible indicators of cadmium pollution for two populations of *Acartia clausi* (Copepoda) living at two differently polluted areas. *Bull. Environ. Contam. Toxicol.*, **23**, 642–649.

Negilski, D. S. (1976). Acute toxicity of zinc, cadmium and chromium to the marine fishes, yellow-eye mullet (*Aldrichetta forsteri* C.&V.) and small-mouthed hardyhead (*Atherinsoma microstoma* Whitley). *Aust. J. Mar. Freshwater Res.*, **27**, 137–149.

Nelson, D. A., Calabrese, A., Nelson, B. A., MacInnes, J. R., and Wenzloff, D. R. (1976). Biological effects of heavy metals on juvenile bay scallops *Argopecten irradians* in short-term exposures. *Bull. Environ. Contam. Toxicol.*, **16**, 275–282.

Nimmo, D. R., Rigby, R. A., Bahner, L. H., and Sheppard, J. M. (1977a). *Mysidopsis bahia*: An estuarine species suitable for toxicity tests to determine the effects of a pollutant. In *Proceedings of the Annual Symposium, Memphis, Tenn., USA*, Mayer, F. L., and Hamelink, J. L. (eds.). American Society for Testing and Materials, Philadelphia, PA, pp. 109–116.

Nimmo, D. W. R., Lightner, D. V., and Bahner, L. H. (1977b). Effects of cadmium on the shrimps, *Penaeus duorarum*, *Palaemonetes pugio* and *Palaemonetes vulgaris*. In *Physiological Response of Marine Biota to Pollutants*, Vernberg, F. J., Calabrese, A., Thurberg, F. P., and Vernberg, W. B. (eds.). Academic, New York, pp. 131–183.

Nimmo, D. R., Rigby, R. A., Bahner, L. H., and Sheppard, J. M. (1978). The acute and chronic effects of cadmium on the estuarine mysid, *Mysidopsis bahia*. *Bull. Environ. Contam. Toxicol.*, **19**, 80–85.

O'Hara, J. (1973). The influence of temperature and salinity on the toxicity of cadmium to the fiddler crab, *Uca pugilator*. *Fish. Bull*, **71**, 149–153.

Paffenhöfer, G.-A., and Knowles, S. C. (1978). Laboratory experiments on feeding, growth, and fecundity of and effects of cadmium on *Pseudodiaptomus*. *Bull. Mar. Sci.*, **28**, 574–580.

Papathanassiou, E., and King, P. E. (1983). Ultrastructural studies on the gills of *Palaemon serratus* (Pennant) in relation to cadmium accumulation. *Aquat. Toxicol.*, **3**, 273–284.

Pavicic, J., and Jarvenpää, T. (1974). Cadmium toxicity in adults and early larval stages of the mussel *Mytilus galloprovincialis* Lam. In *Comparative Studies of Food and Environmental Contamination. Proceedings of a Symposium. Otaniemi, August 27–31, 1973*. International Atomic Energy Agency, Vienna, pp. 179–188.

Pesch, G. G., and Stewart, N. E. (1980). Cadmium toxicity to three species of estuarine invertebrates. *Mar. Environ. Res.*, **3**, 145–156.

Portman, J. E. (1968). Progress report on a programme of insecticide analysis and toxicity-testing in relation to the marine environment. *Helgol. Wiss. Meeresunters.*, **17**, 247–256.

Portman, J. E. (1970). The toxicity of 120 substances to marine organisms. Shellfish information leaflet number 19, United Kingdom Ministry of Agriculture, Fisheries and Food, London.

Portman, J. E., and Wilson, V. W. (1971). Shellfish information leaflet 22, United Kingdom Ministry of Agriculture Fisheries and Food, London.

Price, R. K. J., and Uglow, R. F. (1979). Some effects of certain metals on development and mortality within the moult cycle of *Cragnon cragnon* (L.). *Mar. Environ. Res.*, **2**, 287–299.

Pruell, R. J., and Engelhardt, F. R. (1980). Liver cadmium uptake, catalase inhibition and cadmium thionein production in the killifish (*Fundulus heteroclitus*) induced by experimental cadmium exposure. *Mar. Environ. Res.*, **3**, 101–111.

Ray, S., and McLeese, D. W. (1983). Factors affecting uptake of cadmium and other trace metals from marine sediments by some bottom-dwelling marine invertebrates. In *Wastes in the Ocean*, Vol. 2, Kester, D. R., Ketchum, B. H., Duedall, I. W., and Park, P. K. (eds.). Wiley, New York, pp. 185–197.

Reisch, D. J., and Carr, R. S. (1978). The effect of heavy metals on the survival, reproduction, development and life cycles for two species of polychaetous annelids. *Mar. Pollut. Bull.*, **9**, 24–27.

Reisch, D. J., and Martin, J. M. (1976). The effect of heavy metals on laboratory population of two polychaetes with comparisons to the water quality conditions and standards in southern California marine waters. *Wat. Res.*, **10**, 299–302.

Roch, M., and Maly, E. J. (1979). Relationship of cadmium-induced hypocalcemia with mortality in rainbow trout, *Salmo gardneri*, and the influence of temperature on toxicity. *J. Fish. Res. Board Can.* **36**, 1297–1303.

Rosenberg, R., and Costlow, J. D. (1976). Synergistic effects of cadmium and salinity combined with constant and cycling temperatures on the larval development of two estuarine crab species. *Mar. Biol.*, **38**, 291–303.

Sosnowski, S. L., and Gentile, J. H. (1978). Toxicological comparison of natural and cultured populations of *Acartia tonsa* to cadmium, copper, and mercury. *J. Fish. Res. Bd. Can.*, **35**, 1366–1369.

Soyer, M.-O., and Prevot, P. (1981). Ultrastructural damage by cadmium in a marine dinoflagellate, *Prorocentrum micans. J. Protozool.*, **28**, 308–313.

Sprague, J. B. (1971). Measurement of pollutant toxicity to fish—III. Sublethal effects and "safe" concentrations. *Wat. Res.*, **5**, 245–266.

Sullivan, J. K. (1977). Effects of salinity and temperature on the acute toxicity of cadmium to the estuarine crab *Paragrapsus gaimardii*. (Milne-Edwards). *Aust. J. Mar. Freshwater Res.*, **28**, 739–743.

Sunda, W. G., Engel, D. W., and Thuotte, R. M. (1978). Effect of chemical speciation on toxicity of cadmium to grass shrimp, *Palaemonetes pugio:* Importance of free cadmium ion. *Environ. Sci. Technol.*, **12**, 409–413.

Thurberg, F. P., Dawson, M. A., and Collier, R. S. (1973). Effects of copper and cadmium on osmoregulation and oxygen consumption in two species of estuarine crabs. *Mar. Biol.*, **23**, 171–175.

Thurberg, F. P., Calabrese, A., Gould, E., Greig, R. A., Dawson, M. A., and Tucker, R. K. (1977). Response of the lobster, *Homarus americanus*, to sublethal levels of cadmium and mercury. In *Physiological Response of Marine Biota to Pollutants*, Vernberg, F. J., Calabrese, A., Thurberg, F. P., and Vernberg, W. B. (eds.). Academic, New York, pp. 185–197.

Tucker, R. K. (1979). Effects of *in vivo* cadmium exposure on ATPases in gill of the lobster, *Homarus americanus. Bull. Environ. Contam. Toxicol.*, **23**, 33–35.

Tucker, R. K., and Matte, A. (1980). *In vitro* effects of cadmium and lead on ATPases in the gill of the rock crab, *Cancer irroratus. Bull. Environ. Contam. Toxicol.*, **24**, 847–852.

Vernberg, W. B., De Coursey, P. J., and O'Hara, J. (1974). Multiple environmental factor effects of physiology and behavior of the fiddler crab, *Uca pugilator*. In *Pollution and Physiology of Marine Organisms*, Vernberg, F. J., and Vernberg, W. B. (eds.). Academic, New York, pp. 381–425.

Verriopoulos, G., and Moraitou-Apostolopoulou, M. (1981). Effects of some environmental factors on the toxicity of cadmium to the copepod *Tisbe holothuriae. Arch. Hydrobiol.*, **91**, 287–293.

von Westernhagen, H., and Dethlefsen, V. (1975). Combined effects of cadmium and salinity on development and survival of flounder eggs. *J. Mar. Biol. Assoc. U.K.*, **55**, 945–957.

von Westernhagen, H., Rosenthal, H., and Sperling, K.-R. (1974). Combined effects of cadmium and salinity on development and survival of herring eggs. *Helgol. Wiss. Meeresunters.*, **26**, 416–433.

von Westernhagen, H., Dethlefsen, V., and Rosenthal, H. (1975). Combined effects of cadmium and salinity on development and survival of garpike eggs. *Helgol. Wiss. Meeresunters.*, **27**, 268–282.

von Westernhagen, H., Dethlefsen, V., and Rosenthal, H. (1979). Combined effects of cadmium, copper and lead on developing herring eggs and larvae. *Helgol. Wiss. Meeresunters.,* **32,** 257–278.

Voyer, R. A., Wentworth, C. E., Barry, E. P., and Hennekey, R. J. (1977). Viability of embryos of the winter flounder *Pseudopleuronectes americanus* exposed to combinations of cadmium and salinity at selected temperatures. *Mar. Biol.,* **44,** 117–124.

Voyer, R. A., Heltshe, J. F., and Kraus, R. A. (1979). Hatching success and larval mortality in an estuarine teleost, *Menidia menidia* (Linnaeus), exposed to cadmium in constant and fluctuating salinity regimes. *Bull. Environ. Contam. Toxicol.,* **23,** 475–481.

Voyer, R. A., Cardin, J. A., Heltshe, J. F., and Hoffman, G. L. (1982). Viability of embryos of the winter flounder *Pseudopleuronectes americanus* exposed to mixtures of cadmium and silver in combination with selected fixed salinities. *Aquat. Toxicol.,* **2,** 223–233.

Watling, H. R. (1978). Effect of cadmium on larvae and spat of the oyster, *Crassostrea gigas* (Thunberg). *Trans. Roy. Soc. S. Afr.,* **43,** 125–134.

Watling, H. R. (1981). Effects of metals on the development of oyster embryos. *S. Afr. J. Sci.,* **77,** 134–135.

Weis, J. S. (1976). Effects of mercury, cadmium, and lead salts on regeneration and ecdysis in the fiddler crab, *Uca pugilator*. *U.S. Dept. Comm. Nat. Ocean. Atmos. Admin. Fish. Bull.,* **74,** 464–467.

Weis, J. S. (1978). Interactions of methylmercury, cadmium, and salinity on regeneration in the fiddler crabs, *Uca pugilator, U. pugnax* and *U. minax. Mar. Biol.,* **49,** 119–124.

Weis, J. S. (1980). Effect of zinc on regeneration in the fiddler crab, *Uca pugilator* and its interactions with methylmercury and cadmium. *Mar. Environ. Res.,* **3,** 249–255.

Wright, D. A., and Frain, J. W. (1981). Cadmium toxicity in *Marinogammarus obtusatus:* Effect of external calcium. *Environ. Res.,* **24,** 338–344.

Zaroogian, G. E., and Morrison, G. (1981). Effect of cadmium body burdens in adult *Crassostrea virginica* on fecundity and viability of larvae. *Bull. Environ. Contam. Toxicol.,* **27,** 344–348.

9

BIOLOGICAL CYCLING OF CADMIUM IN MARINE ENVIRONMENT

S. Ray
D. W. McLeese

Fisheries and Environmental Sciences
Fisheries Research Branch
Department of Fisheries and Oceans
Biological Station
St. Andrews, New Brunswick, Canada

9.1. Introduction
9.2. Chemical Form of Cadmium in Marine Environment
9.3. Distribution of Cadmium in Biota
9.4. Bioavailability of Cadmium
    9.4.1. Water
    9.4.2. Sediment
9.5. Bioaccumulation of Cadmium
    9.5.1. Water
        9.5.1.1. Phytoplankton
        9.5.1.2. Mollusks
        9.5.1.3. Crustaceans
        9.5.1.4. Polychaetes
        9.5.1.5. Fish
    9.5.2. Sediment
    9.5.3. Food
9.6. Environmental Factors Affecting Bioaccumulation of Cadmium
    9.6.1. Temperature
    9.6.2. Salinity
    9.6.3. Seasonal Variation

9.7. Storage of Cadmium
    9.7.1. Subcellular Mechanisms
        9.7.1.1. Metallothionein
        9.7.1.2. Membrane-Limited Vesicles
    9.7.2. Extracellular Mechanisms
9.8. Excretion of Cadmium
9.9. Summary
References

9.1. INTRODUCTION

Since the early 1970s, with the increase in sophistication in trace-metal analysis, the reported concentrations of cadmium in unpolluted open-ocean surface waters have steadily decreased (Raspor, 1980; Mart and Nürnberg, 1986; and Chapter 1, this volume). For example, recent values are less than 5 ng/kg for the North Atlantic (Boyle et al., 1981; Bruland and Franks, 1983; Mart et al., 1982; Spivack et al., 1983) and North Pacific (Boyle et al., 1981; Boyle and Huested, 1983; Bruland et al., 1978; Mart et al., 1982). However, cadmium concentration increases with nutrient level in the surface waters. Levels in the Canadian eastern Arctic have been variously reported to be 16–22 ng/kg (Danielsson and Westerlund, 1983); 7–10 ng/kg (Mart et al., 1982, 1984), and 20–50 ng/kg (Campbell and Yeats, 1982). Higher concentrations may be found where water circulation and water mass exchange are limited (Kremling and Petersen, 1978; Mart et al., 1982, 1984; Spivack et al., 1983). High cadmium concentrations, up to 125 µg/L, have been reported in water of certain coastal and estuarine areas due to weathering and/or anthropogenic input of cadmium (Abdullah and Royle, 1974; Bloom and Ayling, 1977; Chan et al., 1974; Holmes et al., 1974; Loring et al., 1980; Mart et al., 1982, 1984; Thornton et al., 1975). However, in view of the recent analytical trend, some of the earlier values may be doubtful.

    The vertical distribution of cadmium is directly related to the nutrient level in the water column (Boyle and Huested, 1983; Bruland and Franks, 1983; Danielsson and Westerlund, 1983; also discussed in Chapter 2, this volume). The low concentration in surface water increases to the vicinity of 120 ng/L at about 1000 m depth and then decreases slightly at greater depths. However, only a small percentage of the cadmium remains in true solution, since the metal may be adsorbed on particulate matter that can remain suspended indefinitely or may settle slowly to the bottom in relatively calm water. Marine clays from unpolluted areas would be expected to contain 0.2–0.5 mg Cd/kg (Chapter 2). In contaminated areas, and especially in areas receiving metal refinery waste or mine effluent, high levels, up to about

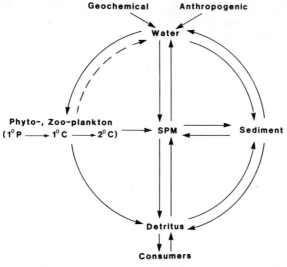

Figure 1. Cadmium cycle in marine environment.

1400 mg Cd/kg (dry wt.) have been reported (Bloom and Ayling, 1977; Bower et al., 1978; Holmes et al., 1974; Loring et al., 1980; Skei et al., 1972).

Marine organisms accumulate cadmium in their tissues to levels considerably higher than those in seawater. The primary concerns regarding the ubiquitous presence of cadmium in the marine environment are that high levels in marine organisms may have detrimental effects on the organisms themselves or on the suitability of commercial species as food for humans. There are three possible sources of cadmium to marine organisms: (1) in solution or in colloidal suspension; (2) in food or on particulate matter that may be ingested; and (3) in sediment. The relative importance of these sources is not understood properly and is difficult to establish since the organisms are usually exposed to more than one source. There is a need to define the role of marine organisms in the cycling of cadmium in the marine environment (Fig. 1) (i.e., accumulation and excretion by biota, effects, transfer, and ultimate fate of the metal).

9.2. CHEMICAL FORM OF CADMIUM IN MARINE ENVIRONMENT

Cadmium may be present in one or more of the three phases of the marine environment (i.e., water, suspended particulate matter, and bottom sediment) and may be in equilibrium between the phases. The processes and mechanisms are discussed in Chapter 1, and only broad features that affect biological cycling are discussed here.

Biological availability (bioavailability) of cadmium to marine organisms is governed largely by its chemical form. Recently, electrochemical techniques have been used to speciate dissolved cadmium in seawater. Chemical forms of cadmium in sediment also have been studied by selective chemical extraction procedures, but these are not specific for different forms and can provide only a qualitative picture.

In seawater cadmium may be either in solution as free hydrated ions or complexed with labile and nonlabile chemical species, both inorganic or organic. The predominant inorganic species is chloride, while the organic ligands may be polyphenols, amino acids, humates, proteins, or other tissue breakdown products. The degree and nature of complexation depends on the alkali earth metals, chelating ligands, and the stability constant ratios of each of the complexes and also are influenced by physiochemical parameters such as salinity and pH.

Cadmium is present in seawater mostly in the "ionic" form, predominantly in a variety of chloride complexes (Baric and Branica, 1967; Lu and Chen, 1977; Nürnberg and Valenta, 1983; Sillen 1961, 1962; Zirino and Yamamoto, 1972). Differences of opinion exist regarding the presence in organic complexes (Batley and Florence, 1976; Duinker and Kramer, 1977; Mart et al., 1984; Nürnberg and Valenta, 1983).

In particulate matter, cadmium concentrations are low compared with those in marine sediments. Suspensions of manganese oxides will absorb cadmium, but most particulate metal is associated with fecal matter. Sodium and other ions also may compete with cadmium or surface sites on the particulates depending on pH and ionic strength of the seawater.

Sediments are the ultimate recipient of nearly all trace metals in the sea, and sediment concentrations of metals, including cadmium, are several orders of magnitude higher than in the water. Sediments may determine the dissolved metal concentrations in associated and overlying waters. Bioaccumulation of cadmium by marine organisms has in several instances been related to the concentration and chemical species of cadmium present in the sediment. Ray and McLeese (1983) determined sorption characteristics of cadmium in a sediment–seawater system using an estuarine sediment separated into four size fractions. The sorption data followed the Freundlich isotherm and indicated that the sorption coefficients were directly related to total organic carbon (TOC) and cation exchange capacity (CEC) of the sediment but not to the particle size of the fractions. Cadmium was strongly bound to particles of high organic carbon content and cation exchange capacity. Fractions with different particle size but similar organic carbon had very similar sorption coefficients for cadmium concentrations varying widely from 9 to 220 ng Cd/mL. If cadmium is strongly bound to sediments with high organic carbon content, it is not bioavailable, but may become so, following physicochemical changes in water and sediments (Malo, 1977).

Chemical extractions have been used to differentiate metals associated with ion-exchangeable, easily reducible, moderately reducible organic and silicate residue fractions of the sediments (Salomons and Förstner, 1980).

9.3. DISTRIBUTION OF CADMIUM IN BIOTA

Cadmium is present in almost all marine organisms that have been examined, from bacteria to vertebrates, although data for phytoplankton and zooplankton are scarce. In uncontaminated areas, the organisms have low concentrations, but some marine invertebrates in contaminated areas can accumulate cadmium rapidly to several orders of magnitude above the levels in organisms from unpolluted areas. Calculated bioconcentration factors (BCF) for plankton, seaweed, mollusk, crustacea, and fish are usually of the order 10^4, 10^2–10^3, 10^3–10^4, 10^3, and 10^2, respectively (Preston, 1973). Bryan (1976) listed concentrations of cadmium in whole marine organisms as follows (in µg/g dry wt.) phytoplankton 2; seaweeds 0.5; zooplankton (copepods) 4; coelenterate 1; echinoderms 2; bivalve mollusks 2; oysters 10; gastropod mollusks 6; cephalopod mollusks 5; decapod crustaceans 1; tunicates 1; and fish 0.2.

Reported values should be considered with caution. Hamilton (1981) pointed out that sample collection and preparation, use of certified reference materials, sample composition and matrix effect, method of analysis and good housekeeping may affect the "quality (accuracy) of data" for "minor trace elements present in a variety of materials from the marine environment." Results may be reported as wet weight or dry weight, making comparison difficult.

The characteristics and habits of different groups of organisms will determine how they obtain cadmium and how much of it. Cadmium content of sediment is likely to have a major influence on uptake by sediment-ingesting animals such as polychaetes. Closely related species can accumulate different concentrations of cadmium from sediment (Jackim et al., 1977). Filter feeders are affected by the metal content in the water and by the suspended particulate matter. Because of such differences, the rates and extent of bioaccumulation may vary significantly among species in the same location. In addition, the extent of bioaccumulation will depend on the relative ability of the animals to depurate the metal. The concentration factors for cadmium in the filter-feeding scallops, oysters, and mussels have been estimated as 2.3×10^6, 3.2×10^5, and 1×10^5, respectively (Ketchum, 1972). These values are much higher than the estimates by Preston (1973) listed above and illustrate the variation in the literature. The great ability of certain mollusks to accumulate cadmium and other pollutants from the environment has led to their use as monitors. Occasionally, organisms are not identified to species. Most literature values are limited to organisms from the Northern Hemisphere.

Concentrations of cadmium in marine organisms reported up to about 1977 have been reviewed by Coombs (1979) and Phillips (1980). The *Journal of Water Pollution Control Federation* (literature issues) is an excellent source for recent values.

9.4. BIOAVAILABILITY OF CADMIUM

The best indication of bioavailability is usually the level of cadmium within an organism because prediction of bioavailability from levels in the environment is a complex matter. Determination of total cadmium concentrations in seawater, suspended matter, and sediments does not provide a reliable picture of bioavailability since different chemical forms of the metal differ in biological availability. Physicochemical factors such as redox potential, pH, sorption, chelation, complexation, precipitation, and hydrolysis may also affect bioavailability.

Thermodynamic considerations lead to the conclusion that bioavailability should depend on free metal ion activity, the driving force for chemical reactions. Bioaccumulation of cadmium involves the transport of the ion or of complexed cadmium to receptor sites, followed by transfer through the animal's integument or membranes. In most cases, only the free metal is expected to interact with the receptor sites and the extent of binding would depend on the ion activity. The action may be in competition with other metal ions and thus controlled by relative metal ion activities and relative stability constants for the various metal–ligand combinations. The surface layer of the sediment is likely to play a significant role in the transport to receptor sites, and its cadmium concentration may differ from concentration in the bulk of the sediment.

9.4.1. Water

It is generally agreed that the ionic form of cadmium in seawater is most bioavailable to marine organisms, so the relative proportions of ionic cadmium and all the other chemical forms will be important in governing bioaccumulation. There are conflicting reports on the importance of uptake of cadmium organically bound with EDTA, NTA, or humic acid complexes compared to the "ionic" form as in $CdCl_2$.

In many cases, organic binding of cadmium makes it less available to marine organisms. Uptake by the diatom *Phaeodactylum tricornutum* in the presence of EDTA is negligible in comparison with exposure to cadmium alone (Cossa, 1976). The phytoplankter *Criscosphaera elongata* exposed to cadmium in the presence of a natural exudate from phytoplankton accumulated 12–24% less metal than in exposures without the exudate (Hardstedt-Roméo and Gnassia-Barelli, 1980). The oyster *Crassostrea virginica* accumulated significantly less cadmium in the presence of EDTA, NTA, and humic acid (Hung, 1982). Ray et al. (1979) observed a reduction of approximately 30% in accumulated cadmium in the polychaete *Nereis virens* and shrimp *Pandalus montagui* exposed to CdEDTA, and there is a reduction of about 45% by *Macoma balthica* compared with exposure to the same concentration of cadmium as $CdCl_2$ (McLeese and Ray, 1984). A similar decrease has been observed for barnacles (*Semibalanus balanoides*) exposed to humate, alginate, and EDTA complexes of cadmium (Rainbow et al., 1980), but cadmium uptake by the crab *Carcinus maenas* was relatively unaffected in the presence of sodium salts of EDTA and phosphate (Gutierrez-Galindo, 1980).

In contrast, increased cadmium uptake in the presence of EDTA was reported for the dinoflagellate *Prorocentrum micans* (Prévot and Soyer, 1978), George and Coombs (1977) found that prior complexation of ionic cadmium with EDTA, humic acid, or alginic acid doubled uptake rate and final tissue concentration of metal in *Mytilus edulis* compared with those exposed to ionic cadmium. Obviously, there is no single simple relationship between uptake of organically complexed cadmium and the ionic form.

9.4.2. Sediments

Marine organisms can accumulate cadmium from marine sediments, but the processes controlling bioavailability are poorly understood. In coastal systems, sediments may become suspended due to turbulence, enhancing metal exchange between sediment, particulate matter, and water. Polychaete worms play an important biogeochemical role in the diagenesis of estuarine sediments (Gordon, 1966; Young, 1968). Sediment texture and composition, salinity, and microbial activity may also affect exchange processes.

The chemical forms of cadmium in sediments are particularly important in determining bioavailability, and indeed, ambiguous relationships observed between the metal contents of animal tissues and sediments drew attention to the diverse forms of metals in sediments. The metals in the crystal lattice of the minerals are not believed to be available to the ecosystem. Adsorption of cadmium on hydrous iron or manganese oxides or on detrital materials also decreases their bioavailability, and this depends on the equilibrium solute concentration in the associated waters due to sorption–desorption and dissolution–precipitation reactions (Jenne and Luoma, 1977). Similarly, the concentration of cadmium in *N. virens* was related to the concentration of the element leached into the water (Ray et al., 1980a).

Attempts have been made to distinguish available and total metal by extracting fractions of sediments with chemical reagents that are neutral, weakly acidic, or basic (Salomons and Förstner, 1980). This seems to be a promising method for assessing bioavailability.

Luoma and Jenne (1976) related bioavailability of cadmium and other metals for the deposit-feeding clam, *M. balthica,* to binding strength between the metals and the sediment particles of different chemical composition to which the metals were sorbed. Uptake of cadmium depended on the physico/chemical form of the natural sediments, that is, whether they were rich in organic detritus, iron oxide, inorganic carbonate or biogenic carbonate, and so on. Extraction of sediment-bound cadmium by 70% ethanol or 1 N ammonium acetate provided a good estimate of bioavailability. The amount of cadmium extracted from sediments with weak acids (0.1 N HCl; 25% CH_3COOH), reducing agents (1 N hydroxylamine hydrochloride, sodium dithionite plus citrate) or oxidizing agents (3% H_2O_2 plus citrate) correlated poorly with bioavailability. Cooke et al. (1979) examined bioavailability of cadmium to the cockle, *Cerastoderma edule,* from four sediments prepared in the laboratory (iron oxide, manganese oxide, calcium carbonate, and biogenic calcium carbonate). Significant uptake occurred with exposure to the biogenic calcium carbonate sediment which, unlike the other three sediments, produced an appreciable amount of cadmium in the overlying water.

9.5. BIOACCUMULATION OF CADMIUM

Bioaccumulation of cadmium from water, food, and sediment by a variety of marine organisms has been studied under field and laboratory conditions, with most of the information coming from laboratory studies. Basic understanding of the bioaccumulation process is still incomplete because its components of uptake, storage, and elimination are not yet fully understood. Reviews have been provided by Phillips (1980) and Ray (1984).

Critical comparison of data is difficult since experimental conditions have varied over a wide range to suit particular needs. Exposures have been static or dynamic, varying from a few hours to periods up to 1 yr. Exposure concentrations have varied from about 5 µg Cd/L to about 100 mg Cd/L. Results have often been expressed on a whole-body (soft-tissue) basis. Even when individual organs were analyzed, some organs were arbitrarily combined as "gut" or "viscera." The tendency now is to express measurements on a dry-weight basis whereas earlier data were usually reported in terms of wet weight. Often, the accuracy of measurements has not been checked by use of standard reference materials. There should be much more standardization of protocols for conducting tests and expressing results if the results are to be useful for comparison.

9.5.1. WATER

Bioaccumulation of cadmium from water can take place either by diffusion through the general body surface or more particularly through the gills. Uptake is generally proportional to both exposure concentration and time. An equilibrium concentration of cadmium in the body should not be assumed in most tests because of limited exposure times.

9.5.1.1. Phytoplankton

Only a few marine planktonic species have been studied for bioaccumulation of cadmium. Cossa (1976) observed that uptake by the diatom *P. tricornutum* varied with growth phase of a culture and was controlled by adsorption processes. The uptake by the alga *Porphyra umbilicalis* was dependent on light conditions and had an initial rapid uptake phase followed by a slow period (McLean and Williamson, 1977).

9.5.1.2. Mollusks

Bioaccumulation has been studied in exposures from 100 h (Brooks and Rumsby, 1967) to 40 weeks (Zaroogian, 1980; Zaroogian and Cheer, 1976) and concentrations from 5µg Cd/L (Zaroogian, 1980; Zaroogian and Cheer 1976) to 50 mg Cd/L (Brooks and Rumsby, 1967). In various laboratory studies of oysters, the concentrations of cadmium in the soft tissues did not reach equilibrium under any of the experimental conditions. Accumulation varied over a wide range up to 292 µg Cd/g dry weight (Zaroogian, 1980) after exposure to 15 mg Cd/L for 40 weeks. Tissue concentrations of cadmium in the oyster *C. virginica* were in the order gill > viscera ≥ mantle, while the order for the total load of cadmium was viscera > gill ≥ mantle (Zaroogian, 1980). Similarly, in *Saccostrea commercialis*, concentrations were in the order viscera ≃ gill > mantle > muscle compared with the order in control oysters viscera > gill ≃ mantle > muscle. This suggests that the gill is most important in the direct accumulation of cadmium from water (Ward, 1982). There are species differences, and *Ostrea edulis* held for 5 months in a floating pontoon accumulated a total body burden of 120 µg Cd, significantly higher than the burden of 74 µg for *Crassostrea gigas* (Boyden, 1975).

The mussel *M. edulis* is the most studied mollusk, and George and Coombs (1977) observed an initial lag period in uptake by mussels exposed to different concentrations of cadmium in seawater. Final tissue concentrations were in the order kidney ≫ viscera > gills ≫ mantle > muscle. Scholz (1980) and Janssen and Scholz (1979) reported that the main body burden was in the midgut gland, and the order of concentrations in the tissues was midgut > gill > kidney > mantle > foot > adductor muscle. Mussels of the genus *Mytilus* have been widely used as indicators of pollution in coastal areas because of their sessile mode of life, high capacity for bioaccumulation of many pollutants, and their presence in both hemispheres (Goldberg et al., 1978). As filter feeders, they would acquire cadmium from the water, food, or other suspended particulate matter.

Cadmium uptake by the filter-feeding clams *Mulinia lateralis, Mya arenaria,* and *M. edulis* was significantly greater than by the deposit-feeding clam *Nucula proxima* (Jackim et al., 1977). *Nucula proxima* did not accumulate cadmium on exposure to 5 µg Cd/L for 14 days, while the filter feeders accumulated 2.0–9.4 µg Cd/g (dry wt.) under the same conditions. However, *N. proxima* accumulated 2.0–4.6 μg Cd/g (dry) when exposed to 20 μg Cd/L for 14 days compared with 10–60 μg Cd/g for the filter feeders. Scallops (*Aquipecten irradians* and *Argopecten irradians*) appear to accumulate particularly high levels of cadmium when exposed to water-borne levels of the metal (Carmichael and Fowler, 1981; Eisler et al., 1972).

9.5.1.3. Crustaceans

Almost all studies of bioaccumulation by crustaceans have been conducted in the laboratory. There have been a few field studies of concentration in animals from clean and naturally contaminated areas. Exposures have varied widely, but in most studies that included uptake kinetics, the levels of cadmium in crustaceans were proportional to both exposure concentration and time, and the levels did not reach equilibrium during the experiment.

Direct proportionality between whole-body concentration and exposure concentration or time has been shown in a number of laboratory studies with shrimp (*Palaemonetes pugio* by Engel and Fowler, 1979b; *P. pugio, P. vulgaris,* and *Penaeus duorarum* by Nimmo et al., 1977; *Palaemon elegans* by White and Rainbow, 1982; *Pandalus montagui* by Ray et al., 1980b; and *Crangon crangon* by Dethlefsen, 1977/78). Similarly, Nimmo et al. (1977) observed that cadmium level in muscle tissue of *P. duorarum* increased from about 0 to 0.25 mg/kg, when the animals were exposed to 2.0 μg Cd/L for 50 days. *Palaemon elegans* also accumulated cadmium when exposed to 2.5 μg Cd/L for 21 days (White and Rainbow, 1982). Cadmium was localized selectively in the tissues of shrimp, and concentrations in the various tissues varied by several orders of magnitude. Several authors have listed the order in which cadmium accumulated in tissues: hepatopancreas > gill > exoskeleton > muscle > serum (Nimmo et al., 1977); hepatopancreas ≫ gill eyestalk > abdominal muscle > gonadal tissues > exoskeleton (Dethlefsen, 1977/78); hepatopancreas ≫ carcass (exoskeleton, gills, head, legs, etc.) > eggs > abdominal muscle (Ray et al., 1980b).

Laboratory studies with euphausiids, barnacles, and crabs indicate that elevated levels of cadmium in water lead to selective accumulation in the gills and hepatopancreas. In the crab *C. maenas,* the order of tissue concentrations was hepatopancreas > gills > carapace > chela muscle (Bjerregaard, 1982) or midgut gland > gill > muscle > exoskeleton (Jennings and Rainbow, 1979). In other crabs, *Callinectes sapidus* showed an order of carapace > hepatopancreas > gills > eye stalk > green glands > claws > heart (Hutcheson, 1974), and *Cancer pagurus* had hepatopancreas > gill ≫ carapace > gonad > claw muscle in exposures from 10 μg/L lasting as long as 300 days (Davies et al., 1981). The order of localization for *Cancer* was similar for animals exposed in the laboratory and in field locations with high cadmium levels. Similar results were obtained for a euphausiid using ^{109}Cd (Benayoun et al., 1974) and a barnacle (Rainbow et al., 1980).

Lobsters (*Homarus americanus*) accumulated cadmium at the lowest concentration of 7 µg/L tested in the laboratory (Ray and McLeese, 1982). The uptake route was primarily through the gills, and tissue concentrations were hepatopancreas > gills > green glands > claw nonmuscle > claw muscle > abdominal muscle. In tissues of cadmium-contaminated lobsters from Belledune Harbour, New Brunswick, Canada, residues roughly paralleled the order determined in laboratory tests, and the hepatopancreas contained more than 90% of the body burden (Ray et al., 1981a). No accumulation of cadmium occurred in 60-day exposures of lobsters to 6 µg Cd/L (Thurburg et al., 1977), a concentration similar to that which had an effect in the above example.

9.5.1.4. Polychaetes

Bioaccumulation of cadmium by deposit-feeding polychaetes also increases proportionally with time and the water-borne concentration [*Nereis diversicolor:* Bryan (1974), Bryan and Hummerstone (1973); *Nereis virens:* Ray et al. (1980a); *Nereis japonica:* Ueda et al. (1976); *Glycera dibranchiata:* Rice and Chien (1979)]. Cadmium concentrations as high as 100 mg/L were used for exposure periods up to 35 days (Bryan and Hummerstone, 1973) with concentrations in the worms reaching up to 4000 µg/g (dry). It is not known whether the accumulation process reached equilibrium with respect to cadmium concentration in water. Smaller worms accumulated higher amounts of cadmium per unit weight and at a faster rate than the larger worms (Ray et al., 1980a).

However, cadmium uptake by *Ophryotrocha diadema* (Klöckner, 1979) was a nonlinear process related to exposure time and cadmium concentration in the seawater. Maximum body loads of up to 1700 µg Cd/g were observed in worms exposed to 1 µg Cd/L for 64 days. Absorption saturation was not attained.

9.5.1.5. Fish

Pentreath (1977) studied cadmium accumulation by plaice *Pleuronectes platessa* and thornback ray *Raja clavata* using ^{115m}Cd as a tracer and determined that the uptake by both fish is very low and that accumulation was linear was time (70 days). Uptake rate for the plaice was about 4 times greater than for the ray. Pentreath also studied uptake by eggs and larvae of plaice. The eggs attained equilibrium in cadmium concentration within 24 h with no further accumulation during the 15-day study period; uptake by larvae followed the pattern for fish.

9.5.2. Sediment

Bioaccumulation of cadmium by marine organisms cannot always be related to concentrations in the sediments (Section 9.4.2.), but this has been done in some cases. Bryan and Hummerstone (1973) collected polychaete worms

(*N. diversicolor*) from 26 estuaries and found the cadmium concentrations in worms to be roughly proportional to those in the sediments. Concentrations in the worms ranged from 0.08 to 3.6 µg/g (dry), and those in the sediments from 0.2 to 9.3 µg/g, both showing 46-fold variation. Ayling (1974) concluded that 1 µg Cd/g (dry) in the mud resulted in about 25 µg/g in tissues of the Pacific oyster (*C. gigas*), based on 15 sites in the Tamar River, Tasmania. Dry weights in mud and oysters were 0.4–5.7 and 4.2–134 µg/g, respectively, and larger oysters tended to have lower concentrations of cadmium. Bryan and Hummerstone (1978) exchanged deposit-feeding bivalves (*Scrobicularia plana*) between two estuaries having different levels of cadmium in the sediments; cadmium concentrations in the transposed animals slowly approached those of the natives, but equality was not achieved in a year.

A larger number of investigations show inconsistent relations between sediment cadmium and levels of the metal in organisms. It appears that metal must leach from the sediment into the water in order to become easily bioavailable. One example of poor sediment–organism correlation has been provided by Neff et al. (1978). Five species were exposed to three natural sediments for periods up to 6 weeks, and availability of cadmium and nine other metals was assessed. Out of 136 combinations, only 49 showed statistically significant relations between metal levels in the sediment and the organisms. Ray and McLeese (1983) collected shrimp (*Crangon septemspinosa*) and the bivalve mollusk *Macoma balthica* from sediments with 100-fold variation in cadmium content, but the corresponding range in tissue levels was only 3-fold for shrimp and 5-fold for *Macoma*. Polychaete worms exposed to cadmium-spiked sediment during 24 days increased their tissue levels linearly with time and concentration in the sediment, but it appeared that the primary relationship was with cadmium leached into the water (Ray et al., 1980a). Clearer evidence was provided in 30-day exposures of three invertebrates to two cadmium-polluted sediments. The polychaete *N. virens* and the bivalve *M. balthica* increased their tissue levels with only one sediment, the one that caused trace increases of cadmium in the overlying water. The shrimp, *Crangon,* did not show an increase (Ray et al., 1981b). The amount of cadmium leached into the water depends on the ion exchange capacity and the organic content of the sediments (Ray and McLeese, 1983). The importance of the release of cadmium from sediments to the associated water, allowing for accumulation by marine invertebrates, has been documented by several other workers (Bryan, 1985; Cooke et al., 1979; Holmes et al., 1974; Jenne and Luoma, 1977; Kneip and Hazen, 1979; Luoma and Jenne, 1976).

9.5.3. Food

Only a few studies have been conducted on bioaccumulation of cadmium by marine organisms from food, and there are conflicting results on the relative importance of uptake from food and water. The only general conclusion seems to be that cadmium intake with food increases the residues in the hepatopancreas and viscera, while uptake from water increases the levels in the gills and exoskeleton.

Some results suggest that food is a relatively unimportant source of cadmium. Nimmo et al. (1977) reported low uptake of cadmium from food by shrimp, *P. pugio,* and estimated that to produce an equivalent total body load in the shrimp, about 15,000 times more cadmium would be required in food than in water. Jennings and Rainbow (1979) estimated that only 10% of the cadmium available in food was accumulated by the crab *C. maenas.* The midgut gland and the exoskeleton contained 16.9 and 22.2%, respectively, of the total absorbed cadmium from food, compared with 10 and 59–80% when it was administered through water. Comparative importance of the two routes was not determined, but it seems, that uptake from water is more important. Similarly, plaice, *P. platessa,* and thornback ray, *R. clavata,* retain only 5 and 17%, respectively, of cadmium administered through food (Pentreath, 1977).

Other results suggest that food intake of cadmium is of primary importance, or support the general conclusion in the first paragraph of differential distribution, with no appraisal of the overall relative importance. Davies et al. (1981) compared uptake and disposition of cadmium in crab, *C. pagurus,* administered through food and in water, and concluded that accumulation of cadmium in midgut gland is primarily from food, while gills and exoskeleton are primary routes for cadmium uptake from water. They also compared laboratory results with those for crabs from Orkney and Shetland Islands and suggested that the dominant route for the uptake for cadmium was from food. Gutierrez-Galindo (1980) compared the visceral concentration of cadmium in the crab *C. maenas* given cadmium through food and water for 15 days and suggested that cadmium accumulation through food was more important. However, the questionable conclusion is based on the analysis of only the viscera of the animals. Ray and McLeese (1982) compared disposition of cadmium in various tissues of American lobster, *H. americanus,* and concluded that the gills are the primary deposition sites for cadmium uptake from water. There also was a sudden increase in cadmium uptake in hepatopancreas of animals exposed to 21 µg Cd/L for 45 days, which might have been related to induction of metallothionein synthesis. In contrast, most of the cadmium administered through food was deposited in the hepatopancreas. No attempt was made to assess the relative importance of the two routes for cadmium uptake.

Benayoun et al. (1974) observed that cadmium was accumulated in the internal organs of *M. norvegica* when administered through either food or water, but they could not assess the relative importance of the two routes of uptake.

Marine mammals provide evidence that food intake of cadmium can be important since these air-breathing creatures must have minimal intake of the metal from water. There are surprising differences in cadmium residues among related species that have not been explained in the published literature.

Arctic ringed seals (*Phoca hispida*) contain extremely high levels of about 48 mg/kg in the liver, while samples of European seals have a maximum of 8 mg/kg with levels down to 1 mg/kg or less (Table 1). Canadian harp seals (*Phoca groenlandica*), which migrate between the arctic and more southerly regions, are intermediate with respect to the cadmium levels in their livers. The arctic narwhal (*Monodon monoceros*) has even higher levels in the liver. Concentrations in kidneys of all those animals were higher than in the livers, but muscle levels were one or two orders of magnitude lower (Table 1). The elevated cadmium levels in arctic marine mammals appear to be natural. Concentrations of cadmium rise quickly during the first 2 yr of the seal's life and then level off (Johansson et al., 1980; Ronald et al., 1984). The primary intake route is probably food and not water, and the leveling off in cadmium concentration of seals could result from change in diet from crustaceans to fish (C. R. Macdonald, University of Guelph, Ontario, Canada, personal communiction).

9.6. ENVIRONMENTAL FACTORS AFFECTING BIOACCUMULATION OF CADMIUM

Accumulation of cadmium by marine organisms in nature may vary over a wide range depending on environmental factors such as temperature, salinity, and season, which represents an interaction of environmental factors with the organism.

9.6.1. Temperature

Bioaccumulation of cadmium usually increases with increasing temperature because of increased metabolic activity by the organisms. For example, concentrations of cadmium in animals, and therefore presumably uptake rates, have been shown to increase with increasing temperature in the clams *Mya arenaria, Mulinia lateralis,* and *Nucula proxima* (Jackim et al., 1977). Cadmium concentration in *M. arenaria* exposed to 5 µg Cd/L for 14 days increased from 2.2 to 4.2 µg/g (dry) when temperature increased from 10 to 20 °C, while those exposed to 20 µg Cd/L increased from 16.8 to 29.0 µg/g under the same conditions. The corresponding values for *M. lateralis* were 3.6 to 8.9 µg/g and 10.3 to 20.5 µg/g. For *N. proxima* exposed to 20 µg Cd/L, cadmium concentration rose from 2.6 to 4.6 µg/g (dry) for such a temperature change. Average cadmium in the oyster *C. virginica* exposed to 45 µg Cd/L increased from 86 to 123 µg Cd/g (dry wt.) at temperatures of 15 and 20 °C (Hung, 1982). Several individual tissues were analyzed, and these tissues doubled or tripled in concentration at the higher temperature. Concentra-

Table 1. Averages or Ranges of Cadmium Concentrations in Tissues of Selected Marine Mammals[a]

Species	Location	Number of Individuals	Average Cadmium (mg/kg dry wt.)			Reference
			Muscle or Whole Body	Liver	Kidney	
Ringed seal	West Greenland	29–48	0.43[a]	48[a]	190[a]	Johansen et al., 1980
Harp seal	Canada, East Arctic					
	Pups	3–6	0.18[a]	5.4[a]	28[a]	Ronald et al., 1984
	Juveniles	8–21	0.40[a]	14[a]	81[a]	Ronald et al., 1984
	Adults	15	1.4[a]	30[a]	110[a]	Ronald et al., 1984
Harbor seal	Holland, North Sea	11	—	0.5[a]	—	Koeman et al., 1972
	Germany, North Sea	70	0.01–0.4[a]	0.05–1.1[a]	0.30–5.0[a]	Harms et al., 1978
		16–58	—	0.15[a]	0.85[a]	Drescher et al., 1977
	Scotland, East Coast	4	—	6.5[a]	—	Holden, 1978
	Scotland, East Coast	9	—	8.5[a]	—	Holden, 1978
Gray seal		20	—	6.5[a]	16[a]	McKie et al., 1980
	United Kingdom					
	Farne Islands	64	—	3.3[a]	22[a]	Caines, 1978
	Outer Hebrides	4	—	<5.0[a]	58[a]	Heppelston and French, 1973
Narwhal	Canada, East Arctic	38–58	0.69	133	298	Wagemann et al., 1983
		—	1.9–18[a]	63–590[a]	—	Hatfield and Williams, 1975

[a]Converted from wet weight using a factor of 5.0.

tions of cadmium in adductor muscle, mantle, gill, and hepatopancreas were 14, 42, 70, and 59 µg/g, respectively, at 10 °C and 23, 126, 144, and 145 µg/g at 20 °C. Uptake rates of cadmium in an oyster (*Saccostrea echinata*) increased from 0.27 to 0.68 µg Cd/g/day for a temperature change of 20 to 30 °C at high salinity (36‰). At a lower salinity of 20‰, there was an even greater change in uptake from 0.48 to 2.0 µg Cd/g/day (Denton and Burdon-Jones, 1981). Uptake rates for individual tissues were also computed, but actual tissues concentrations of cadmium were not given.

Effect of temperature on toxicity and uptake of cadmium has been studied for various crustaceans. Uptake rates of cadmium by the crab *C. sapidus* exposed to 10 mg/L for 96 h at 10, 20, and 33 °C at 5‰ salinity showed slopes for the gill tissues of 0.78, 1.20, and 3.13, respectively; the corresponding values for the hepatopancreas were 0.56, 1.05, and 1.07, respectively (Hutcheson, 1974). The actual metal concentrations were not given. Bengtsson (1977) found that the accumulated level of cadmium for the shrimp *Leander adspersus* exposed to various levels of cadmium up to 320 µg/L for 35 days at 15 °C had values that were about twice the level compared to exposures at 6 °C. Similarly, Fowler and Benayoun (1974) observed increased cadmium uptake by the shrimp *Lysmata seticaudata* at higher temperatures.

The coelenterate hydrozoan *Laomedea loveni* exposed to 1.2 µg Cd/L for 7 days at 15‰ salinity had cadmium concentrations of 0.35 µg/g at 5 °C and 0.83 µg/g at 20 °C (Theede et al., 1979).

The general conclusion is that uptake rates and concentration of cadmium increase in marine invertebrates at higher temperature. The only exception seems to be the mussel *M. edulis*. There are several reports that this species shows no significant change in either concentration or uptake with increase in temperature (Carpene and George, 1981; Fowler and Benayoun, 1974; Jackim et al., 1977).

9.6.2. Salinity

Salinity is an important environmental variable for estuarine and coastal organisms. In most cases studied, accumulation of cadmium increased with decrease in salinity.

Jackim et al. (1977) reported increased cadmium uptake by *M. edulis, M. lateralis,* and *N. proxima* with decreases in salinity. When the animals were exposed to 20 µg Cd/L at 21 °C for 21 days, the cadmium levels were 86.6, 8.8, and 2.6 µg Cd/g (dry), respectively, at 30‰ salinity but increased to 108.0, 36.9, and 5.4 µg Cd/g at 20‰ salinity. Increases at 10 °C were from 32.1, 24.4, and 0.6 µg/g at 30‰ salinity to 83.2, 52.3, and 2.1 µg/g at the lower salinity. These were similar increases in uptake in *M. edulis* exposed to 40 µg Cd/L at 10 or 18 °C and 15 or 35‰ salinity for 21 days (Phillips, 1976) and in oysters when the salinity was reduced from 36 to 20‰ salinity (Denton and Burdon-Jones, 1981; see Section 9.6.1.). Parallel increases in accumulation and uptake rates for cadmium have been reported for three crabs (Hutcheson, 1974; O'Hara, 1973; Wright, 1977a), a shrimp (Engel and Fowler, 1979b; Vernberg et al., 1977), the common goby *Pomatoschistus minutus* (Bengtsson, 1977), and flounder eggs (Westernhagen and Dethlefsen, 1975).

In contrast to the above observations, *M. edulis* had reduced cadmium levels in various tissues when exposed to 0.1 mg/L for 544 h at 11‰ salinity rather than at 30‰ (Briggs, 1979). It was suggested that the discrepancy may have resulted from valve closure by the mussels in response to stress. Change in salinity apparently did not affect the cadmium levels in the hydrozoan *L. loveni* (Theede et al., 1979). Several reasons have been postulated for increased uptake of cadmium at lower salinity:

(a) Oxygen consumption in organisms may increase in dilute seawater (Bass, 1977; Taylor, 1977), resulting in increased gill irrigation, which may permit greater metal uptake.

(b) The total amount of biologically available Cd^{2+} is higher at lower salinities (Engel et al., 1981; Mantoura et al., 1978). Free cadmium varies from 100% of the total in solution at 0‰ salinity to only 6% at 35‰ salinity (Mantoura et al., 1978).

(c) Antagonism between cadmium and calcium as suggested by Wright (1977a,b). Higher concentration of calcium at higher salinities decreases cadmium uptake.

9.6.3. Seasonal Variation

Most studies of seasonal variation of cadmium concentration have involved filter-feeding bivalve mollusks. There are seasonal changes in various species, but no single pattern can be given as a generalization. Similarly, there does not seem to be any single cause, and the explanations have included weight changes of the organism, gametogenesis, feeding, age, and other factors.

Phytoplankton in Monterey Bay, California, showed a major yearly variation from <1 to about 7 µg Cd/g (dry). The lowest levels were found during the low upwelling period, which coincided with periods of high phytoplankton productivity, which presumably diluted the amount of the metal per unit mass of phytoplankton (Knauer and Martin, 1973). Such a change would seem to have the potential for affecting cadmium levels in filter feeders ingesting the plankton.

A single population of limpets (*Patella vulgata*) showed an increase from 400 µg Cd/g dry weight in March to 720 µg/g in January (Boyden, 1977). Body weights were lowest in January since this species spawns in the winter. To explain the higher cadmium in terms of the loss of body weight requires that relatively little cadmium was passed out with the gametes.

Julshamn (1981) reported slightly higher cadmium content in *O. edulis* collected during the winter months when little feeding occurred. Frazier (1975, 1976) determined that in the oysters *C. virginica* held in suspended trays in Chesapeake Bay for 21 months, the total soft-tissue body burden of cadmium increased throughout the winter and spring, attaining maximum levels in the first week of July, followed by a 50% reduction over the next 11 weeks. Frazier suggested that the uptake might be controlled by a combination of four factors: feeding, spawning, shell metabolism, and environmental chemistry in the various seasons. Most laboratory studies are too short to show seasonal changes, but the oysters *C. virginica* exposed to 5, 10, or 15 μg Cd/L at ambient temperatures over a 40-week period showed cadmium accumulation that followed a seasonal pattern similar to that in the field studies mentioned above for Chesapeake Bay (Zaroogian, 1980; Zaroogian and Cheer, 1976). Quahog (*Mercenaria mercenaria*) varied from 6.5 μg Cd/g dry weight in August to 33.5 μg/g in April and July for whole-body soft tissues. Genest and Hatch (1981) attributed this to monthly variations in environmental factors affecting the organism. The mussel *Choromytilus meridionalis* changed from 0.3 μg Cd/g in June to 1.0 μg/g in November, and Orren et al. (1980) discounted explanations based on gonad development or food availability. Cossa et al. (1980) found gradual decreases in cadmium content until maturity in mussel (*M. edulis*) from the Gulf of St. Lawrence. This was not explained by changes in body weight with maturation but appeared to be physiological variation associated with reproduction and seasonal adaptation.

In contrast to the above studies, Goldberg et al. (1978) concluded that seasonal variability in cadmium level was not significant in oysters (*C. virginica* and *Ostrea equestris*) and mussels (*M. edulis* and *M. californis*) collected from the U.S. east and west coasts. The variations could be accounted for by differences in age, sexual state, food, detritus uptake, and other factors, some of which are, of course, related to season.

9.7. STORAGE OF CADMIUM

Cadmium is not distributed uniformly in the body of an organism but is localized in certain organs, and tissues may differ by several orders of magnitude. Body distribution of cadmium from laboratory exposure normally parallels that found under natural conditions.

Details of relative concentrations in different organs have been given in Section 9.5, but some major differences between phyla deserve mention here. In mollusks, cadmium is mostly accumulated in kidney, gill, viscera, or mantle and to a lesser degree in adductor muscle, foot, or the shell. In crustaceans, the midgut gland (hepatopancreas) is usually a major storage site, although the exoskeleton is sometimes as important, and the gills and green gland also contain appreciable concentrations. The liver and the kidney are the primary sites of accumulation in fish and aquatic mammals. In most organisms, whether vertebrates or invertebrates, the muscle tissue contain only negligible amounts of cadmium.

9.7.1. Subcellular Mechanisms

9.7.1.1. Metallothionein

Most organisms that accumulate cadmium from the environment have evolved mechanisms that sequester or otherwise handle this nonessential and potentially harmful trace metal, and marine organisms are no exception. One of the main mechanisms is synthesis of low-molecular-weight metal-binding proteins called *metallothioneins* (Kägi and Nordberg, 1979). These are cytoplasmic, low-molecular-weight proteins with characteristic amino acid composition and high metal content.

Metalloproteins isolated from marine organisms contain Cd and other metals such as copper, zinc, and mercury. Cadmium-binding proteins have been found in all the major groups of marine vertebrates, many crustaceans and mollusks, an annelid worm, an echinoderm, and marine bacteria. In some cases, the proteins have been definitely characterized as metallothioneins (see below), and in other cases they should be described as metallothioneinlike proteins, that is, proteins of unknown amino acid composition but having similar elution properties in gel permeation chromatography and similar spectral properties or proteins that may differ in their amino acid composition, especially cysteine content.

The apoprotein, thionein, is an inducible protein. The synthesis of metallothionein occurs when intake of essential trace metals like copper and zinc exceeds metabolic requirements. It is also synthesized in response to the nonessential metals cadmium and mercury, possibly as a protective detoxifying mechanism. It is assumed that metals bound to the protein are not available for binding to enzymes or for causing damage to intracellular membranes.

The proteins induced in several marine organisms by exposure to cadmium have been purified and characterized, and properties are similar to those for human and equine metallothionein (crab, Olafson et al., 1979; plaice, Overnell and Coombs, 1979; a bacterium, Olafson et al., 1980). Besides metallothionein, varying proportions of apparent dimers and trimers of metallothionein have also been found in several cases (Carpene et al., 1980, 1983; George et al., 1979; Scholz, 1980). Several high-molecular-weight proteins have been reported (Carpene et al., 1983; Nolan and Duke, 1983; Pruell and Engleherdt, 1980; Rainbow and Scott, 1979; Ray and White, 1981; Siewicki et al., 1983) as well as low-weight ones (Frankenne et al., 1980; George, 1983a,b; Howard and Nickless, 1977; Ray and White, 1981; Siewicki et al., 1983; Thompson et al., 1982).

The widely accepted term metallothionein was first introduced by Kägi and Vallee (1960). But as the number of investigators in the field increased, so did confusion with terminology. A liberal interpretation of the term accompanied by a variety of claims of finding another new metallothionein from yet another organism has been occurring over the past few years. An international committee has now established a definition, nomenclature, and the required biochemical and physiological properties for metalloproteins that are considered to be true metallothioneins (Kägi and Nordberg, 1979). Vallee (1979) has provided further clarification by suggesting nine criteria for identification of metallothionein. These include: cysteine constituting 30–35% of the total amino acid content; 7 g atom Me^{2+} (Cd, Zn, or Cu) per 20–21 SH groups of the protein; absence of aromatic amino acids, histidine and adsorption band near 280 nm; and molecular weights of 6000–7000 daltons. He also cautioned against the practice of identifying a "new, unknown" metallothionein by comparing its gel chromatography elution pattern with that of a known metallothionein. Unfortunately, authors continue to claim the isolation of metallothioneins from other marine organisms based only on the similarity of elution patterns of the proteins on gel filtration. As suggested earlier, these proteins should preferably be called "metallothionein-like protein," as has been done by several authors (Jenkins et al., 1982; Olafson et al., 1979; Ray and White, 1981; Scholz, 1980) or simply cadmium-binding proteins.

9.7.1.2. Membrane-Limited Vesicles

Many organisms can accumulate cadmium and other metals in intracellular electron-dense granules within membrane-limited vesicles (Brown, 1982). Vesicles are formed by surrounding and trapping cadmium and other metals within a membrane within the cells. Such a process appears to serve as a detoxification mechanism for the cell since the metal is rendered chemically inert within the vesicle. The metal composition of these granules is variable. Commonly, calcium occurs along with other metals like iron, zinc, and manganese. Cadmium is normally associated with sulfur and frequently with phosphorus in the vesicles. George and Pirie (1979) and George et al. (1980) have shown that granules isolated from mussels and scallops are composed mainly of calcium, manganese, and zinc phosphates, accounting for about 50% of the granules. Other elements detected are potassium, magnesium, copper, iron, cesium, sulfur, and chlorine. The molar ratio of the sum of the metals to that of the phosphate is usually close to unity. Such granules have been isolated from a variety of marine invertebrates like the mussel *M. edulis* (George, 1983a; George and Pirie, 1979; George et al., 1982; Janssen and Scholz, 1979, Marshall and Talbot, 1979); the scallops *Argopecten irradians* and *Argopecten gibbus* (Carmichael et al., 1979); and the crab *C. maenas* (Chassard-Bouchaud, 1982). Some results have suggested that metallothioneins may be associated with the particulate structure within the cell and not be freely available within the cytoplasm (George, 1983b; George and Pirie, 1979).

9.7.2. Extracellular Mechanisms

The mucous layer and cuticles of marine organisms that are exposed to water may be considered as extracellular traps for cadmium because of their high affinity for metals (Cossa, 1976; Wright, 1977a,b). In several species of marine fish, white mucous corpuscles can be seen in the intestinal lumen of unfed fish. This material is evacuated regulary and is termed *intestinal corpuscles*. The corpuscles contain high concentrations of calcium and magnesium, probably in the form of carbonate precipated from seawater contained in the intestine. In cadmium-contaminated fish, they may contain most of the body burden of cadmium, although the relative weights of corpuscle to total weight may be very small. Cadmium-containing "corpuscles" have been reported to occur in intestines of a variety of fish (*Anguilla anguilla, Myoxocephatus scorpius, Servanus cabrilla, Moena chryselis,* and *Scorpaena* sp.) and were shown to be involved in uptake as well as in excretion of cadmium (Noël-Lambot, 1981).

9.8. EXCRETION OF CADMIUM

In many marine invertebrates, the biological half-life of cadmium is estimated as several months, and experiments often indicate little excretion, if any.

The excretion of cadmium by the mussels *Mytilus galloprovincialis* (Majori and Petronio, 1973) and *M. edulis* (George and Coombs, 1977; Scholz, 1980) and the oyster *C. virginica* (Greig and Wenzloff, 1978) was reported to be very slow or even not detectable. Most experiments were short, and in only a few were the kinetics of excretion determined. Scholz (1980) determined that the half-life for cadmium in *M. edulis* was between 14 and 29 days, while George and Coombs (1977) suggested that the excretion rate is 18 times slower than the uptake rate. Similarly, there was no decrease in cadmium content in the crab *C. sapidus* (Hutcheson, 1974), or shrimps *L. seticaudata* (Fowler and Benayoun, 1974) and *Pandalus montagui* (Ray et al., 1980b). Lobsters (*H. americanus*) excreted cadmium extremely slowly or not at all (McLeese et al., 1981). Cadmium-contaminated lobsters were maintained for 8 months in clean water, and although there was a detectable decrease of cadmium concentration in the hepatopancreas and gill, there was no decrease in eight other tissues. The calculated half-lives for cadmium in the hepatopancreas and gill were 500 and 200 days, respectively. There was no detectable cadmium excretion by the polychaete *N. virens* (Ray et al., 1980a) during 80 days. The loss form plaice *P. platessa* and thornback ray *R. clavata* was very slow (Pentreath, 1977) and half-lives could not be ascertained definitely because of large variance but appeared to range from 98 to 204 days.

In contrast to the above findings, several authors have reported substantial loss of cadmium from contaminated organisms held in clean seawater. Denton and Burdon-Jones (1981) and Mowdy (1981), have suggested appreciable excretion from oysters (*C. virginica* and *S. echinata*) depending on salinity and temperature. Rapid excretion of cadmium also has been claimed for the crab *C. maenas* (Jennings and Rainbow, 1979; Wright, 1977a,b). However, most of the initial loss seems to have taken place from cadmium adsorbed on the exterior surface of the exoskeleton. *Nereis japonica* (Ueda et al., 1976) was reported to excrete 30% of total body burden of cadmium within 7 days. Similarly, there was 50% Cd loss within 2–3 days in the mummichog *Fundulus heteroclitus* (Eisler, 1974). It is not known whether the loss was only of surface-adsorbed cadmium.

In most organisms, cadmium is stored in the form of metallothionein or membrane-limited vesicles or both (Sections 9.7.1.1 and 9.7.1.2) and is immobilized and prevented from interfering with cellular mechanisms (Fowler et al., 1981; George and Viarengo, 1985). It is thus possible that such detoxification processes contribute to effective retention and consequently a very long half-life of cadmium in the marine organisms. Futhermore, the observed linearity of increase of cadmium concentration with time of exposure in bioaccumulation studies must also be related to effective retention or very low rates of excretion of cadmium by the organisms.

9.9 SUMMARY

Cadmium occurs in surface ocean water in only trace amounts but may be found at higher concentrations at greater depths. The vertical distribution of cadmium is controlled by biogeochemical cycles and normally parallels the distribution of nutrients like phosphates, silicates, and nitrates in the water column. The detailed mechanism is not known, but this behavior indicates that cadmium is bioaccumulated by aquatic organisms at the surface and regenerated from the sinking biological debris in deeper waters.

Marine organisms bioaccumulate cadmium not only from the aqueous media but also from bottom and suspended sediments and from food. The process may be controlled by physicochemical factors like salinity and temperature. The chemical form of cadmium in the environment is of prime importance in bioaccumulation by marine organisms. The ultimate level of cadmium in the organism is controlled not only by biotic and abiotic processes but also by metabolism of the metal by the organisms.

Most of our knowledge regarding biological cycling of cadmium comes from laboratory studies; consequently, the results cannot be easily applied to field situations without some danger of misinterpretation. Tremendous progress has been made over the years, but basic understanding is still nebulous and will remain so until the mobilization, storage, and uptake mechanisms are fully understood.

REFERENCES

Abdullah, M. I., and Royle, L. G. (1974). A study of the dissolved and particulate trace elements in the Bristol Channel. *J. Mar. Biol. Assoc. U.K.,* **54,** 581–597.

Ayling, G. M. (1974). Uptake of cadmium, zinc, copper, lead and chromium in the Pacific oyster, *Crassostrea gigas,* grown in the Tamar River, Tasmania. *Wat. Res.,* **8,** 729–738.

Baric, A., and Branica, M. (1967). Ionic state polarography of sea water. I. Cadmium and zinc in sea water. *J. Polarogr. Soc.,* **13,** 4–8.

Bass, L. E. (1977). Influence of temperature and salinity on oxygen consumption of tissues in the American oyster (*Crassostrea virginica*). *Comp. Biochem. Physiol.,* **58B,** 125–130.

Batley, G. E., and Florence, T. M. (1976). Determination of chemical forms of dissolved Cd, Pb, and Cu in sea water. *Mar. Chem.,* **4,** 347–363.

Benayoun, G., Fowler, S. W., and Oregioni, B. (1974). Flux of cadmium through euphausiids. *Mar. Biol.,* **27,** 205–212.

Bengtsson, B.-E. (1977). Accumulation of cadmium in some aquatic animals from the Baltic Sea. *Ambio. Special Report* No. 5, pp. 69–73.

Bjerregaard, P. (1982). Accumulation of cadmium and selenium and their mutual interaction in the shore crab *Carcinus maenas* (L). *Aquat. Toxicol.,* **2,** 113–125.

Bloom, H., and Ayling, G. M. (1977). Heavy metals in the Derwent estuary. *Environ. Geol.,* **2,** 3–22.

Bower, P. M., Simpson, J. H., Williams, S. C., and Li, Y. H. (1978). Heavy metals in the sediments of Foundry Cove, Cold Spring, New York. *Environ. Sci. Technol.,* **12,** 683–687.

Boyden, C. R. (1975). Distribution of some trace metals in Poole Harbour, Dorset, *Mar. Pollut. Bull.,* **6,** 180–187.

Boyden, C. R. (1977). The effect of size upon metal content of shellfish. *J. Mar. Biol. Assoc. U.K.,* **57,** 675–714.

Boyle, E. A., and Huested, S. S. (1983). Aspects of the surface distributions of copper, nickel, cadmium and lead in the North Atlantic and North Pacific. In *Trace Metals in Sea Water,* Wong, C. S., Boyle, E., Bruland, K. W., Burton, J., and Goldberg, E. D. (eds.). Plenum, New York, pp. 379–394.

Boyle, E. A., Huested, S. S., and Jones, S. (1981). On the distributions of Cu, Ni and Cd in the surface waters of the North Atlantic and North Pacific ocean. *J. Geophys. Res.,* **86**(C9), 8048–8066.

Briggs, L. B. R. (1979). Effects of cadmium on the intracellular pool of free amino acids in *Mytilus edulis. Bull. Environ. Contam. Toxicol.,* **22,** 838–845.

Brooks, R. R., and Rumsby, M. G. (1967). Studies on the uptake of cadmium by the oyster, *ostrea sinuata* (Lamarck). *Aust. J. Mar. Freshwater Res.,* **18,** 53–61.

Brown, B. E. (1982). The form and function of metal containing "granules" in invertebrate tissues. *Biol. Rev.,* **37,** 621–668.

Bruland, K. W., and Franks, R. P. (1983). Mn, Ni, Cu, Zn and Cd in the western North Atlantic. In *Trace Metal in Sea Water,* Wong, C. S., Boyle, E., Bruland, K. W., Burton, J. D., and Goldberg, E. D. (eds.). Plenum, New York, pp. 395–414.

Bruland, K. W., Knauer, G. A., and Martin, J. H. (1978). Cadmium in Northeast Pacific waters. *Limnol. Oceanogr.*, **23**, 618–625.

Bryan, G. W. (1974). Adaptation of an estuarine polychaete to sediments containing high concentrations of heavy metals. In *Pollution and Physiology of Marine Organisms,* Vernberg, F. J., and Vernberg, W. B. (eds.). Academic, New York, pp. 123–135.

Bryan, G. W. (1976). Heavy metal contamination in the sea. In *Marine Pollution,* Johnston, R. (ed.). Academic, New York, pp. 185–302.

Bryan, G. W. (1985). Bioavailability and effects of heavy metals in marine deposits. In *Wastes in the Ocean,* Vol. 6, Ketcham, B. H., Capuzzo, J. M., Burt, W. V., Duedall, I. W., Park, P. K., and Kester, D. R. (eds.). Wiley, New York, pp. 41–79.

Bryan, G. W., and Hummerstone, L. G. (1973). Adaptation of the polychaete *Nereis diversicolor* to estuarine sediments containing high concentrations of zinc and cadmium. *J. Mar. Biol. Assoc. U.K.,* **53**, 839–857.

Bryan, G. W., and Hummerstone, L. G. (1978). Heavy metals in the burrowing bivalve *Scrobicularia plana* from contaminated and uncontaminated estuaries. *J. Mar. Biol. Assoc. U.K.,* **58**, 401–419.

Caines, L. A. (1978). Heavy metal residues in grey seals (*Halichoerus grypus*) from the Farne Islands. International Council for Exploration Sea. Report. C.M.1978/E.40.

Campbell, J. A., and Yeats, P. (1982). The distribution of manganese, iron, nickel, copper and cadmium in the waters of Baffin Bay and the Canadian Arctic Archepelago. *Oceanol. Acta,* **5**, 161–168.

Carmichael, N. G., and Fowler, B. A. (1981). Cadmium accumulation and toxicity in the kidney of the bay scallop *Argopecten irradians. Mar. Biol.,* **65**, 35–43.

Carmichael, N. G., Squibb, K. S., and Fowler, B. A. (1979). Metals in the molluscan kidney: A comparison of two closely related bivalve species (*Argopecten*) using X-ray microanalysis and atomic absorption spectroscopy. *J. Fish. Res. Bd. Can.,* **36**, 1149–1155.

Carpene, E., and George, S. G. (1981). Absorption of cadmium by gills of *Mytilus edulis* (L.). *Mol. Physiol.,* **1**, 23–34.

Carpene, E., Cortesi, P., Crisetig, G., and Serrazanetti, G. P. (1980). Cadmium-binding proteins from the mantle of *Mytilus edulis* (L.) after exposure to cadmium. *Thalassia Jugosl.,* **16**, 317–323.

Carpene, E., Cattani, O., Hakim, G., and Serrazanetti, G. P. (1983). Metallothionein from foot and posterior adductor muscle of *Mytilus galloprovincialis. Comp. Biochem. Physiol.,* **74C**, 331–336.

Chan, J. P., Cheung, M. T., and Li, F. P. (1974). Trace metals in Hong Kong waters. *Mar. Pollut. Bull.,* **5**, 171–174.

Chassard-Bouchaud, C. (1982). Ultrastructural study of cadmium concentration by the digestive gland of the crab *Carcinus maenas* (Crustacea Decapoda). Electorn probe X-ray microanalysis. *C. R. Sci. Acad. Sci. Ser. 3,* **294**, 153–157.

Cooke, M., Nickless, G., Lawn, R. E., and Roberts, D. J. (1979). Biological availability of sediment-bound cadmium to the edible cockle, *Cerastoderma edule. Bull. Environ. Contam. Toxicol.,* **23**, 381–386.

Coombs, T. L. (1979). Cadmium in aquatic organism. In *The Chemistry, Biochemistry and Biology of Cadmium,* Webb, M. (ed.). Elsevier/North Holland Biomedical Press, Amsterdam, pp. 93–139.

Cossa, D. (1976). Sorption du cadmium par une population de la diatomée *Phaeodactylum tricornutum* en culture. *Mar. Biol.,* **34**, 163–167.

Cossa, D., Bourget, E., Pouliot, D., Piuze, J., and Chamut, J. P. (1980). Geographical and seasonal variations in the relationship between trace metal content and body weight in *Mytilus edulis. Mar. Biol.,* **58**, 7–14.

Danielsson, L.-G., and Westerlund, S. (1983). Trace metals in the Arctic Ocean in *Trace Metals in Seawater,* Wong, C. S., Boyle, E., Bruland, K. W., Burton, J. D., and Goldberg, E. D. (eds.). Plenum, New York, pp. 85–95.

Davies, I. M., Topping, G., Graham, W. C., Falconer, C. R., McIntosh, A. D., and Saward, D. (1981). Field and experimental studies on cadmium in the edible crab *Cancer pagurus*. *Mar. Biol.,* **64,** 291–297.

Denton, G. R. W., and Burdon-Jones, C. (1981). Influence of temperature and salinity on the uptake, distribution and depuration of mercury, cadmium and lead by the black-lip oyster *Saccostrea echinata. Mar. Biol.,* **64,** 317–326.

Dethlefsen, V. (1977/78). Uptake, retention and loss of cadmium by brown shrimp (*Crangon crangon*). *Meeresforsch.,* **26,** 137–152.

Drescher, H. E., Harms, U., and Huschenbeth, E. (1977). Organochlorines and heavy metals in the harbour seal *Phoca vitulina* from the German north sea coast. *Mar. Biol.,* **41,** 99–106.

Duinker, J. D., and Kramer, C. J. M. (1977). An experimental study on the speciation of dissolved zinc, cadmium, lead and copper and River Rhine and North Sea water by differential pulse anodic stripping voltametry. *Mar. Chem.,* **5,** 207–228.

Eisler, R. (1974). Radiocadmium exchange with seawater by *Fundulus heteroclitus* (L.) (Pisces: Cyprinodontidae). *J. Fish. Biol.,* **6,** 601–612.

Eisler, R., Zaroogian, G. E., and Hennekey, R. H. (1972). Cadmium uptake by marine organisms. *J. Fish. Res. Bd. Can.,* **29,** 1367–1369.

Engel, D. W., and Fowler, B. A. (1979a). Copper- and cadmium-induced changes in the metabolism and structure of molluscan gill tissue. In *Marine Pollution: Functional Responses,* Vernberg, W. B., Thurberg, F. P., Calabrese, A., and Vernberg, F. (eds.). Academic, New York, pp. 239–256.

Engel, D. W., and Fowler, B. A. (1979b). Factors influencing cadmium accumulation and its toxicity to marine organisms. *Environ. Health Pespect.,* **28,** 81–88.

Engel, D. W., Sunda, W. G., and Fowler, B. A. (1981). Factors affecting trace metal uptake and toxicity to estuarine organisms. I. Environmental parameters. In *Biological Monitoring of Marine Pollutants,* Vernberg, J., Calabrese, A., Thurberg, F. P., and Vernberg, W. B. (eds.). Academic, New York, pp. 127–144.

Fowler, B. A., Carmichael, N. G., Squible, K. S., and Engel, D. W. (1981). Factors affecting trace metal uptake and toxicity to estuarine organisms. II. Cellular mechanisms. In *Biological Monitoring of Marine Pollutants,* Vernberg, J., Calabrese, A., Thurberg, F. P., and Vernberg, W. B. (eds.). Academic, New York, pp. 145–163.

Fowler, S. W., and Benayoun, G. (1974). Experimental studies on cadmium flux through marine biota. In *Comparative Studies of Food and Environmental Contamination IAEA-SM 175/10,* International Atomic Energy Agency, Vienna, pp. 159–178.

Frankenne, F., Noël-Lambot, F., and Disteche, A. (1980). Isolation and characterization of metallothioneins from cadmium-loaded mussel *Mytilus edulis. Comp. Biochem. Physiol.,* **66C,** 179–182.

Frazier, J. M. (1975). The dynamics of metals in the American oyster, *Crassostrea virginica*. I. Seasonal effects. *Chesapeake Sci.,* **16,** 162–171.

Frazier, J. M. (1976). The dynamics of metals in the American oyster, *Crassostrea virginica*. II. Environmental effects. *Chesapeake Sci.,* **17,** 188–197.

Genest, P. E., and Hatch, W. I. (1981). Heavy metals in *Mercenaria mercenaria* and sediments from the New Bedford Harbour region of Buzzard's Bay, Massachusetts. *Bull. Environ. Contam. Toxicol.,* **26,** 124–138.

George, S. G. (1983a). Heavy metal detoxification in the mussel *Mytilus edulis*. Composition of Cd-containing kidney granules (tertiary lysosomes). *Comp. Biochem. Physiol.,* **76C,** 53–57.

George, S. G. (1983b). Heavy metal detoxification in *Mytilus* kidney. An *in vitro* study of Cd- and Zn-binding to isolated tertiary lysosomes. *Comp. Biochem. Physiol.*, **76C**, 59–65.

George, S. G., and Coombs, T. L. (1977). The effects of chelating agents on the uptake and accumulation of cadmium by *Mytilus edulis*. *Mar. Biol.*, **39**, 261–268.

George, S. G., and Pirie, B. J. S. (1979). The occurrence of cadmium in sub-cellular particles in the kidney of the marine mussel, *Mytilus edulis*, exposed to cadmium. *Biochim. Biophys. Acta*, **580**, 234–244.

George, S. G., and Viarengo, A. (1985). A model for heavy metal homeostatis and detoxification in mussels. In *Marine Pollution and Physiology: Recent Advances*, Vernberg, F. J., Thurberg, F. P., Calabrese, A., and Vernberg, W. B. (eds.). University of South Carolina Press, Columbia, SC, pp. 125–143.

George, S. G., Carpene, E., Coombs, T. L., Overnell, J., and Youngson, A. (1979). Characterization of cadmium-binding proteins from mussels, *Mytilus edulis* (L.) exposed to cadmium. *Biochim. Biophys. Acta*, **580**, 225–233.

George, S. G., Pirie, B. J. S., and Coombes, T. L. (1980). Isolation and elemental analysis of metal rich granules from the kidney of the scallop, *Pecten maximus* (L.). *J. Exp. Mar. Biol. Ecol.*, **42**, 143–156.

George, S. G., Coombs, T. L., and Pirie, B. J. S. (1982). Characterization of metal containing granules from the kidney of the common mussel, *Mytilus edulis*. *Biochim. Biophys. Acta*, **716**, 61–71.

Goldberg, E. D., Bowen, V. T., Farrington, J. W., Harvey, G., Martin, J. H., Parker, P. L., Risebrough, R. W., Robertson, W., Schneider, E., and Gamble, E. (1978). The mussel watch. *Environ. Conserv.*, **5**, 101–126.

Gordon, D. C., Jr. (1966). The effects of the deposit feeding polychaete *Pectinaria gouldii* on the intertidal sediments of Barnstable Harbour. *Limnol. Oceanogr.*, **11**, 327–332.

Greig, R. A., and Wenzloff, D. R. (1978). Metal accumulation and depuration by the American oyster, *Crassostrea virginica*. *Bull. Environ. Contam. Toxicol.*, **20**, 499–504.

Gutierrez-Galindo, E. A. (1980). Etude comparee des roles de la nourriture et de l'eau dans l'accumulation du cadmium par le crabe *Carcinus maenas* en presence d'EDTA et de phosphate. *Rev. Int. Oceanogr. Med.*, **58**, 69–79.

Hamilton, E. I. (1981). The analysis of elements—Quality factors. *Mar. Pollut. Bull.*, **12**, 393–394.

Härdstedt-Roméo, M., and Gnassia-Barelli, M. (1980). Effect of complexation by natural phytoplankton exudates on the accumulation of cadmium and copper by the haptophyceae *Cricosphaera elongata*. *Mar. Biol.*, **59**, 79–84.

Harms, U., Drescher, H. E., and Huschenbeth, E. (1978). Further data on heavy metals and organochlorines in marine mammals from German coastal waters. *Reports on Marine Research 26. Sonderdr. Bd.*, **26**, 153–161.

Hatfield, C. T., and Williams, G. L. (1975). A summary of possible environmental effects of disposing mine tailings into Strathcona Sound, Baffin Island. C. T. Hatfield Association Ltd., Report to Canada Department of Indian Affairs and Northern Development.

Heppelston, P. B., and French, M. C. (1973). Mercury and other metals in British seals. *Nature*, 302–304.

Holden, A. V. (1978). Pollutants and seals—a review. *Mamm. Rev.*, **8**, 53–66.

Holmes, C. W., Slade, E. A., and McLerran, C. J. (1974). Migration and redistribtuion of zinc and cadmium in marine estuarine system. *Environ. Sci. Technol.*, **8**, 255–259.

Howard, A. G., and Nickless, G. (1977). Heavy metal complexation in polluted molluscs. 1. Limpets (*Patella vulgata* and *Patella intermedia*). *Chem. Biol. Interac.*, **16**, 107–114.

Hung, Y.-W. (1982). Effects of temperature and chelating agents on the cadmium uptake in the American oyster. *Bull. Environ. Contam. Toxicol.*, **28**, 546–551.

Hutcheson, M. S. (1974). The effect of temperature and salinity on cadmium uptake by the blue crab, *Callinectes sapidus*. *Chesapeake Sci.*, **15**, 237–241.

Jackim, E., Morrison, G., and Steele, R. (1977). Effects of environmental factors on radiocadmium uptake by four species of marine bivalves. *Mar. Biol.*, **40**, 303–308.

Janssen, H. H., and Scholz, N. (1979). Uptake and cellular distribution of cadmium in *Mytilus edulis*. *Mar. Biol.*, **55**, 133–141.

Jenkins, K. D., Brown, D. A., Oshida, P. S., and Perkins, E. M. (1982). Cytosolic metal distribution as an indicator of toxicity in sea urchins from the southern California Bight. *Mar. Pollut. Bull.*, **13**, 413–421.

Jenne, E. A., and Luoma, S. N. (1977). Forms of trace elements in soils, sediments, and associated waters: An overview of their determination and biological availability. In *Biological Implications of Metals in the Environment*, Wildung, R. E., and Drucker, H. (eds.). Conf.-750929. NTIS Springfield, VA, pp. 110–143.

Jennings, J. R., and Rainbow, P. S. (1979). Studies on the uptake of cadmium by the crab *Carcinus maenas* in the laboratory. I. Accumulation from seawater and a food source. *Mar. Biol.*, **50**, 131–139.

Johansen, P., Kapel, F. O., and Kraul, I. (1980). Heavy metals and organochlorines in marine mammals from Greenland. International Council for Exploration Sea. Report C.M. 1980/E:32.

Julshamn, K. (1981). Studies on major and minor elements in Molluscs in western Norway. II. Seasonal variation in the contents of 10 elements in oysters (*Ostrea edulis*) from three oyster farms. *Fisk. Dir. Skr. Ser. Ernaering.*, **1**, 183–197.

Kägi, J. H. R., and Nordberg, M. (eds.) (1979). *Metallothionein*. Birkhauser-Verlag, Basel.

Kägi, J. H. R., and Vallee, B. L. (1960). Metallothionein: A cadmium- and zinc-containing protein from equine Renal cortex. *J. Biol. Chem.*, **235**, 3460–3465.

Ketchum, B. H. (ed.) (1972). *Waters Edge: Critical Problems of the Coastal Zone*. MIT Press, Cambridge, MA.

Klöckner, K. (1979). Uptake and accumulation of cadmium by *Ophryotrocha diadema* (Polychaeta). *Mar. Ecol. Prog. Ser.*, **1**, 71–76.

Knauer, G. A., and Martin, J. (1973). Seasonal variations of cadmium, copper, manganese, lead and zinc in water and phytoplankton in Monterey Bay, California. *Limnol. Oceanogr.*, **18**, 597–604.

Kneip, T. J., and Hazen, R. E. (1979). Deposit and mobility of cadmium in a marsh-cove ecosystem and the relation to cadmium concentration in biota. *Environ. Health Perspect.*, **28**, 67–74.

Koeman, J. H., Peeters, W. H. M., Smit, C. J., Tjioe, P. S., and De Goeij, J. J. M. (1972). Persistent chemicals in marine mammals. *Overdruk vit "Tno-Neiuws"*, **27**, 570–578.

Kremling, K., and Petersen, H. (1978). The distribution of Mn, Fe, Zn, Cd and Cu in Baltic seawater; A study on the basis of one anchor station. *Mar. Chem.*, **6**, 155–177.

Loring, D. H., Bewers, J. M., Seibert, G., and Kranck, K. (1980). A preliminary survey of circulation and heavy metal contamination in Belledune Harbour and adjacent areas. In *Cadmium Pollution of Belledune Harbour, New Brunswick, Canada*, Uthe, J. F., and Zitko, V. (eds.). Dept. of Fisheries & Oceans. St. Andrews, N. B., Canada. *Can. Tech. Rep. Fish. Aquat. Sci.*, **963**, 35–47.

Lu, J. C. S., and Chen, K. Y. (1977). Migration of trace metals in interfaces of seawater and polluted artificial sediments. *Environ. Sci. Technol.*, **11**, 174–179.

Luoma, S. N., and Jenne, E. A. (1976). Estimating bioavailability of sediment-bound trace metals with chemical extractants. In *Trace Substances in Environmental Health—X*, Hemphill, D. D. (ed.). University of Missouri, Columbia, pp. 343–351.

Majori, L., and Petronio, F. (1973). Accumulation phenomenon which takes place in a mussel

(*Mytilus galloprovincialis* LMK) grown in an artificially polluted environment. Verification of a simplified model of the dynamic equilibrium of metal distribution between mussel and sea water. Note II—Pollution from cadmium. *Ig. Mod.*, **66**, 39–63. Fish. Mar. Serv. Transl. Ser. No. 3143 (1974).

Malo, B. A. (1977). Partial extraction of metals from aquatic sediments. *Environ. Sci. Technol.*, **11**, 277–282.

Mantoura, R. F., Dickson, A., and Riley, J. P. (1978). The complexation of metals with humic materials in natural waters. *Est. Coast. Mar. Sci.*, **6**, 387–408.

Marshall, A. T., and Talbot, V. (1979). Accumulation of cadmium and lead in the gills of *Mytilus edulis:* X-ray microanalysis and chemical analysis. *Chem. Biol. Interac.*, **27**, 111–123.

Mart, L., and Nürnberg, H. W. (1986). Distribution of Cd in the sea. In *Cadmium in the Environment*, Mislin, H., and Ravera, O. (eds.), Experientia Supplementum, Vol. 50 Birkhäuser Verlag, Basel, pp. 28–40.

Mart, L., Rützel, H., Klahre, P., Sipos, L., Platzek, U., Valenta, P., and Nürnberg, H. W. (1982). Comparative studies on the distributions of heavy metals in the oceans and coastal waters. *Sci. Tot. Environ.*, **26**, 1–17.

Mart, L., Nürnberg, H. W., and Rützel, H. (1984). Comparative studies on cadmium levels in the North Sea, Norwegian Sea, Barents Sea and the Eastern Arctic Ocean. *Fresen. Z. Anal. Chem.*, **317**, 201–209.

McKie, J. C., Davies, I. M., and Topping, G. (1980). Heavy metals in grey seals (*Halichoerus grypus*) from the east coast of Scotland. International Council for Exploration Sea. Report C.M. 1980/E:41.

McLean, M. W., and Williamson, F. B. (1977). Cadmium accumulation by marine red alga *Porphyra umbilicalis*. *Physiol. Plant.*, **41**, 268–272.

McLeese, D. W., and Ray, S. (1984). Uptake and excretion of cadmium, CdEDTA, and zinc by *Macoma balthica*. *Bull. Environ. Contam. Toxicol.*, **32**, 85–92.

McLeese, D. W., Ray, S., and Burridge, L. E. (1981). Lack of excretion of cadmium from lobsters. *Chemosphere*, **10**, 775–778.

Mowdy, D. E. (1981). Elimination of laboratory-acquired cadmium by the oyster *Crassostrea virginica* in the natural environment. *Bull. Environ. Contam. Toxicol.*, **26**, 345–351.

Neff, J. W., Foster, R. S., and Slowey, J. F. (1978). Availability of sediment-absorbed heavy metals to benthos with particular emphasis on deposit feeding infauna. Dredged material research program. Technical Report D-78-42, Office Chief of Engineers, U.S. Army, Washington, DC.

Nimmo, D. W. R., Lightner, D. V., and Bahner, L. H. (1977). Effects of cadmium on shrimps, *Panaeus duorarum*, *Palaemonetes pugio* and *Palaemonetes vulgaris*. In *Physiological Responses of Marine Biota to Pollutants*, Vernberg, F. J., Calabrese, A., Thurburg, F. P., and Vernberg, W. B. (eds.). Academic, New York, pp. 131–183.

Noël-Lambot, F. (1981). Presence in the intestinal lumen of marine fish of corpuscles with a high cadmium-zinc and copper-binding capacity: A possible mechanism of heavy metal tolerance. *Mar. Ecol. Prog. Ser.*, **4**, 175–181.

Nolan, C. V., and Duke, E. J. (1983). Cadmium-binding proteins in *Mytilus edulis:* Relation to mode of administration and significance in tissue retention of cadmium. *Chemosphere*, **12**, 65–74.

Nürnberg, H. W., and Valenta, P. (1983). Potentialities and applications of voltametry in chemical speciation of trace metals in the sea. In *Trace Metals in Sea Water*, Wong, C. S., Boyle, E., Bruland, K. W., Burton, J. D., and Goldberg, E. D. (eds.). Plenum, New York, pp. 671–697.

O'Hara, J. (1973). Cadmium uptake by Fiddler crabs exposed to temperature and salinity stress. *J. Fish. Res. Bd. Can.*, **30**, 846–848.

Olafson, R. W., Kearns, A., and Sim, R. G. (1979). Heavy metal induction of metallothionein synthesis in the hepatopancreas of the crab *Scylla Serrata*. *Comp. Biochem. Physiol.*, **62B**, 417–424.

Olafson, R. W., Loya S., and Sim, R. G. (1980). Physiological parameters of prokaryotic metallothionein induction. *Biochem. Biophys. Res. Commun.*, **95**, 1495–1503.

Orren, M. J., Eagle, G A., Hennig, H. F-K.O., and Green, A. (1980). Variations in trace metal content of the mussel *Choromytilus meridionalis* (Kr) with season and sex. *Mar. Pollut. Bull.*, **11**, 253–257.

Overnell, J., and Coombs, T. L. (1979). Purification and properties of plaice metallothionein, a cadmium-binding protein from the liver of the plaice (*Pleuronectes platessa*). *Biochem. J.*, **183**, 277–283.

Pentreath, R. J. (1977). The accumulation of cadmium by the plaice, *Pleuronectes platessa* L. and the thornback ray, *Raja clavata* L. *J. Exp. Mar. Biol. Ecol.*, **30**, 223–232.

Phillips, D. J. H. (1976). The common mussel *Mytilus edulis* as an indicator of pollution by zinc, cadmium, lead and copper. 1. Effect of environmental variables on uptake of metals. *Mar. Biol.*, **38**, 59–69.

Phillips, D. J. H. (1980). Toxicity and accumulation of cadmium in marine and estuarine biota. In *Cadmium in the Environment*, Part 1, Nriagu, J. O. (ed.). Wiley-Interscience, New York, pp. 425–569.

Preston, A. (1973). Cadmium in the marine environment of the United Kingdom. *Mar. Pollut. Bull.*, **4**, 105–107.

Prévot, P., and Soyer, M. (1978). Action du cadmium sur un Dinoflagellé libre: *Prorocentrum micans* E.: Croissance, absorption du cadmium et modifications cellulaires. *C. R. Acad. Sci., Paris.* 287, (Série D), 833–836.

Pruell, R. J., and Engelherdt, F. R. (1980). Liver cadmium uptake, catalase inhibition and cadmium thionein production in the killifish (*Fundulus heteroclitus*) induced by experimental cadmium exposure. *Mar. Environ. Res.*, **3**, 101–111.

Rainbow, P. S., and Scott, A. G. (1979). Two heavy metal-binding proteins in the midgut gland of the crab *Carcinus maenas*. *Mar. Biol.*, **55**, 143–150.

Rainbow, P. S., Scott, A. G., Wiggins, E. A., and Jackson, R. W. (1980). Effect of chelating agents on the accumulation of cadmium by the barnacle *Semibalanus balanoides*, and complexation of soluble Cd, Zn and Cu. *Mar. Ecol. Progr. Ser.*, **2**, 143–152.

Raspor, B. (1980). "Distribution and speciation of cadmium in natural waters. In *Cadmium in the Environment, Part 1, Ecological Cycling*, Nriagu, J. O. (ed.). J Wiley, New York, pp. 147–236.

Ray, S. (1984). Bioaccumulation of cadmium in marine organisms. *Experientia*, **40**, 14–23.

Ray, S., and McLeese, D. W. (1982). Notes on bioavailability of cadmium and lead to a marine crustacean through food and water. Environment Canada Surveillance Report EPS-5-AR-82-1, pp. 19–24.

Ray, S., and McLeese, D. W. (1983). Factors affecting uptake of cadmium and other trace metals from marine sediments by some bottom-dwelling marine invertebrates. In *Wastes in the Ocean*, Vol. II, *Dredged Material Disposal in the Ocean*, Kester, D., Duedall, I. W., Ketchum, B. H., and Park, P. K. (eds.). J Wiley, New York, pp. 185–197.

Ray, S., and White, M. (1981). Metallothionein-like protein in lobster (*Homarus americanus*). *Chemosphere*, **10**, 1205–1213.

Ray, S., McLeese, D. W., and Pezzack, D. (1979). Chelation and interelemental effects on the bioaccumulation of heavy metals by marine invertebrates. In Proceedings of the International Conference on Management Control, Heavy Metal Environment, CEP Consultants LTD, Edinburgh, U.K., pp. 35–38.

Ray, S., McLeese, D. W., and Pezzack, D. (1980a). Accumulation of cadmium by *Nereis Virens*. *Arch. Environ. Contam. Toxicol.*, **9**, 1–8.

Ray, S., McLeese, D. W., Waiwood, B. A., and Pezzack, D. (1980b). Disposition of cadmium and zinc in *Pandalus montagui*. *Arch. Environ. Contam. Toxicol.*, **9**, 675–681.

Ray, S., McLeese, D. W., and Burridge, L. E. (1981a). Cadmium in tissues of lobsters captured near a lead smelter. *Mar. Pollut. Bull.*, **12**, 383–386.

Ray, S., McLeese, D. W., and Peterson, M. R. (1981b). Accumulation of copper, zinc, cadmium and lead from two contaminated sediments by three marine invertebrates—A laboratory study. *Bull. Environ. Contam. Toxicol.*, **26**, 315–322.

Rice, M. A., and Chien, P. K. (1979). Uptake, binding and clearance of divalent cadmium in *Glycera dibranchiata* (Annelida: polychaeta). *Mar. Biol.*, **53**, 33–39.

Ronald, K., Frank, R. J., Dougan, J., Frank, R., and Bran, H. H. (1984). Pollutants in harp seals (*Phoca groenlandica*). II. Heavy metals and selenium. *Sci. Tot. Environ.*, **38**, 153–166.

Salomons, W., and Förstner, U. (1980). Trace metal analysis on polluted sediments. Part II. Evaluation of environmental impact. *Environ. Technol. Lett.*, **1**, 506–517.

Scholz, N. (1980). Accumulation, loss and molecular distribution of cadmium in *Mytilus edulis*. *Helogöl. Meeresunters.*, **33**, 68–78.

Siewicki, T. C., Sydlowski, J. S., and Webbe, E. S. (1983). The nature of cadmium binding in commercial eastern oysters (*Crassostrea virginica*). *Arch. Environ. Contam. Toxicol.*, **12**, 299–304.

Sillen, L. G. (1961). The physical chemistry of sea water. In *Oceanography*, Sears, M. (ed.). American Association for the Advancement of Science, Publication 67, Washington, DC, pp. 549–582.

Sillen, L. G. (1962). *Chemical Equilibrium in Analytical Chemistry*. Wiley-Interscience, New York.

Skei, J. M., Price, N. B., and Calvert, S. E. (1972). The distribution of heavy metals in sediments of Sörfjord, West Norway. *Wat. Air Soil Pollut.*, **1**, 452–461.

Spivack, A. J., Huested, S. S., and Boyle, E. A. (1983). Copper, nickel, cadmium in the surface waters of the Mediterranean. In *Trace Metals in Sea Water*, Wong, C. S., Boyle, E., Bruland, K. W., Burton, J. D., Goldberg, E. D. (eds.). Plenum, New York, pp. 505–512.

Taylor, A. C. (1977). The respiratory responses of *Carcinus maenas* (L.) to changes in environmental salinity. *J. Exp. Mar. Biol. Ecol.*, **29**, 197–210.

Theede, H., Scholz, N., and Fischer, H. (1979). Temperature and salinity effects on the acute toxicity of cadmium to *Laomedea loveni* (Hydrozoa). *Mar. Ecol. Prog. Ser.*, **1**, 13–19.

Thompson, K. A., Brown, D. A., Chapman, P. M., and Brinkhurst, R. O. (1982). Histopathological effects and cadmium-binding protein synthesis in the marine oligochaete *Monopylephorus cuticulatus* following cadmium exposure. *Trans. Am. Microsc. Soc.*, **101**, 10–26.

Thornton, I., Watling, H., and Darracott, A. (1975). Geochemical studies in several rivers and estuaries used for oyster rearing. *Sci. Tot. Environ.*, **4**, 325–345.

Thurberg, F. P., Calabrese, A., Gould, E., Greig, R. A., Dawson, M. A., and Tucker, R. K. (1977). Response of the lobster, *Homarus americanus*, to sublethal levels of cadmium and mercury. In *Physiological Responses of Marine Biota to Pollutants*, Vernberg, F. J., Calabrese, A., Thurberg, F. P., and Vernberg, W. B. (eds.). Academic, New York, pp. 185–197.

Ueda, T., Nakamura, R., and Suzuki, Y. (1976). Comparison of ^{115m}Cd accumulation from sediments and sea water by polychaete worms. *Bull. Jap. Soc. Sci. Fish.*, **42**, 299–306.

Vallee, B. L. (1979). Metallothionein: Historical review and perspectives. In *Metallothionein*, Kägi, J. H. R., and Nordberg, M. (eds.). Birkhauser-Verlag, Basel, pp. 19–40.

Vernberg, W. B., DeCoursey, P. J., Kelly, M., and Johns, D. M. (1977). Effects of sublethal concentrations of cadmium on adult *Palaemonetes pugio* under static and flow through conditions. *Bull. Environ. Contam. Toxicol.*, **17,** 16–24.

von Westernhagen, H., and Dethlefsen, V. (1975). Combined effects of cadmium and salinity on development and survival of flounder eggs. *J. Mar. Biol. Assoc. U.K.*, **55,** 945–957.

Wagemann, R., Snow, N. B., Lutz, A., and Scott, D. P. (1983). Heavy metals in tissues and organs of the narwhal (*Monodon monoceros*). *Can. J. Fish. Aquat. Sci.*, **40** (Suppl. 2), 206–214.

Ward, T. J. (1982). Laboratory study of the accumulation and distribution of cadmium in the Sydney rock oyster *Saccostrea commercialis* (I & R). *Aust. J. Mar. Freshwater Res.*, **33,** 33–44.

White, S. L., and Rainbow, P. S. (1982). Regulation and accumulation of copper, zinc and cadmium by the shrimp *Palaemon elegans*. *Mar. Ecol. Progr. Ser.*, **8,** 95–101.

Wright, D. A. (1977a). The effect of salinity on cadmium uptake by the tissues of the shore crab *Carcinus maenas*. *J. Exp. Biol.*, **67,** 137–146.

Wright, D. A. (1977b). The effect of calcium on cadmium uptake by the shore crab *Carcinus maenas*. *J. Exp. Biol.*, **67,** 163–173.

Young, D. K. (1968) Chemistry of southern Chesapeake sediments. *Chesapeake Sci.*, **9,** 254–260.

Zaroogian, G. E. (1980). *Crassostrea virginica* as an indicator of cadmium pollution. *Mar. Biol.*, **58,** 275–284.

Zaroogian, G. E., and Cheer, S. (1976). Accumulation of cadmium by the American oyster, *Crassostrea virginica*. *Nature*, **261,** 408–410.

Zirino, A., and Yamamoto, S. (1972). A pH dependent model for the chemical speciation of copper, zinc, cadmium, and lead in seawater. *Limnol. Oceanogr.*, **17,** 661–671.

10

METHODS OF CADMIUM DETECTION

R. W. Dabeka

Food Research Division
Food Directorate, Health Protection Branch,
Health and Welfare Canada,
Ottawa, Ontario, Canada

M. Ihnat*

Plant Research Centre, Research Branch,
Agriculture Canada,
Ottawa, Ontario, Canada

10.1. Introduction
10.2. Methods of Analysis
10.3. Contamination Control
    10.3.1. Laboratory and Analyst
    10.3.2. Laboratory Ware
    10.3.3. Reagents
    10.3.4. Dealing with Blanks
    10.3.5. Monitoring and Controlling Contamination
10.4. Project Design and Analytical Input
10.5. Sample Handling
    10.5.1. Sampling
    10.5.2. Sample Decomposition
    10.5.3. Sample Storage

*Contribution No. 1003 from Plant Research Centre.
Present address: Land Resource Research Centre, Research Branch, Agriculture Canada, Ottawa, Ont. K1A0C6.

232   Methods of Cadmium Detection

10.6. Concentration Techniques
10.7. Determination of Cadmium
    10.7.1. Flame Atomic Absorption Spectrometry
    10.7.2. Graphite Furnace Atomic Absorption Spectrometry
    10.7.3. Electroanalytical Methods
    10.7.4. Inductively Coupled Plasma Atomic Emission Spectrometry
    10.7.5. Neutron Activation Analysis
    10.7.6. Spark Source Mass Spectrometry
    10.7.7. Inductively Coupled Plasma Mass Spectrometry
10.8. Reliability of Literature Data and Reference Materials
10.9. Analytical Recommendations
Appendix: An Example of Contamination Control Development
References

10.1. INTRODUCTION

A variety of analytical methodologies of varying complexity and applicability pertinent to the determination of cadmium in organic and inorganic environmental matrices are discussed in this chapter: flame atomic absorption spectrometry (FAAS), flame atomic absorption spectrometry with solvent extraction (SEFAAS), electrothermal (graphite furnace) atomic absorption spectrometry (EAAS), inductively coupled plasma atomic emission spectrometry (ICPAES), anodic stripping voltammetry (ASV), instrumental neutron activitation analysis (INAA), inductively coupled plasma mass spectrometry (ICPMS), and spark source mass spectrometry (SSMS). Factors such as sampling, sample storage, and the role of laboratory-ware and reagents and the analyst as sources of contamination are of underlying importance to all analytical methods; in addition, sample decomposition and/or analyte concentration are essential steps for all techniques with the exception of INAA and perhaps SSMS and ICPMS. In selecting a particular technique, consideration of the following technical and operational characteristics can be useful: multielement capability: detection limit; sample preparation requirements and hence analytical throughout; spectral, physical, and chemical interferences; precision and accuracy; and operational expertise and cost, with weighting and judgmental factors at the discretion of the analyst and establishment. Detection limits on a sample basis range from 0.05 and 0.5 µg/g for ASV and EAAS, respectively, to 100 µg/g for ICPAES, permitting most of the techniques to be applicable to the measurement of the cadmium in solid aquatic materials with the exception of freshwater fish. After clean-up and concentration of water samples when necessary EAAS with preconcentration and ASV can be used to quantitate cadmium in natural-water matrices. Requirements for sample preparation range from minimal for INAA through moderate for FAAS, ICPAES, ICPMS, and ASV to extensive for SEFAAS, EAAS, and SSMS. The time required for complete processing of a sample ranges from typically 1 h for FAAS, SEFAAS, EAAS, ICPAES, and ASV to possibly days for INAA and SSMS. Precision and accuracy are generally excellent for all of the techniques with the exception of SSMS, which exhibits moderate precision and accuracy. Operational and method developmental expertise required for technical and scientific personnel vary from low or moderate for FAAS, SEFAAS, and ASV to moderate or high for the techniques of EAAS, ICPAES, ICPMS, INAA, and SSMS.

The main problems currently encountered in the determination of cadmium are traceable to the knowledge and experience of the analyst. For accurate determination of the low levels of cadmium found in the aquatic environment, the analyst should have a detailed knowledge of both the analytical chemistry of cadmium and the instruments used for its determination. Too often, the assumption is made that the only prerequisites to performing analyses are purchase of the analytical instrument and application of a published or "cook-book" method. This view point is sometimes promoted by instrument manufacturers, and, indeed, with the widespread use of microprocessors, little skill is needed to run instruments. Unfortunately, even correct application of an instrument does not guarantee accurate results, and it is vital that the analyst understand the capabilities and limitations of the instrument. Every instrumental technique is subject to interferences, and application of common validation methods, such as standard additions and recovery studies, will not reveal all problem areas. One can use standard additions or have 100% recovery and still obtain a result deviating by over an order of magnitude from the true value.

Accurate cadmium determinations require a high degree of expertise and judgment in choosing an appropriate method and suitable sampling, storage, manipulation, and contamination and quality control procedures. Few published methods are capable of conveying to the inexperienced analyst all the knowledge needed to avoid problems, and there is a need for researchers to make sure that their analytical staff receive proper training. With proper procedures and training, satisfactory measurements of endogenous cadmium is possible.

Concentrations of cadmium in natural materials vary widely, as can be surmised from information presented in the other chapters in this volume. Reported cadmium levels range from about 0.001–0.05 μg/L in fresh water and seawater to be in the vicinity of 1000 μg/kg in marine biota and estuarine sediments (Chapters 9; 4 and Bender and Gagner, 1976; Bewers et al., 1976; Bruland et al., 1978; Chan, 1977; Chester and Stoner, 1974; Eaton, 1976; Florence and Batley, 1977; Hume, 1975; Ihnat, 1978, 1982a; Knauer and Martin, 1981; Kopp, 1970; Martin et al., 1976; Nurnberg et al., 1976; Wahlgren et al., 1971; D. J. MacGregor, unpublished report). Materials uncontaminated by anthropogenic sources of cadmium contain the element at levels typically below 1 µg/kg, often substantially lower. Quite different, elevated levels occur in heavily contaminated materials. Whereas these higher concentrations present less of a problem for the determination of cadmium, the challenge to the analyst lies in accurately measuring the low endogenous levels in environmental materials.

10.2. METHODS OF ANALYSIS

This chapter considers, for the most part, the measurement of total cadmium in samples. In water, for example, there are various forms of cadmium (ionic, complexed, colloidal) as outlined in Chapters 5; and 7 (physical cycling and freshwater uptake). Not all forms are easily available to biota, and it may sometimes be appropriate to measure the easily available forms of cadmium rather than total cadmium in the water. This is best done by pretreatment of the sample in some appropriate way such as filtration prior to application of the techniques of determination discussed in this chapter.

Whether a particular analytical technique can be applied to the quantitative measurement of cadmium in a given material depends on several considerations, the prime one being the detectivity of the technique in comparison to the level of the element in the sample. Detection limits of the various techniques are summarized in Table 1 and are given as concentrations in the solution or material presented to the measuring instrument as well as these values converted to detection limits with respect to the original sample. Comparison of the latter with native cadmium levels in various materials, recalling the criterion that the elemental concentration must be at least about 10 times the detection limit for satisfactory determination, will give an indication of method potential. Most of the techniques can be applied to most of the solid aquatic materials, a possible exception being freshwater fish for which FAAS, ICPAES, INAA, and SSMS may be hard-pressed. The techniques of ICPMS, EAAS, and ASV have the potential for direct determination of low native levels of cadmium in fresh water, whereas the technique of FAAS can be used provided a preconcentration step using evaporation, solvent extraction, or ion exchange precedes the measurement. The very high salt content of seawater dictates a clean-up of the matrix and concentration of the element by ion exchange or solvent extraction prior to its measurement by SEFAAS, EAAS, ICPAES, or ASV. Application of SSMS to water analysis necessitates concentration by evaporation to yield solid residue for the preparation of conducting electrodes.

Two points should be borne in mind when making such an evaluation. First, the detection limits quoted refer to typically good data usually on clean standard solutions under ideal operating conditions. Matrix-laden solutions of actual samples will tend to exhibit a deterioration of detection limit. On the other hand, the use of larger sample sizes where feasible can

Table 1. Limits of Detection of Analytical Methods Used to Measure Cadmium in Aquatic Materials

Technique	Detection Limit (ng/mL or ng/g) On Sample		Reference
	As Presented To Instrument[a]	Original Basis[b]	
Flame atomic absorption spectrometry (direct aspiration) (FAAS)	1	50	Robinson, 1974; Kirkbright and Sargent, 1974; Slavin, 1968; Varian Techtron, 1972; Dean and Rains, 1975
Flame atomic absorption spectrometry with solvent extraction (SEFAAS)	1	5	de Vries et al., 1975; Posma et al., 1975; Environment Canada, 1974
Electrothermal atomization atomic absorption spectrometry (EAAS)	0.01[c]	0.5	Fuller, 1977
Inductively coupled plasma atomic emission spectrometry (ICPAES)	2	100	Boumans, 1980; Boumans and Bosveld, 1979; Scott et al., 1974; Dickinson and Fassel, 1969; Winge et al., 1979; McQuaker et al., 1979; Boumans and de Boer, 1972
Anodic stripping voltammetry (ASV)	0.001	0.05	Batley, 1983
Instrumental neutron activation analysis (INAA)	40[d]	40	Guinn, 1971
Spark source mass spectrometry	0.3[d]	1–100	Elser, 1976
Inductively coupled plasma mass spectrometry (ICPMS)	0.1	0.1–5	Boorn et al., 1985

[a] Detection limits are typical, good limits reported in the literature usually defined as the concentration required to produce a signal equal to twice the standard deviation of the blank (ng/mL) unless otherwise stated.
[b] Based on 1 g solid sample digested and made up to 50 mL for FAAS, EAAS, ICPAES, and ASV; similar treatment for SEFAAS but extraction into 5 mL of organic solvent to effect 10-fold concentration of element. Based on 1 g solid sample in INAA. Rough estimate for SSMS.
[c] Based on 10 μL volume introduced into the atomizer.
[d] Absolute detection limit in nanograms.

reduce the detection limit on a sample basis, enabling the technique to be applied to materials with far lower levels of the element. An example of the latter is using 10 g instead of 1 g in INAA, leading to a detection limit on a sample basis of 4 ng/g instead of 40 ng/g, and concentrating natural stream water by a factor of 50 yields a detection limit of 0.02 ng/mL for FAAS.

Although limit of detection is an important criterion for method selection, other technical and operational characteristics have a bearing on the choice of method. A number of such parameters are summarized in Table 2 for the different methods touched upon in this chapter. Further discussion of advantages and disadvantages are included in Section 10.7 dealing with the individual methods.

In spite of a plethora of published analytical methods for cadmium, interlaboratory studies (where each laboratory uses the method of its own choice) on cadmium in water (Pszonicki et al., 1982), cod (Slabyj et al., 1983), milk powder (Dybczynski et al., 1980), and marine biologicals (Berman and Boyko, 1985) demonstrate that accuracy is still a problem for many laboratories. For the water sample, containing 4 ng/mL cadmium, results averaged 3.9 ng/mL and varied from 0.76 to 10 ng/mL, indicating that fair agreement can be obtained for simple matrices. For the cod study, results obtained for an unspiked sample of unknown cadmium concentration ranged from 8 to 2067 ng/g, over two orders of magnitude. For the sample spiked at 78 ng/g (probably over three orders of magnitude greater than the natural level) reported values ranged 14–85 ng/g. For the milk powder study, results ranged from 1.1 to 1660 ng/g, with two out of eight results reporting 1.1 ng/g, and the next closest result being 20 ng/g. Subsequent to the study, two other laboratories using INAA found 1 ng/g cadmium in the sample. For the marine biological samples Berman and Boyko found that three-quarters of the participating laboratories produced valid results at the 300–400-ng/g level, and that about one-third did well at the 80-ng/g level. Thus, for low levels of cadmium in complex matrices, poor interlaboratory agreement is the rule rather than the exception.

Of a more optimistic nature, recent collaborative studies (where each laboratory uses the same well-defined method) on foods and seawater have revealed good agreement between results at cadmium levels as low as 0.2 ng/mL for seawater (Lamathe et al., 1982) and solutions of digested marine samples (Sperling and Bahr, 1981) using graphite-furnace atomic absorption spectrometry, and as low as 30 ng/g for foods using anodic stripping voltammetry (Capar et al., 1982). For such studies, however, only a single method was used, and greater controls were placed on the participating laboratories. For two of the above studies, some of the participating laboratories were requested to repeat the analyses, and in one instance, the organizer of the study actually visited some of the laboratories to detect and eliminate problems (Sperling and Bahr, 1981). In instances such as these, reporting of the initial results obtained from poorer laboratories would provide a better means of evaluating the true reliability of the method and the analyst apply-

Table 2. Comparison of Technical and Operational Characteristics of Analytical Methods for Measurement of Cadmium

	FAAS	SEFAAS	EAAS	ICPAES	ASV	INAA	SSMS	ICPMS
Multielement capability	No	No	No	Yes	No	Yes	Yes	Yes
Detection limit for Cd on sample basis, ng/g	50	5	0.5	100	0.05	40	1–100	0.1–5
Requirement for sample preparation	Moderate	Extensive	Moderate to extensive	Moderate	Moderate	Minimal	Extensive	Minimal to moderate
Determination rate, h^{-1a}	200	200	20	400	6–30	0	0	200
Time for complete processing of one sample, hr	1–	1–	1–	1–	2–24	1 day to 1 month	2 h to 2 days	1–
Precision	High	High	Moderate to high	High	Moderate to high	Moderate to high	Moderate	High
Spectral, physical, chemical interferences	Low to moderate	Low	Low to high	Low to high	Low to moderate	Low to high	Low	Moderate
Accuracy	High	High	High	High	High	Moderate to high	Moderate	Moderate to high
Operational/developmental expertise	Low to moderate	Low to moderate	Moderate to high	Moderate to high	Moderate to High	High	High	Moderate to high
Purchase cost of equipment ($000)	10–50	10–50	20–100	50–250	20	50 + reactor (1000)	100	300

^{a}Excluding sample preparation time and in multielement mode where applicable. Numbers include cadmium as well as other elements.

ing it and the effects of using different instruments. For all of the above studies, the levels determined are high compared to natural levels in environmental samples, and in the case of the anodic stripping voltammetric method, unspiked samples were not sent to collaborators.

Since many toxicological, epidemiological, and geophysical studies rely heavily on the accuracy of analytical results, the focal point of this section is devoted to analytical accuracy rather than to a review of the methods themselves. Among the factors affecting analytical accuracy are instrument application, use of standard reference materials (SRMs) during method development, and laboratory reference materials during method application, contamination control, storage, concentration techniques, and digestion or ashing methods. These factors will be discussed with a view to providing the analyst with selected but not comprehensive practical information on evaluating and assuring accuracy when analyzing samples in the aquatic environment: fresh water, seawater, sediments, and aquatic biota.

10.3. CONTAMINATION CONTROL

10.3.1. Laboratory and Analyst

Laboratory designs for rigid contamination control are well known (Murphy, 1976; Tschopel et al., 1980; Moody, 1982). However, with few exceptions, atmospheric cadmium levels are low enough so that most analyses of fresh and salt water, biological materials, and soil samples for cadmium can be performed without use of a clean room supplied with HEPA-filtered air. Forty-eight percent of air samples from Canadian localities contained less than the detection limit of 1 ng/m^3 cadmium, while only 3% contained greater than 10 ng/m^3 (Air Pollution Control Directorate, Environment Canada, unpublished data, 1975). Atmospheric cadmium contamination in a laboratory has been reported to be about 2 ng/m^3 (Maienthal, 1970), and over an 18-h period, 5% nitric acid in open 250-mL Erlenmeyer flasks in perchloric acid hoods and in a circulating oven have yielded blank standard deviations of less than 0.2 ng for cadmium in the absence of a clean room (Dabeka, 1979).

Sources of cadmium contamination do exist in laboratories, however, and can often be traced to the analytical history of the laboratory (determination of high cadmium levels in ores powdered prior to analysis), tobacco containing more than 1 µg/g cadmium (Lewis et al., 1972) and the fingers of smokers, zinc products (antirust paints and galvanized materials), and plastics with cadmium pigments. The latter have presented particular problems for blood sampling (Nackowski et al., 1977) and instrumental pipetting (Stoeppler and Brandt, 1980; Sperling and Bahr, 1981). Cadmium contamination has been traced to the break-point for acid ampoules used for water sampling (Calabrese et al., 1979) and to sample homogenization (Bunker et al., 1982).

An important source of laboratory contamination is the inexperience of the analyst in trace and microtrace techniques of handling, washing, and manipulations. Often, the cadmium concentrations of sample and standard solutions used in a laboratory can vary by as much as seven orders of magnitude. Under such circumstances, cross-contamination can occur when automatic pipet washers are used, splashing from sinks occurs, and strict segregation of dilute and concentrated solutions within the laboratory is not maintained. A single drop of a concentrated cadmium standard solution on the outer lip of a flask or the stopcock of a separatory funnel can ruin an analysis. Furthermore, due to the possibility of adsorption of cadmium from inadequately acidified solutions (Singh, 1977), conventional washing techniques are not applicable and an initial wash in concentrated acid should be directly followed with a deionized water rinse.

10.3.2. Laboratory Ware

Most materials [Pyrex, Vycor, high-purity quartz, polymethyl pentene (TPX), polypropylene, conventional polyethylene, and Teflon-FEP] used for labware contribute no detectable cadmium contamination to samples (Dabeka et al., 1976); however, since zinc can be leached from linear polyethylene, its use should be avoided. Cleaning of these materials can be affected by immersion in concentrated acids followed directly by deionized water rinse. The porosity of some organic base materials can cause extensive leaching of metals from the plastic, and care should be exercised in not using plastic bottles that have contacted concentrated cadmium solutions in the past.

A time-saving approach to sample collection or storage is application of commercial sampling tubes that do not require cleaning. In this respect, at least two laboratories are using 15-mL polystyrene graduated centrifuge tubes with screw caps for collection and storage of water and other samples (G. C. Gillette-Welling, personal communication, 1983; Dabeka and McKenzie, 1986). The tubes (Falcon No. 2095) yield a cadmium blank standard deviation of less than 0.05 ng.

Commercial lint-free cellulose wipes are useful for wiping labware and covering it during drying or storage. Provided no acids are contacted with the wipes, no detectable cadmium contamination results at the intermediate picogram level (Dabeka and McKenzie, 1986).

10.3.3. Reagents

The Analytical Chemistry Section, Division of Chemistry, National Research Council of Canada, has obtained high-purity water with about 0.0002 ng/mL cadmium (Dabeka et al., 1976) by passing building distilled water through an adsorber (Illinois Water Treatment Company, IWT) and 2 IWT

Research Model demineralizers. The current cost of the system, including mountings, is less than $500 (Canadian).

Regarding inorganic acids, reported levels of cadmium in reagent-grade acids are sufficiently low to enable determination of trace levels of cadmium in most samples. For microtrace levels, subboiling distillation is probably the most inexpensive and reliable method to obtain acids with cadmium levels less than 0.01 ng/mL (Murphy, 1976; Dabeka et al., 1976).

Purification of salt solutions, such as ammonium dihydrogen phosphate, citric acid, and buffers, which are commonly used for cadmium determinations, can be effected by extracting the solution with chloroform or methylisobutyl ketone in the presence of a chelating agent. Dithizone can be used for extraction from basic media, but it is ineffective for acidic solutions. Ammonium pyrrolidine carbodithioate is an effective chelating agent over a wider pH range (1.5–9); however, it, as well as some other carbodithioates, are water soluble in neutral and basic solutions and decompose with time. The analyst must ascertain that the decomposition product or residues of the chelating agent or organic phase do not affect the analytical method to which the solution is being applied. Aqueous solutions of chelating agents can usually be purified by filtration through 0.3-μm cellulose acetate membrane filters. Cadmium can be removed from water-immiscicle organic solvents by storing over 0.5–1.0 M nitric or hydrochloric acids. Concentrated ammonia solutions containing less than 0.02 ng/mL cadmium can be prepared by saturating cooled deionized water with commercial-grade ammonia gas passed through a 0.3-micrometer membrane filter.

10.3.4. Dealing with Blanks

For natural level cadmium determinations, blanks will almost always be positive. The most common errors related to blanks are variability and validity. Analysts dealing with positive blanks should always run multiple blanks in order to assess detection limits and accuracy for low-level determinations. Unfortunately, most automated analytical instruments (atomic absorption and plasma emission spectrometry) are not equipped to handle multiple blanks, and hence, their use is passively discouraged because of the need to recalculate results manually.

Regarding blank validity, blank measurements are usually subtracted from those of the samples. As long as the sample signals are large compared with those of the blanks, few errors can be expected for this approach. For sample signals of the same order of magnitude as the blank, the validity of this approach is based on the assumption that blank and sample recoveries are identical. Often, particularly if the preparation steps for samples and sample or standard blanks differ, this assumption is not valid (and is rarely validated when the method is developed).

As a subtle example of this, some elements form glasses with pyrex or quartz at moderate ashing temperatures. When dry ashing methods are used for decomposing organic matter, the blanks for vessels containing no sample will differ from those with sample because in the first instance the majority of the elemental blank will be in contact with the walls of the vessels, whereas in the second case, the bulk of the sample matrix will keep the blank off the walls. Errors of this kind can often be identified by checking recovery of pure standards (in the absence of sample) added to the vessel before the ashing step.

10.3.5. Monitoring and Controlling Contamination

The most systematic and efficient way to define sources of contamination and improve technique on contamination control is to run multiple blanks at various stages of the methodology, beginning with instrumental determination, proceeding to washing techniques and laboratory sources, and ending with stepwise reverse progression through the methodology until the total method, including the sampling step, is encompassed. The approach is summarized for a solvent extraction graphite furnace atomic absorption method (Appendix).

10.4. PROJECT DESIGN AND ANALYTICAL INPUT

For moderate or large projects requiring chemical analyses (e.g., monitoring surveys), the analytical chemist is sometimes the last professional to be consulted and, in some instances, first learns about the project after the samples have been collected and stored for analysis. This approach can result in inappropriate sample collection, contamination of samples, unavailability of adequate analytical methods, poor quality assurance, and a complete waste of resources put into collection and analysis of the samples. It is vital that an experienced analyst be consulted when such projects are first designed because the goals of such projects will directly govern the approach taken for sampling and analysis.

The role of the analyst lies in defining the accuracy, precision, and sensitivity of the analytical method; the techniques necessary to assure that no contamination or losses occur during collection and processing of samples; and the quality control measures that will be necessary throughout the survey steps to assure maximum accuracy of results. The analyst should also be prepared to give accurate estimates on the concentrations of element expected in the samples and their expected homogeneity based on sample size and method and location of collection. A small preliminary study should be considered a necessary prerequisite to any major study. A statistician acquainted with chemical analysis and the goals of the study should be consulted to approve the final survey design.

10.5. SAMPLE HANDLING

10.5.1. Sampling

One of the most complex issues facing analysts is proper sampling procedures. The reasons for any analyses will directly govern the approach taken for sampling.

In the case of natural waters, cadmium can exist in the ionic form, or may bind to either soluble organics or insoluble particulates. Analysis of the whole sample can usually be effected by acidification before or after collection; however, problems arise if a distinction between soluble and insoluble forms of cadmium is required. For instance, if the sample is acidified prior to filtration, cadmium will be leached from the particulate matter, and the extent of leaching will depend on the nature of the insoluble particulates, the storage time, and the acidity. If acidication is not done prior to filtration, cadmium can adsorb on the walls of storage vessels (Massee et al., 1981; Tschopel et al., 1980) and on filter paper during filtration; furthermore, the adsorption can be expected to differ for ionic cadmium and for that bound to soluble organics. As a result of the above, there is no absolute way to segregate insoluble and soluble forms of cadmium for surveys in which the nature of soluble organics and insoluble particulates will vary from sample to sample. Also, interpretation of results from any attempt to do so will be ambiguous. Several publications demonstrate good sampling and procedural techniques for natural waters (Patterson and Settle, 1975; Spencer et al., 1982; Ihnat, 1978, 1981, 1982a; Berman et al., 1983; Tschopel et al., 1980).

In the case of biological materials, cross-contamination is the major sampling problem. Since cadmium levels in different tissues or organs of biological materials can vary over two to three orders of magnitude or more, cross-contamination during dissection can significantly affect the results obtained, and even in cases when the whole sample is analyzed, the approach taken in washing, sealing, or skinning is of extreme importance. An illustration of excellent sample procedures for biological samples has been published using lead as an example (Patterson and Settle, 1975).

10.5.2. Sample Decomposition

Instrumental, separation, and concentration techniques often require that the cadmium in environmental materials be converted to solutions of the chemically labile form. For water, this is most simply effected by acidification and/or addition of complexing agents.

For biological materials, dry ashing or wet digestion methods, the latter usually requiring strong oxidants (Gorsuch, 1970; Analytical Methods Committee, 1960) must be used. The key elements affecting choice between the two are that wet digestions demand significant analyst time and, in the presence of perchloric acid, can be dangerous; however, cadmium recoveries are generally quantitative (Gorsuch, 1970). Dry ashing methods are less time consuming and, as a result, have gained wide usage and have even been collaboratively studied (Capar et al., 1982). Cadmium is, however, a relatively volatile element, and cadmium losses can occur from the surface of a graphite tube at temperatures as low as 300 °C. Thus, the success of dry ashing methods depends to a great degree on the method of ashing and the specific matrix ashed. Cadmium losses from fish (Feinberg and Ducauze, 1980; Slabyj et al., 1983) and from seaweed (van Raaphorst et al., 1978) have been reported. Also, the validity of dry ashing methods reported in the literature will to some degree depend on whether the spiked cadmium standard simulates the cadmium indigenous to the sample both in chemical form and physical distribution. Considering the foregoing as well as the additional question of blank validity discussed in Section 10.3.4, dry ashing methods should never be used when high accuracy is required.

For wet digestions, an accuracy consideration is the ability of the methodology to handle the relatively high and often variable concentrations of acids remaining after the digestion. Variable acid concentrations, even after neutralization and buffering, will affect the ionic strength of the resulting solution, influencing the accuracy of electrochemical determination. Also, variable acid concentrations will affect the accuracy of EAAS determinations, and some ions such as sulfate and phosphate contribute to nonspecific background during atomization.

Recent advancements in wet digestions include safety improvements when digesting with nitric and perchloric acids (Dabeka, 1979), pressure dissolution using hydrochloric (Kuennen et al., 1982) or nitric acid (Williams, 1978), and, respectively, polyethylene and polystyrene tubes and UV irradiation of digests (Lobel, 1978). Fish samples of up to 5 g have been routinely digested with nitric and sulfuric acids using test tubes in an aluminum hot block (Agemian et al., 1980), and a similar system has been used to compare dry ashing and wet digestion methods for soils (Ritter et al., 1978).

10.5.3. Sample Storage

Although sample collection and storage often occur outside of the analytical laboratory and are viewed as adjunct operations to chemical analysis, they constitute integral components of the overall scheme of analysis and must be properly carried out to ensure reliable and representative data. Storage of solid samples for cadmium determination presents no unusual problems. Dried or freeze-dried biological and sediment materials can be kept at room temperature in tightly covered glass containers. Wet materials must be stored frozen to preserve sample integrity.

The storage of fresh natural water and seawater samples presents considerably more difficulty with the need to pay attention to parameters such as pH, type of storage container and its cleaning, and particulate matter content of the solution. Although solutions containing the element at hundreds or thousands of micrograms per milliliter do not present particular storage difficulties, the ppb to ppt levels of endogenous cadmium in natural waters can easily suffer positive or negative alterations during storage. An excellent treatment of the trace (less than 1 µg/mL) chemistry of aqueous solutions has been presented by Benes and Majer (1980), and a review of contamination problems in trace element analysis has been published by Robertson (1972). The stability of model aqueous solutions of cadmium in various types of containers and as a function of pH has been reported (Struempler, 1973; Smith, 1973; King et al., 1974). The behavior of the element in actual freshwater (Rattonetti, 1974) and seawater (Berman et al., 1983) matrices has been studied. Struempler (1973) observed model solutions containing cadmium at 1 µg/L to be stable at pH 2 in borosilicate glass containers or at pH 2 and 6 in polyethylene containers for about 20 days. Shorter term (1–2-day) studies (Smith 1973; King et al., 1974) indicated stability at high concentrations (up to 1000 µg/L) in plastic containers at pH 3–10 and borosilicate glass at pH 1–3. Long term ($\geq$1 yr) studies by Berman et al. (1983) with seawater acidified with nitric acid to pH 1.6 and stored at room temperature in cleaned linear polyethylene bottles showed excellent stability of native cadmium at the 0.03 µg/L level. It would appear that acidification of natural water samples to about pH 1.5 immediately upon collection and storage in acid-cleaned polyethylene containers (Moody and Lindstrom, 1977) is a satisfactory approach for analysis and storage of cadmium-containing water samples. For measurement of other than total cadmium pretreatment of the sample such as filtration would precede acidification.

10.6. CONCENTRATION TECHNIQUES

With the exception of sediments and contaminated waters, the cadmium levels in most aquatic environmental samples are sufficiently low, or the matrices sufficiently complex, to warrant, depending on the sensitivity and specificity of the instrumentation, consideration of concentration and separation techniques, such as solvent extraction, coprecipitation, and ion exchange (Leyden and Wegscheider, 1981).

Evaporation is the simplest of concentration methods; however, it is only applicable to relatively pure aqueous samples since simultaneous concentration of the accompanying matrix often leads to matrix interferences during instrumental determination. Two important factors affecting accuracy during evaporation techniques are (1) assuring that the solution is acidified prior to evaporation and (2) avoiding evaporation to dryness. If these factors are not considered, adsorption of cadmium onto the walls of the container are possible, and for some materials, the adsorption is irreversible. Since standard solutions are usually acidified, recovery studies may not reveal that a problem exists.

Accuracy considerations for solvent extraction are volume effects, colloid formation, and quantitation. Volume effects are caused by the mutual solubility of the two phases, a function of ionic strength, pH, and solvent affinity of the ions in solution. As a result, the solubility of the organic phase in standard and sample aqueous solutions will differ, often by as much as 1%. While a 1% change in solubility is minor if equal volumes of the two phases are used, decreasing the volume of the organic phase in order to concentrate the cadmium can result in serious errors. Colloid formation in the organic phase can be sample dependent and can also affect the volume. Quantitative extraction, simply defined as a distribution ratio of 10^2 or more, is important from the viewpoint that in many situations of substoichiometric extraction, the method will be insufficiently rugged (Youden and Steiner, 1975) with respect to changes of sample matrix, analyst technique, pH, and reagent concentrations. In addition to the above considerations, the accuracy and usefulness of solvent extraction methods can be judged on details such as extract and chelating agent stability, extraction and, if applicable, stripping times, and loading factors, particularly in the presence of copper or iron, which may be abundant in environmental samples and also may bind with the chelating agent. The stability of cadmium in the organic phase is often variable, and the organic phase cannot be reliably analyzed by some instrumental techniques. Back extraction (stripping) into a low-pH aqueous phase overcomes these problems and is gaining increasing acceptance among analysts.

For ion exchange concentration techniques, accuracy and applicability factors are loading capacity, long-term resin stability (both shelf-life and under repeated regeneration), and assurance that the elution or regeneration step is also removing more strongly binding ions, such as copper, the accumulation of which with time reduces the loading capacity. Ion exchange methods are particularly amenable to freshwater and salt water analyses, where a large concentration factor can be effected rapidly and conveniently. Since ion exchange methods usually require at least 5–30 mL elution solution to quantitatively remove the cadmium from the column, they are not amenable to low-level cadmium determinations in soils and biological materials, the digestion of which is a limiting factor on sample size. This limitation can be overcome by direct analyses of the resin using NAA (Greenberg and Kingston, 1983) or X-ray fluorescence spectrometry (Clanet et al., 1981). As in other methods of concentration, elution of sample or resin constituents may interfere with instrument application, as in the case of Chelex-100 using EAAS (Danielsson et al., 1982).

Good coprecipitation methods should be fully quantitative, enabling application of synthetic standards rather than precipitated ones, the latter being acceptable only if the instrumental technique demands it. Assurance should be made that other coprecipitating elements, including the reagent or its decomposition products, do not cause instrumental interferences.

The success of quantitative precipitation often depends on details in the experimental stage, including ionic strength, final volume of solution, time factors in reagent addition, manipulations and filtration, and order of reagent addition. Evaluation of published methods should take all the above factors into consideration. Since the solubility of the precipitate may depend on the presence of excess coprecipitant, care must be taken to determine and specify whether or not the precipitation flask or the filtered precipitate itself should be washed. Also, where chelating agents are used as precipitants, the nature of samples analyzed should be scrutinized to avoid situations where the amount of other metals that bind with the chelating agent will reduce the concentration of chelating agent available for the element of interest to ineffectual levels.

10.7. DETERMINATION OF CADMIUM

10.7.1. Flame Atomic Absorption Spectrometry

In-depth treatments of the principles and applications of FAAS are available in excellent comprehensive treatises edited by Mavrodineanu (1970) and Dean and Rains (1969, 1971, 1975) and written by Kirkbright and Sargent (1974). AAS attributes of relative simplicity of operation, high element specificity and good detection limits, adaptability to virtually all of the metallic elements and low to moderate cost of basic instrumentation has led to widespread usage of this technique of analysis. In a typical analysis of biological materials with 0.5 µg Cd/g, several grams of the material is digested with nitric and perchloric acids, brought to a suitable small volume, and aspirated into a flame. With suitably careful experimental technique, very good precision and accuracy can be realized. With sediments and soil samples, decomposition should include hydrofluoric acid to dissolve siliceous material in order to measure total cadmium (Ihnat, 1981, 1982b).

More often than not, the low natural concentrations of cadmium in environmental materials will necessitate incorporation of some technique of preconcentration prior to measurement by FAAS. This can be achieved by ion exchange or solvent extraction (SEFAAS) or chelate coprecipitation to effect typically 100-fold concentration. An added advantage to these sample pretreatment steps is separation of the sought-for element from the remainder of the bulk matrix, substantially reducing interelement interferences in the subsequent FAAS step (Chakrabarti et al., 1978; Ihnat, 1982c, and references therein). One approved procedure (Hanson, 1973; Analytical Methods Committee 1975) uses an Amberlite LA-2/MIBK (methylisobutyl ketone) mixture to extract the cadmium in the form of the iodocadmate ion from the sulfuric acid–hydrogen peroxide digest of the biological sample. The most popular chelation–solvent extraction combinations for cadmium detection are NDDC (sodium diethyldithiocarbamate)/MIBK and APDC (ammonium pyrrolidine carbodithioate)/MIBK, with the latter combination being the most popular (McClellan, 1975; Horwitz, 1980). Application of preconcentration techniques to the challenging analysis of seawater have been reviewed (Riley, 1975). Most often, these are used in combination with electrothermal atomic spectrometry to reliably detect exceeding low levels of cadmium in matrix-rich seawater.

10.7.2. Graphite Furnace Atomic Absorption Spectrometry (GFAAS)

GFAAS, a popular form of EAAS, is probably one of the most sensitive, convenient, and rapid techniques currently available for the determination of trace and microtrace levels of cadmium. Solution detection limits or less than 0.05 ng/mL are common, and instrumental precision is usually 0.5–5% relative standard deviation (RSD). Unfortunately, GFAAS is susceptible to many chemical and physical interferences which, to a great degree, depend on the model of commercial atomizer; the surface structure of the tubes (which change with use); the drying, ashing, and atomization parameters; and even on the time constant of the spectrometer used. Also, unlike other instrumental techniques, new designs, innovations, and modifications continue to appear frequently. As a result, most GFAAS methods published to date are not easily transferable to other instruments and, occasionally, not even to other laboratories with the same instrumentation. The only exception to this is for the analysis of water or standard solutions, for which good (15–30% RSD) interlaboratory precision has been achieved for solutions containing as little as 0.2 ng/mL cadmium (Lamathe et al., 1982; Sperling and Bahr, 1981).

Factors significantly affecting accuracy are drying and ashing temperatures, aliquot volume, matrix effects, background correction, measurement mode, and furnace design. Drying temperature is the most unreliable parameter in conventional furnaces and is a function of cooling water temperature, electrode contact stability, temperature gradient along the tube length, and tube age and surface. Depending on analyte composition and aliquot volume, inappropriate choice of drying temperature can cause running of analyte within the tube during the drying step, resulting in sensitivity changes exceeding 300% for the same cadmium solution during an analytical run (R. W. Dabeka, unpublished data, 1984). The only way of overcoming this effect is to choose an aliquot volume, the sensitivity of which will be independent of drying temperature over a wide range (110–180 °C) and to avoid using solutions containing more than 1–2% nitric or any other acid. For direct pipetting onto the walls of tube-type furnaces, the aliquot volume (in microliters should not exceed $0.06 \times l \times d$, where l and d are, respectively, the length and diameter of the tube in millimeters. The above volume can be increased 2–3 times for tubes with wells (not grooved edges) and, depending on the platform dimensions, 2–6 times, for platforms within tubes. It should be noted that the volumes calculated using the above formula are significantly less than those usually recommended by the manufacturers of the instrumentation.

For cadmium, choice of appropriate ashing temperature can significantly improve accuracy. Cadmium losses during the ashing step strongly depend on the nature of the analyte matrix and can occur at ashing temperatures ranging from 300 to 1000 °C. Irrespective of matrix, the ashing temperature chosen should be roughly 50–100 degrees lower than that at which cadmium losses first appear for standards or samples, so that changes in tube temperature with tube age will not affect accuracy.

With the exception of pure water, the constituents of most environmental samples can cause chemical interferences for cadmium GFAAS determinations (Belayev et al., 1980; Matousek, 1981; Slavin, 1984; Slavin and Manning, 1982). Since the effects vary with furnace design, temperature program, tube surface, and application of platforms or other modifications, all furnace results reported for environmental samples should include recovery studies on the "dirtiest" samples to evaluate the extent of interferences, particularly when separation steps are avoided. For cadmium, as well as for some other elements, perchloric acid should be avoided (Koirtyohann et al., 1981) because it suppresses cadmium absorbance and rapidly degrades tube performance. Chloroform residues can occasionally cause background correction problems (Sperling, 1982). Sample and standard matrices should be well matched with respect to acidity and organic and ionic composition to assure best results.

Addition of matrix modifiers to analyte solutions is commonly used to provide a consistent physical medium from which samples are atomized and to either increase or decrease the volatility of the element being determined. Ammonium phosphates (0.01–0.2%) decreased cadmium volatility, enabling application of higher ashing temperatures (600–900 °C) to remove more easily vaporized organic materials prior to atomization. EDTA (0.1%), on the other hand, has been used to increase cadmium volatility, enabling atomization at a temperature of 1000 °C, which, in the case of seawater, leaves many of the less volatile salts on the tube surface, effectively reducing interelemental interferences and nonspecific background (Guevremont et al., 1980).

Probably the most common error made in GFAAS is the assumption that spectrometers equipped with simultaneous background correctors are adequately correcting for nonspecific background, generated by complex matrices during the atomization step. The reason for this situation is the failure of less experienced analysts to realize that neither standard addition techniques nor recovery studies, used frequently to validate results, will reveal the presence of uncorrected background. A recent collaborative study of GFAAS method for lead in milk revealed that most laboratories using conventional deuterium-continuum simultaneous background correction were unable to correct for a nonspecific background of about 0.2 absorbance units (R. W. Dabeka, 1984), significantly lower than the claims usually made by spectrometer manufacturers. Since the nonspecific absorption at the cadmium wavelength is usually greater for the same matrix than that at the 283.3-nm lead line, it would appear that the only widely applicable and reliable analytical methods are those involving separation of cadmium from the bulk of any complex sample matrix. Indeed, the only instance when cadmium determinations in biological and seawater matrices should be attempted by any but the most experienced of analysts is in the presence of more effective background correction techniques such as Zeeman or the Smith–Hieftje system (Sotera and Kahn, 1982). Changes by instrument manufacturers to the background correctors of spectrometers made in 1985 and 1986 have significantly improved their capabilities. Irrespective of the method of background correction used, the only definitive way of assuring adequate background correction is to use a completely independent method of analysis.

A simple practical relationship exists between peak-height and peak-area measurements using GFAAS. If peak-height measurements give good results when the method is tested, it is usually an indication of a good method, and in all likelihood, peak-area measurements will also yield accurate results. If only peak-area measurements give good results, then the applicability of the method will be limited to those instruments capable of measuring area, and interelemental or physical interferences can be expected for some models of atomizers, irrespective of the measurement mode. Peak-area measurements are less precise for solutions containing low levels of analyte (Sperling, 1983).

For cadmium determinations, the major advance in GFAAS has been atomization from the L'vov platform (Ottaway, 1982), which attempts to achieve atomization in an isothermal environment, reducing both chemical and physical interferences. The low atomization temperatures necessary for atomization of cadmium enables application of platforms to most commercial tube-type atomizers; however, optimal benefits are derived when rapid heating rates coupled with low atomization temperatures (1400–1800 °C) are used. The disadvantages of platforms are that the longer cooling time required for platforms increases the time between consecutive atomizations by 20–40 s, and the lower atomization temperatures used result in carryover effects caused by deposition of cadmium on cooler areas of the tube or, for high-salt matrices, interferences caused by accumulation of residues within the tube. Both effects can be eliminated by adding a high-temperature cleaning step to the furnace program. Regardless of the atomizer used, application of platforms will generally yield more accurate cadmium results.

10.7.3. Electroanalytical Methods

Wide variations of elecroanalytical methods exist and have been covered in recent publications (Batley, 1983; Brihaye and Duyckaerts, 1983; Fleet, 1980; Ryan and Wilson, 1982). For cadmium, electroanalytical methods are convenient and sensitive, attaining solution detection limits, depending on the technique, as low as 0.5 pg/mL (Mart et al., 1980). Their advantage over furnace methods is the comparative lack of interference by environmental macroelements, such as sodium, calcium, and magnesium, which severely interfere with GFAAS measurements; thus, there is little need for separation techniques. Furthermore, stripping methods, by their very application, serve as excellent concentrators of cadmium from bulk solutions. The disadvantages of electroanalytical methods, particularly for the low natural levels of cadmium in environmental samples, are poorer specificity and precision, slow analytical throughout, and the need for high technical expertise. The latter is particularly evident from the results of a collaborative study of an anodic stripping voltammetric method for the determination of cadmium in foods (Capar et al., 1982). Although reasonable accuracy and interlaboratory precision (17% RSD) were obtained for fish at the 50-ng/g level, the detection limit of the method, 5–10 ng/g, was insufficient for background-level measurements of all the tested samples, and only spiked samples were sent to collaborators. Based on sample size, aliquot volume, and solution detection limits for the instrumentation, the detection limit for the method should have been an order of magnitude lower.

The main accuracy considerations for electroanalytical methods are interferences by organic materials and electrochemically active metals and the analytical skills needed in maintaining reproducible results on a long-term basis, a necessity for routine analytical work. Some electroanalytical methods have been applied to cadmium speciation studies and the validity of some of this research is questionable and deserves special attention.

Inorganic interferences reported for electrochemical cadmium determinations are Sn(II) (Gajan et al., 1982) and Cu(II) (Schulze and Frenzel, 1983). The nature of the tin interference is postulated as formation of a Sn(II) tartrate complex that is oxidized at the same potential as cadmium. For copper, the interference is due to intermetallic compound formation between copper and cadmium in the electrode mercury phase, and such interferences are common when thin-film mercury electrodes, which generally increase analytical speed and sensitivity, are used.

Organic interferences, usually caused by organic material common to almost all environmental samples, can be of two types. First, organic compounds can complex with cadmium, reducing the concentration of electrochemically active free cadmium ions. The effect can be reduced by acidification of the analyte solution, UV irradiation at elevated temperatures, and addition of mercuric ion to the solution to preferentially bind with the organic material and exchange with bound cadmium (Mart et al., 1980; Schonberger and Pickering, 1980). The mercuric ions thus added may also be used for *in situ* coating of thin-film electrodes for stripping analysis (Jagner et al., 1981; Mart et al., 1980). The second interference caused by organics is variations of peak heights and potentials caused by adsorption of organics on the electrode and participation by the organics in electrochemical processes. The effects vary with structure of the organics, chemical environment, and type of electrode used (Brihaye and Duyckaerts, 1983). Application of standard addition methods will only correct for the interference if the organics do not have adsorption peaks or waves and if their action is independent of metal ion concentration. Application of a rotating mercury film ring-disk electrode with linear anodic stripping voltammetry was found to be superior to a disk electrode for separating organic adsorption peaks from those of the metal ion (Brihaye and Duyckaerts, 1983); computerized potentiometric stripping was found to be superior to anodic stripping for reducing interferences caused by organics (Jagner et al., 1981). Also, for stripping processes, complexing agents added just prior to stripping can be used to shift cadmium peaks.

The above discussion on organic interferences has implications with respect to digestion methods for biological samples and aqueous samples containing organics, and complete destruction of organic matter is strongly recommended for accurate analytical results. Conversely, results obtained using partial digestions (including more rigorous treatments such as Teflon pressure bombs) should be viewed with caution, particularly if the purpose of the analysis is to be obtain a total cadmium concentration.

Analyst technique in electrochemical methods is an important aspect in assuring accuracy and improving detection limits. This is illustrated by specific examples of instrument modification, electrode preparation, care and storage, and contamination control (Batley, 1983; Green et al., 1981; Jagner and Aren, 1979; Jagner et al., 1981; Mart, 1979, 1982; Mart et al., 1980; Poldoski and Glass, 1978).

Because electroanalytical methods measure cadmium ions in solution as opposed to the total cadmium concentration, they have found widespread use in speciation studies. Unfortunately, some investigators have not paid due attention to the characteristics of the electroanalytical technique chosen for such studies. While polarographic methods measure equilibrium concentrations of free metal ions, anodic stripping voltammetry is not generally suitable to such studies. In crude terms, anodic stripping voltammetry is not an equilibrium technique, and during the deposition step, free cadmium ions are removed from the environment studied. If original complexes dissociate as the deposition step progresses, the errors occurring from applying the technique in this situation include not only deposition of more than the free ionic metal present in the original solution but also uncertainty in the stripping step due to purturbation of the stripping curve by uncomplexed ligands present in the solution (Schonberger and Pickering, 1980).

10.7.4. Inductively Coupled Plasma Atomic Emission Spectrometry

In inductively coupled plasma atomic emission spectrometry (ICPAES) (Fassel, 1979), the liquid sample is vaporized and atomized in a high-temperature argon plasma. The plasma is further relied upon to excite ground-state atoms (and ions) to higher energy levels from which radiation is emitted as the atoms revert to the ground state. The measurement of emitted radiation is central to ICPAES. This measurement is proportional to excited-state atomic concentration, and thus to the total number of atoms in the region of observation and in turn to the concentration of the element in the sample. As for FAAS, solid materials must be brought into solution prior to aspiration into the plasma. One distinct advantage of this technique is its ability to simultaneously measure a large number of elements and not only the usual metallic elements but important nonmetals such as boron, and phosphorus as well, permitting 400 determinations per hour. Chemical interferences in the hot plasma are minimal compared to FAAS, but the multitude of emission lines excited give rise to moderate spectral interferences. ICPAES is capable of rapid, precise, and reliable analyses. The detection limit for cadmium by ICPAES is similar to that for FAAS, suggesting that the technique may be applied to samples containing the element at the microgram-per-gram level. Indications from a recent report by McLaren et al. (1981) are that due to spectral interferences from arsenic and iron and the insufficient detectivity of the technique for cadmium in marine sediments, preconcentration of the element or its separation from the matrix might be necessary for reliable application of the technique to such materials.

10.7.5. Neutron Activation Analysis

Neutron activation analysis (NAA) entails irradiation of the sample with neutrons resulting in the production of radioactive isotopes whose concentrations are measured by monitoring γ or β radiation of specific energies (Guinn, 1971; Parsons, 1976). The instrumental version in conjunction with γ-ray spectrometry (INAA) involves no chemical operations and is one of the very few techniques not requiring sample decompositon, a decided advantage. INAA is a multielement technique capable of providing concentration data for a number of elements in one irradiation and counting operation. To realize good limits of detection, however, access to a nuclear reactor is required to produce the high flux of thermal neutrons required. Compared to impressive detection limits for several of the rare-earth elements, the detection limit for cadmium is a fair 40 ng, corresponding to 40 ng/g for a 1-g sample, permitting measurement of the element in aquatic materials containing ppm levels. Although excellent analyses are possible with INAA, performance depends on the element and matrix. A variety of sources of random and systematic errors, including irradiation, counting, and interfering reactions, leads to precision and systematic errors of 2–20%.

10.7.6. Spark Source Mass Spectrometry

For spark source mass spectrometry (SSMS), a sample in solid form is mixed with powdered graphite to form conducting electrodes. A high-energy source, commonly a pulsed RF spark, imparts energy to the sample to volatilize and ionize it with the shower of ions being directed to a mass spectrometer for analysis using photographic or ion multiplier detection. A freshwater sample of suitable volume can be concentrated by freeze drying to yield a residue which, after ashing to remove organic components, is incorporated into the electrodes (Wahlgren et al., 1971). Alternatively, an aqueous sample without residue can be concentrated by evaporation and a drop of the concentrate placed on a silver or graphite electrode for analysis. Water samples have also been mixed directly with the graphite and analysis performed after evaporation. Biological samples such as lyophilized liver have been molded with graphite into electrodes. SSMS is a multielement technique with quite an impressive absolute limit of detection for cadmium of 0.3 ng corresponding to perhaps 0.001–0.1 µg/g, suitable for some, but not all, endogenous levels of cadmium. The technique, however, does not offer the good precision and reliability of the other methods of analysis, requires a good deal of effort on the part of trained personnel, and is not suited for routine processing of large numbers of samples.

10.7.7. Inductively Coupled Plasma Mass Spectrometry (ICPMS)

ICPMS is one of the most sensitive multielement methods of analysis available, and solution detection limits as low as 70 pg/mL have been given for cadmium by instrument manufacturers. In natural water, however, Longerich and Kantipuly (1985) reported a detection limit of 0.23 ng/mL, which is about 5 times greater than the detection limit for GFAAS. Boorn et al. (1985) reported a solution detection limit of 0.08 ng/mL for biological samples pressure digested in nitric acid. Cadmium mass interferences are unlikely to be encountered in any water or biological samples; however, potential interferences for $^{106}Cd^+$ and $^{108}Cd^+$ in marine sediment solutions are zirconium oxides (McLaren et al., 1985). The number of natural cadmium isotopes, while reducing the detection limit, make appropriate choice of cadmium isotope as well as isotope dilution determinations quite easy.

10.8. RELIABILITY OF LITERATURE DATA AND REFERENCE MATERIALS

It is generally difficult to assess the validity of the analytical data published in the literature. Occasionally, details are lacking with respect to preanalysis considerations such as description of the material, location, and method of collection. Furthermore, descriptions of sample storage and treatment and method of analysis are sometimes incompletely recorded. With respect to the actual measurement of cadmium, lack of a laboratory data quality assurance program can lead to analytical information of unknown or poor quality, as attested to by a number of recent interlaboratory comparisons mentioned in an earlier section. The reliable measurement of cadmium in environmental materials is a challenge to the analyst.

An important facet of analytical data quality assurance is the incorporation into the scheme of analysis of a reference material certified with respect to the concentration of the element of interest and compositionally similar to the materials under analysis (Seward, 1975; Uriano and Cali, 1977). Such an approach constitutes an important mechanism of transfer of accuracy from definitive methods of analysis to the methods employed in the laboratory on a daily basis. A list is given in Table 3 of some of the currently available quality control reference materials pertinent to the measurement of cad-

Table 3. Analytical Quality Control Reference Materials Pertinent to Measurement of Cadmium in Aquatic Matrices

Material and Source[a]	Cadmium Concentration[b]
Lobster hepatopancreas TORT-1 (NRCC)	26.3 ± 2.1 µg/g
River sediment SRM 1645 (NBS)	10.2 ± 1.5 µg/g
Oyster tissue SRM 1566 (NBS)	3.5 ± 0.4 µg/g
Tomato leaves SRM 1573 (NBS)	(3 µg/g)
Aquatic plant (*Lagarosiphon major*) CRM 60 (BCR)	2.20 ± 0.10 µg/g
Spinach SRM 1570 (NBS)	(1.5 µg/g)
Aquatic plant (*Platihypnidium riparioides*) CRM 61 (BCR)	1.07 ± 0.08 µg/g
Marine sediment reference material MESS-1 (NRCC)	0.59 ± 0.10 µg/g
Bovine liver SRM 1577a (NBS)	0.44 ± 0.06 µg/g
Estuarine sediment SRM 1646 (NBS)	0.36 ± 0.07 µg/g
Marine sediment reference material BCSS-1 (NRCC)	0.25 ± 0.04 µg/g
Soil sample	
SO-1 (CCRMP)	(0.15 µg/g)
SO-2 (CCRMP)	(0.18 µg/g)
SO-3 (CCRMP)	(0.14 µg/g)
SO-4 (CCRMP)	(0.42 µg/g)
Orchard leaves SRM 1571 (NBS)	0.11 ± 0.01 µg/g
Olive leaves (*Olea europaea*) CRM 62 (BCR)	0.10 ± 0.02 µg/g
Citrus leaves SRM 1572 (NBS)	0.03 ± 0.01 µg/g
Trace elements in water SRM 1643a (NBS)	10 ± 1 µg/kg
Seawater reference material NASS-1 (NRCC)	0.029 ± 0.004 µg/L

[a] References are National Bureau of Standards Certificates of Analyses; Berman et al. (1983); McLaren et al. (1981); Bowman et al. (1979); and Community Bureau of Reference (1982). Sources of supply are Office of Standard Reference Materials, National Bureau of Standards, Gaithersburg, MD 20899 (NBS); Marine Analytical Chemistry Standards Program, Chemistry Division, National Research Council, Ottawa, Ontario Canada K1A OR9, (NRCC); Canadian Certified Reference Materials Project, Canada Centre for Mineral and Energy Technology, Energy Mines and Resources Canada, Ottawa, Ontario Canada K1A OG1, (CCRMP); Community Bureau of Reference BCR, Directorate General XII, Commission of the European Communities, 200 rue de la Loi, B-1049 Brussels, Belgium (BCR).

[b] Uncertainties are generally 95% confidence limits; values in parentheses are uncertified.

mium in aquatic materials. Available materials cover matrices such as plant and animal tissues, sediments, soils, water, and seawater with levels of cadmium ranging from 10 µg/g down to 0.03 µg/L. Perhaps the most challenging of these materials has been the preparation and certification of the seawater reference material, NASS-1 by NRC (Berman et al., 1983), providing a reliable reference for oceanographic trace analysis. A number of aquatic matrices (e.g., fresh natural water, fish, fresh water, and marine plants) are not represented in the current repertoire of available reference materials. If one accepts the premise that for adequate quality control, the entire composition of the reference material must approach that of the sample, the number and types of currently available materials is limited. Analytical laboratories should be encouraged to avail themselves of appropriate existing reference materials for data quality control and report results in publications.

10.9. ANALYTICAL RECOMMENDATIONS

Suitable methods for the determination of cadmium in most environmental matrices exist, but considerable care and expertise is required in their application. Areas deserving specific attention are outlined below:

1. The detection limits and precisions of methods should be improved in the case of the measurement of cadmium in natural waters and uncontaminated biological materials since current detection limits are only a small factor (less than 10) below natural environmental levels.
2. When developing or testing a method, greater emphasis should be placed on verification, using, in addition to recovery studies, certified standard reference materials and application of at least one other completely independent method of analysis to typical samples at typical concentration levels.
3. The need exists to thoroughly define the goals of research projects and integrate them with analytical requirements, beginning with the sampling procedures and ending with final calculations. The cooperation of professional analysts at all phases of the study is strongly recommended.
4. Good analytical quality control should be mandatory for all research, monitoring, and regulatory projects. It should include daily evaluation of replicate blanks and recoveries and analysis of laboratory or certified standard reference materials.
5. There is the need to assure proper inorganic analytical training of professional and technical staff. This can be done by allocating sufficient time for proper method development and testing, promoting conference and workshop attendance, and by facilitating an exchange of ideas via visits to and from other laboratories.

6. In order to update analysts with useful analytical methods, confirm accuracy and precision of methods, and detect problems within one's own laboratory, participation in national and international quality assurance programs, including both collaborative and check-sample studies, is strongly recommended.
7. There is a current shortage of certified standard reference materials for aquatic environmental samples containing cadmium and other elements at uncontaminated levels. To alleviate this situation, there is a need for organizations to financially support preparation and analysis of such materials by qualified personnel.

APPENDIX: AN EXAMPLE OF CONTAMINATION CONTROL DEVELOPMENT

A. Proper Washing Technique

One liter of a blank solution (0.5% nitric acid) containing about 0.05 ng/mL cadmium is prepared to yield an instrumental signal just above the detection limit. The solution is poured into 50 autosampler cups, originating from four cleaning batches, and cadmium is determined in each cup. If cadmium contamination is detected for even one of the cups, all the cups are immersed in a 1-ppm cadmium solution, rewashed, and the procedure repeated until cadmium contamination of the cups is completely controlled. Following this, about 1 mL of the test solution is taken up into each of 20 pipets of different volumes, swirled around within the pipet volume, transferred to 20 sampling cups, and the cadmium determined. Again, if any cadmium contamination is detected, the pipets are then soaked in cadmium solution washed with greater care, and tested again for contamination. The procedure is repeated for flasks, separatory funnels, sample bottles, and all other labware used routinely.

B. Laboratory, Reagent, and Method Contamination Control

To test laboratory air contamination, 5 mL of the test solution is pipetted into 10 open Erlenmeyer flasks or beakers, left in the fume hood or on the laboratory bench overnight, and analyzed the following day. To test for sporadic contamination from organic solvents, 5.00 mL test solution is shaken with 40 mL organic solvent, dispensed in the usual manner, in 10 separatory funnels, and the resulting aqueous phases are analyzed for cadmium variability. Contamination from other reagents is evaluated by varying the amount of each reagent added at different steps of the analytical method and remedying problems by either purifying the specific reagent or by increasing control on the amount of reagent added. Finally, the whole method is applied on two separate days to 10 blanks to assure complete contamination control.

REFERENCES

Agemian, H., Sturtevant, D. P., and Austen, K. D. (1980). Simultaneous acid extraction of six trace metals from fish tissue by hot-block digestion and determination by atomic absorption spectrometry. *Analyst,* **105,** 125–130.

Analytical Methods Committee. (1960). Methods of the destruction of organic matter. *Analyst,* **85,** 643–656.

Analytical Methods Committee. (1975). The determination of small amounts of cadmium in organic matter Part II. Determination of amounts down to the submicrogram level. *Analyst,* **100,** 761–763.

Batley, G. E. (1983). Electroanalytical techniques for the determination of heavy metals in seawater. *Mar. Chem.,* **12,** 107–117.

Belayev, U. I., Shcherbakov, V. I., and Karyakin, A. V. (1980). Studying the effects of macroconstituents on atomic absorption analysis with impulse electrothermal atomization on a graphite rod. *Zh. Anal. Khim.,* **35,** 2074–2079.

Bender, M. L., and Gagner, C. (1976). Dissolved copper, nickel and cadmium in the Sargasso Sea. *J. Mar. Res.,* **34,** 327–339.

Benes, P., and Majer, V. (1980). *Trace Chemistry of Aqueous Solutions, General Chemistry and Radiochemistry.* Elsevier, Amsterdam.

Berman, S. S., and Boyko, V. (1985). First intercomparison exercise for trace metals in marine biological tissues. NRC BT1/TM. National Reseach Council of Canada, Report II NRCC No. 25324, pp. 1–48.

Berman, S. S., Sturgeon, R. E., Desaulniers, J. A. H., and Mykytiuk, A. P. (1983). Preparation of the sea water reference material for trace metals, NASS-1. *Mar. Pollut. Bull.,* **14,** 69–73.

Bewers, J. M., Sundby, B., and Yeats, P. A. (1976). The distribution of trace metals in the western North Atlantic off Nova Scotia. *Geochim. Cosmochim. Acta,* **40,** 687–696.

Boorn, A., Fulford, J. E., and Wegschceider, W. (1985). Determination of trace elements in organic material by inductively coupled plasma-mass spectrometry. *Mikrochim. Acta,* **1985 II,** 171–178.

Boumans, P. W. J. M. (1980). *Line Coincidence Tables for Inductively Coupled Plasma Atomic Emission Spectrometry,* Vols. I and II. Pergamon, Toronto, Ontario.

Boumans, P. W. J. M., and Bosveld, M. (1979). A tentative listing of the sensitivities and detection limits of the most sensitive ICP lines as derived from the fitting of experimental data for an argon ICP to the intensities tabulated for the NBS copper arc. *Spectrochim. Acta,* **34A,** 59–72.

Boumans, P. W. J. M., and de Boer, F. J. (1972). Studies of flame and plasma torch emission for simultaneous multielement analysis—I. Preliminary investigations. *Spectrochim. Acta,* **27B,** 391–414.

Bowman, W. S., Faye, G. H., Sutarno, R., McKeague, J. A., and Kodama, H. (1979). Soil samples S0-1, S0-2, S0-3 and S0-4—Certified reference materials. CANMET Report 79-3, Canada Centre for Mineral and Energy Technology, Energy, Mines and Resources, Canada.

Brihaye, C., and Duyckaerts, G. (1983). Determination of traces of metals by anodic stripping voltammetry at a rotating glassy carbon ring-disc electrode. Part II. Comparison between linear anodic stripping voltammetry with ring collection and various other stripping techniques. *Anal. Chim. Acta,* **146,** 37–43.

Bruland, K. W., Knauer, G. A., and Martin, J. H. (1978). Cadmium in Northeast Pacific waters. *Limnol. Oceanogr.,* **23,** 618–625.

Bunker, V. W., Delves, H. T., and Fautley, R. F. (1982). A system to minimize trace metal contamination. *Ann. Clin. Biochem.,* **19,** 444–445.

Calabrese, E. J., Tuthill, R. W., Sieger, T. L., and Klar, M. J. (1979). Lead and cadmium contamination during acid preservation of water samples. *Bull. Environ. Contam. Toxicol.,* **23,** 107–111.

Capar, S. G., Gajan, R. J., Madzsar, E., Albert, R. H., Sanders, M., and Zyren, J. (1982). Determination of lead and cadmium in foods by anodic stripping voltammetry: II. Collaborative Study. *J. Assoc. Off. Anal. Chem.,* **65,** 978–986.

Chakrabarti, C. L., Subramanian, K. S., and Nakahara, T. (1978). The application of atomic absorption spectrometry to the analysis of trace metals in non-biological marine samples. Marine Analytical Standards Program, NRCC Report No. 6, Ottawa.

Chan, C. H. (1977). Water quality survey on the Niagara River. 1974. Report Series No. 48, Inland Water Directorate, Water Quality Branch.

Chester, R., and Stoner, J. H. (1974). The distribution of zinc, nickel, manganese, cadmium, copper and iron in some surface waters from the world ocean. *Mar. Chem.,* **2,** 17–32.

Clanet, F., Deloncle, R., and Popoff, G. (1981). Chelating resin catcher for capture, preconcentration and determination of toxic metal traces (Zn, Cd, Hg, Pb) in waters. *Wat. Res.,* **15,** 591–598.

Community Bureau of Reference-BCR (1982). Reference materials for elemental analysis. Information Handout No. 25, Brussels, Belgium.

Dabeka, R. W. (1979). Graphite furnace atomic absorption spectrometric determination of lead and cadmium in foods after solvent extraction and stripping. *Anal. Chem.,* **51,** 902–907.

Dabeka, R. W. (1984). Collaborative study of a graphite-furnace atomic absorption screening method for the determination of lead in infant formulas. *Analyst,* **109,** 1259–1263.

Dabeka, R. W., and McKenzie, A. D. (1986). Graphite-furnace atomic absorption spectrometric determination of lead and cadmium in food after nitric-perchloric acid digestion and coprecipitation with ammonium pyrrolidine dithiocarbamate. *Can. J. Spectrosc.,* **31,** 44–52.

Dabeka, R. W., Mykytiuk, A., Berman, S. S., and Russell, D. S. (1976). Polypropylene for the sub-boiling distillation and storage of high purity acids and water. *Anal. Chem.,* **48,** 1203–1207.

Danielsson, L. G., Magnusson, B., and Zhang, K. (1982). Matrix interference in the determination of trace metals by graphite furnace AAS after Chelex-100 preconcentration. *At. Spectrosc.,* **3,** 39–40.

Dean. J. A., and Rains, T. C. (eds.). *Flame Emission and Atomic Absorption Spectrometry,* Vol. I, *Theory* (1969); Vol. 2, *Components and Techniques* (1971); Vol. 3, *Elements and Matrices* (1975). Marcel Dekker, New York.

de Vries, M. P. C., Tiller, K. G., and Beckwith, R. S. (1975). Sources of error in the determination of Cd and Pb in Plant Material by Atomic Absorption. *Commun. Soil Sci. Plant Anal.,* **6,** 629–640.

Dickinson, G. W., and Fassel, V. A. (1969). Emission spectrometric detection of the elements at the nanogram per milliliter level using induction-coupled plasma excitation. *Anal. Chem.,* **41,** 1021–1024.

Dybczynski, R., Veglia, A., and Suschny, O. (1980). Report on the intercomparison run A-11 for the determination of inorganic constituents of milk powder. IAEA/RL/68, International Atomic Energy Agency, Austria.

Eaton, A. (1976). Marine geochemistry of cadmium. *Mar. Chem.*, **4**, 141–154.

Elser, R. C. (1976). Spark source mass spectrometry. In *Trace Analysis, Spectroscopic Methods for Elements*, Winefordner, J. D. (ed.). Wiley, New York, pp. 383–417.

Environment Canada. (1974). *Analytical Methods Manual*. Inland Waters Directorate, Water Quality Branch, Ottawa.

Fassel, V. A. (1979). Simultaneous or sequential determination of the elements at all concentration levels—the renaissance of an old approach. *Anal. Chem.*, **51**, 1290A–1308A.

Feinberg, M., and Ducauze, C. (1980). High temperature dry ashing of foods for atomic absorption spectrometric determination of lead, cadmium and copper. *Anal. Chem.*, **52**, 207–209.

Fleet, B. (1980). Variations in extraction efficiency of aqueous cadmium (II) using the APDC-MIBK procedure. *Analytical Techniques in Environmental Chemistry*. Pergamon Series on Environmental Science, Pergamon, Oxford, Vol. 3, pp. 621–628.

Florence, T. M., and Batley, G. E. (1977). Determination of the chemical forms of trace metals in natural waters, with special reference to copper, lead, cadmium and zinc. *Talanta*, **24**, 151–158.

Fuller, C. W. (1977). *Electrothermal Atomization for Atomic Absorption Spectrometry*. The Chemical Society, London, United Kingdom.

Gajan, R. J., Capar, S. G., Subjoc, C. A. and Sanders, M. (1982). Determination of lead and cadmium in foods by anodic stripping voltammetry I. Development of method. *J. Assoc. Off. Anal. Chem.*, **65**, 970–977.

Gorsuch, T. T. (1970). *The Destruction of Organic Matter*. Pergamon New York.

Green, D. G., Green. L. W., Page, J. A., Poland, J. S., and Van Loon, G. (1981). The determination of copper, cadmium and lead in sea water by anodic stripping voltammetry with a thin film mercury electrode. *Can. J. Chem.*, **59**, 1476–1486.

Greenberg, R. R., and Kingston, H. M. (1983). Trace element analysis of natural water samples by neutron activation analysis with chelating resin. *Anal. Chem.*, **55**, 1160–1165.

Guevremont, R., Sturgeon, R. E., and Berman, S. S. (1980). Application of EDTA to direct graphite-furnace atomic absorption analysis for cadmium in sea water. *Anal. Chim. Acta*, **115**, 163–170.

Guinn, V. P. (1971). Activation analysis. In *Treatise on Analytical Chemistry*, Part I, Kolthoff, I. M., Elving, P. J., and Sandell, E. B. (eds.). Wiley-Interscience, New York, Chapter 98.

Hanson, H. W. (ed.). (1973). *Official, Standardised and Recommended Methods of Analysis*, 2nd ed. The Society for Analytical Chemistry, London, pp. 3–23.

Horwitz, W. (Ed.). (1980). *Official Methods of Analysis of the Association of Official Analytical Chemists*. 13th ed., Section 33.089-33.094, Washington, DC.

Hume, D. N. (1975). Fundamental problems in oceanographic analysis. In *Analytical Methods in Oceanography*, Gibb, T. R. P. Jr. (Ed.). Advances in Chemistry Series 147, American Chemical Society, Washington, DC, pp. 1–9.

Ihnat, M. (1978). *Agricultural Sources, Transport and Storage of Metals: Copper, Zinc, Cadmium and Lead in Waters of Selected Southern Ontario Agricultural Watershed Studies*. Task Group C (Canadian Section) International Reference Group on Great Lakes Pollution from Land Use Activities. International Joint Commission, Windsor, Ontario.

Ihnat, M. (1981). Analytical approach to the determination of copper, zinc, cadmium and lead in natural fresh waters. *Int. J. Environ. Anal. Chem.*, **10**, 217–246.

Ihnat, M. (1982a). Copper, zinc, cadmium and lead in waters of selected southern Ontario agricultural watersheds. *Int. J. Environ. Anal. Chem.*, **11**, 189–210.

Ihnat, M. (1982b). Importance of acid insoluble residue in plant analysis for total macro and micro elements. *Commun. Soil Sci. Plant Anal.*, **13**, 969–979.

Ihnat, M. (1982c). Application of atomic absorption spectrometry to the analysis of foodstuffs.

In *Atomic Absorption spectrometry,* Cantle, J. E. (ed.). Vol. 5 of *Techniques and Instrumentation in Analytical Chemistry,* Elsevier, Amsterdam, pp. 139–210.

Jagner, D., and Aren, K. (1979). Potentiometric stripping analyses for zinc, cadmium, lead and copper in sea water. *Anal. Chim. Acta,* **107,** 29–35.

Jagner, D., Josefson, M., and Westerlund, S. (1981). Determination of zinc, cadmium, lead and copper in sea water by means of computerized potentiometric stripping analysis. *Anal. Chim. Acta,* **129,** 153–161.

King, W. C., Rodriquez, J. M., and Wai, C. M. (1974). Losses of trace concentrations of cadmium from aqueous solution during storage in glass containers. *Anal. Chem.,* **46,** 771–773.

Kirkbright, G. F., and Sargent, M. (1974). *Atomic Absorption and Fluorescence Spectroscopy.* Academic, New York.

Knauer, G. A., and Martin, J. H. (1981). Phosphorus-cadmium cycling in northeast Pacific waters. *J. Mar. Res.,* **39,** 65–76.

Koirtyohann, S. R., Glass, E. D., and Lichte, F. E. (1981). Some observations on perchloric acid interferences in furnace atomic absorption. *Appl. Spectrosc.,* **35,** 22–26.

Kopp, J. F. (1970). The occurrence of trace elements in water. In *Trace Substances in Environmental Health,* Vol. 3, Hemphill, D. D. (Ed.), University of Missouri, Columbia, pp. 59–73.

Kuennen, R. W., Wolnik, K. A., Fricke, F. L., and Caruso, J. A. (1982). Pressure dissolution and real sample matrix calibration for multielement analysis of agricultural crops by inductively coupled plasma atomic emission spectrometry. *Anal. Chem.,* **54,** 2146–2150.

Lamathe, J., Magurno, C., and Equel, J. C. (1982) Essais interlaboratoires: Dosage du cadmium, du cuivre et du plomb dans l'eau de mer par spectrometrie d'absorption atomique electrothermique. *Anal. Chim. Acta,* **142,** 183–188.

Lewis, G. P., Coughlin, L., Jusko, W., and Hartz, S. (1972). Contribution of cigarette smoking to cadmium accumulation in man. *Lancet,* **1,** 291–292.

Leyden, D. E., and Wegscheider, W. (1981). Preconcentration for trace element determination in aqueous samples. *Anal. Chem.,* **53,** 1095A–1060A.

Lobel, P. B. (1978). Rapid digestion procedure for use in metal analysis. *Mar. Pollut. Bull.,* **9,** 22–23.

Longerich, H. P., and Kantipuly, C. J. (1985). Progress in inductively coupled plasma-mass spectrometry. Presented at the Spectroscopy Society of Canada 1985 Workshop on Applications of Inductively-Coupled-Plasma/Mass Spectrometry, Toronto, Canada, October 3–4.

MacGregor, D. J. Unpublished report, Environment Canada, Hull, Quebec.

McClellan, B. E. (1975). Water. In *Flame Emission and Atomic Absorption Spectrometry,* Vol. 3, *Elements and Matrices,* Dean, J. A., and Rains, T. C. (eds.). Marcel Dekker, New York, Chapter 23, pp. 548–576.

McLaren, J. W., Berman, S. S., Boyko, V. J., and Russell, D. S. (1981). Simultaneous determination of major, minor and trace elements in marine sediments by inductively coupled plasma atomic emission spectrometry. *Anal. Chem.,* **53,** 1802–1806.

McLaren, J. W., Beauchemin, D., Mykytiuk, A. P., and Berman, S. S. (1985). Applications of inductively coupled plasma mass spectrometry in marine analytical chemistry. Presented at the Spectroscopy Society of Canada 1985 Workshop on Applications of Inductively-Coupled-Plasma/Mass Spectrometry, Toronto, Canada, October 3–4.

McQuaker, N. R., Kluchner, P. D., and Chang, G. N. (1979). Calibration of an inductively coupled plasma-atomic emission spectrometer for the analysis of environmental materials. *Anal. Chem.,* **51,** 888–895.

Maienthal, E. J. (1970). U. S. National Bureau of Standards Technical Note 545, Taylor, J. K. (ed.). U. S. Government Printing Office, Washington, DC, pp. 53–54.

Mart, L. (1979). Prevention of contamination and other accuracy risks in voltammetric trace metal analysis of natural waters. Part I. Preparatory steps, filtration and storage of water samples. *Fresen. Z. Anal. Chem.*, **296**, 350–357.

Mart, L. (1982). Minimization of accuracy risks in voltammetric ultratrace determination of heavy metals in natural waters. *Talanta*, **29**, 1035–1040.

Mart, L., Nuernberg, H. W., and Valenta, P. (1980). Prevention of contamination and other accuracy risks in voltammetric trace metal analysis of natural waters. Part II. Voltammetric ultratrace analysis with a multicell system designed for clean bench working. *Fresen. Z. Anal. Chem.*, **300**, 350–362.

Martin, J. H., Bruland, K. W., and Broenkow, W. W. (1976). Cadmium transport in the California current. In *Marine Pollutant Transfer*, Windom, H. L., and Duce, R. A. (eds.). Lexington Books, Lexington, MA, pp. 159–184.

Massee, R., Maessen, F. J. M. J., and De Goeij, J. J. M. (1981). Losses of silver, arsenic, cadmium, selenium and zinc traces from distilled water and artificial sea-water by sorption on various container surfaces. *Anal. Chim. Acta*, **127**, 181–193.

Matousek, J. P. (1981). Interferences in electrothermal atomic absorption spectrometry, their elimination and control. *Prog. Anal. At. Spectrosc.* **4**, 247–310.

Mavrodineanu, R. (ed.). (1970). *Analytical Flame Spectroscopy-Selected Topics*. Springer-Verlag, New York.

Moody, J. R. (1982). NBS clean laboratories for trace element analysis. *Anal. Chem.*, **54**, 1358A–1376A.

Moody, J. R., and Lindstrom, R. M. (1977). Selection and cleaning of plastic containers for storage of trace element samples. *Anal. Chem.*, **49**, 2264–2267.

Murphy, T. J. (1976). The role of the analytical blank in accurate trace analysis. In *Accuracy in Trace Analysis: Sampling, Sample Handling, Analysis*, Vol. 1, LaFleur, P. D. (ed.). U. S. National Bureau of Standards Special Publication 442, Washington, DC, pp. 509–539.

Nackowski, S. B., Putnam, R. D., Robbins, D. A., Varner, M. D., White, L. D., and Nelson, K. W. (1977). Trace metal contamination of evacuated blood collection tubes. *Am. Indus. Hygiene Assn. J.*, **38**, 503–508.

National Bureau of Standards Certificate of Analysis SRM 1566 Oyster Tissue; SRM 1570 Spinach; SRM 1571 Orchard Leaves; SRM 1572 Citrus Leaves; SRM 1573 Tomato Leaves; SRM 1577a Bovine Liver; SRM 1643a Trace Elements in Water; SRM 1645 River Sediment; SRM 1646 Estuarine Sediment, Washington, DC.

Nurnberg, N. W., Valenta, P., Mart, L., Raspor, B., and Sipos, L. (1976). Applications of polarography and voltammetry to marine and aquatic chemistry, II. The polarographic approach to the determination and speciation of toxic trace metals in the marine environment. *Z. Anal. Chem.*, **282**, 357–367.

Ottaway, J. M. (1982). A revolutionary development in graphite furnace atomic absorption. *At. Spectrosc.*, **3**, 89–92.

Parsons, M. L. (1976). Nuclear methods, In *Trace Analysis, Spectroscopic Methods for Elements*, Winefordner, J. D. (ed.). New York, pp. 279–343.

Patterson, C. C., and Settle, D. M. (1975). *The Reduction of Orders of Magnitude Errors in Lead Analyses of Biological Materials and Natural Waters by Evaluating and Controlling the Extent and Sources of Industrial Lead Contamination Introduced During Sample Collecting and Analysis*. Division of Geological and Planetary Sciences, California Institute of Technology, Pasedena, CA, Publication No. 2547, pp. 1–38.

Poldoski, J. E., and Glass, G. E. (1978). Anodic stripping voltammetry at a mercury film electrode: Baseline concentrations of cadmium, lead and copper in selected natural waters. *Anal. Chim. Acta,* **101,** 79–88.

Posma, F. D., Balke, J., Herber, R. F. M., and Stulk, E. J. (1975). Microdetermination of cadmium and lead in whole blood by flameless atomic absorption spectrometry using carbon tube and carbon-cup as sample cell and comparison with flame studies. *Anal. Chem.,* **47,** 834–838.

Pszonicki, L., Veglia, A., and Suschny, O. (1982). *Report on the Intercomparison W-3/1 of the Determination of Trace Elements in Water.* IAEA/RL/94. International Atomic Energy Agency, Austria.

Rattonetti, A. (1974). Stability of metal ions in aqueous environmental samples. In *Accuracy in Trace Analysis: Sampling, Sample Handling, Analysis,* Vol. 1, LaFleur, P. D. (ed.). U.S. National Bureau of Standards Special Publication 422, Washington, DC, pp. 633–648.

Riley, J. P. (1975). Analytical chemistry of sea water. In *Chemical Oceanography,* Vol. 3, 2nd ed., Riley, J. P., and Skirrow, G. (eds.). Academic, New York, pp. 193–514.

Ritter, C. J., Bergman, S. C.,, Cothern, C. R., and Zamierowski, E. E. (1978). Comparison of sample preparation techniques for atomic absorption analysis of sewage sludge and soil. *At. Absorpt. Newsl.,* **17,** 70–72.

Robertson, D. E. (1972). Contamination problems in trace element analysis and ultrapurification, In *Ultrapurity, Methods and Techniques,* Zief, M., and Speights, R. (eds.). Marcel Dekker, New York, pp. 207–253.

Robinson, J. W. (ed.). (1974). *Handbook of Spectroscopy,* Vol. 1. CRC Press, Cleveland, OH.

Ryan, M. D., and Wilson, G. S. (1982). Analytical electrochemistry: Methodology and application of dynamic techniques. *Anal. Chem.,* **54,** 20R–27R.

Schonberger, E. A., and Pickering, W. F. (1980). The influence of pH and complex formation in the ASV peaks of Pb, Cu and Cd. *Talanta,* **27,** 11–18.

Schulze, G., and Frenzel. W. (1983). Interferences by copper in the potentiometric stripping analysis for zinc and cadmium. *Fresen. Z. Anal. Chem.,* **314,** 459–462.

Scott, R. H., Fassel, V. A., Kniseley, R. N., and Nixon, D. E. (1974). Inductively coupled plasma-optical emission analytical spectrometry, a compact facility for trace analysis of solutions. *Anal. Chem.,* **46,** 75–80.

Seward, R. W. (ed.). (1975). *Standard Reference Materials and Meaningful Measurements.* National Bureau of Standards Special Publication 408, Washington, DC.

Singh, S. S. (1977). Consideration on sorption of cadmium during laboratory operations. *Can. J. Soil Sci.,* **57,** 217–219.

Slabyj, B. M., Koons, R. D., Bradbury, H. E., and Martin, R. E. (1983). Cadmium determination of frozen cod: An interlaboratory comparison. *J. Food Prot.,* **46,** 122–125.

Slavin, W. (1968). *Atomic Absorption Spectroscopy.* Interscience Publishers, New York.

Slavin, W. (1984). Graphite furnace AAS, a source book. Perkin-Elmer Corporation, Spectroscopy Division, Ridgefield. CT, pp. 83–90.

Slavin, W., and Manning, D. C. (1982). Graphite furnace interferences. A guide to the literature. *Prog. Anal. At. Spectrosc.,* **5,** 243–340.

Smith. A. E. (1973). A study of the variation with pH of the solubility and stability of some metal ions at low concentrations in aqueous solution. Part II. *Analyst,* **98,** 209–212.

Sotera, J. J., and Kahn, H. L. (1982). Background correction in AAS. *Am. Lab.,* **14,** 100–104, 106, 108.

Spencer, M. J., Betzer, P. R., and Piotrowicz, S. R. (1982). Concentrations of cadmium, copper, lead and zinc in surface waters of the northwest Atlantic Ocean—a comparison of Go-flo and Teflon water samplers. *Mar. Chem.,* **11,** 403–410.

Sperling, K. R. (1982). Determination of heavy metals in sea water and in marine organisms by flameless atomic absorption spectrophotometry. XIV. Comments on the usefulness of organohalides as solvents for the extraction of heavy metal (Cadmium) complexes. *Fresen. Z. Anal. Chem.,* **310,** 254–256.

Sperling, K. R. (1983). Determination of heavy metals in sea water and in marine organisms by flameless atomic absorption spectrophotometry. XVII. On the usefulness of signal integration—a warning. *Fresen. Z. Anal. Chem.,* **314,** 417–418.

Sperling, K. R., and Bahr, B. (1981). Determination of heavy metals in sea water and in marine organisms by flameless atomic absorption spectrophotometry. XIII. Correspondence and some possible sources of error in an intercalibration of cadmium. *Fresen. Z. Anal. Chem.,* **306,** 7–12.

Stoeppler, M., Brandt, K. (1980). Contribution to automated trace analysis. *Fresen. Z. Anal. Chem.,* **300,** 372–380.

Struempler, A. W. (1973). Adsorption characteristics of silver, lead, cadmium, zinc, and nickel on borosilicate glass, polyethylene and polypropylene container surfaces. *Anal. Chem.,* **45,** 2251–2254.

Tschopel, P., Kotz, L., Schulz, W., Veber, M., and Tolg, G. (1980). Causes and elimination of systematic errors in the determination of elements in aqueous solutions in the ng/mL and pg/mL range. *Fresen. Z. Anal. Chem.,* **302,** 1–14.

Uriano, G. A., and Cali, J. P. (1977). Role of reference materials and reference methods in the measurement process. In *Validation of the Measurement Process,* Devoe, J. R. (ed.). ACS Symposium Series 63, American Chemical Society, Washington, DC. pp. 140–161.

van Raaphorst, J. G., van Weers, A. W., and Haremaker, H. M. (1978). On the loss of cadmium, antimony and silver during dry ashing of biological material. *Fresen. Z. Anal. Chem.,* **293,** 401–403.

Varian Techtron. (1972). *Analytical Methods for Flame Spectroscopy.* Varian Techtron, Springvale, Australia.

Wahlgren, M. A., Edgington, D. N., and Rawlings, F. F. (1971). Determination of selected trace elements in water samples using spark source mass spectroscopy and neutron activation analysis. In *Nuclear Methods in Environmental Research.* Vogt, J. R., Parkinson, T. F., and Carter, R. R. (eds.). University of Missouri at Columbia, Columbia, MO, pp. 97–103.

Williams, E. V. (1978). New technique for the digestion of biological materials—application to the determination of tin, iron and lead in canned foods. *J. Food Technol.,* **13,** 367–384.

Winge, R. K., Peterson, V. J., and Fassel, V. A. (1979). Inductively coupled plasma-atomic emission spectroscopy–prominent lines. *Appl. Spect.,* **33,** 206–219.

Youden, W. J., and Steiner, E. H. (1975). *Statistical Manual of the Association of Official Analytical Chemists.* Association of Official Analytical Chemists, Washington, DC, pp. 33–35.

Zief, M., and Horvath, J. (1976). High-purity reagents for trace analysis. In *Accuracy in Trace Analysis: Sampling, Sample Handling, Analysis,* Vol. 1, LaFleur, P. D. (ed.). U. S. National Bureau of Standards Special Publication 422, Washington, DC, pp. 363–375.

INDEX

Acclimation, 150
Acid rain, 64
Acute toxicity:
   annelids, 186–187
   crustaceans, 177–186
   echinoderms, 187
   fish, 172–174
   hardness effects, 164–165
   mollusks, 176–181
   testing, invertebrates, 122–125
Adsorption, 76, 79, 96
   onto container walls, 244
   sediment, 8–9, 60
Adsorption–desorption, 72
ALA-D activity, 175
Algae, sensitivity, 121
Amphipoda, acute toxicity, 124
Anaerobic sediments, 74
Analysis methods, 234–238
   contamination control, 238–241
   detection limits, 235
   project design and analytical input, 241
   standard reference materials, 238, 253–255
   technical and operational characteristics, 237
Analytical recommendations, 255–256
Annelids, acute toxicity, 186–187
   subadult and adult, 178
Anthropogenically mobilized cadmium, 29–30
Anthropogenic sources, 3–4, 29–31, 47
   concentration, 23–24
   freshwater, 21
Ashing temperature, 247–248
Associations, under different physiochemical conditions, 70–73
Atmosphere:
   concentration, 3–5
   deposition, ocean, 26
   natural sources, 3
   precipitation, concentration, 5
   transport, 48

Bacteria:
   distribution, 94
   effects on, 118–119
   elimination, 95
   remobilization influence, 80–81
   uptake, 94–95
Behavior, changes and toxicity, 158–159
Binding capacity, organic matter, 149
Bioaccumulation, 206–212
   aquatic insects, 98–103
   bacteria, 94
   crayfish, 102–103
   fish, 103–110
   from food, 210–212
   insect larvae, 99
   macrophytes, 91–92
   marine organisms, 203
   mollusks, 101–102
   phytoplankton, 95–97, 120–121
   salinity, 214–215
   seasonal variation, 215–216
   from sediment, 92
   and sediment levels, 209–210
   temperature and, 212, 214
   from water, 206–209
   see also specific organisms
Bioavailability, 90, 110, 204–206
   dissolved, 90
   marine environment, 202
   nonlabile, 90
   potentially available, 90
   readily available, 90
   sediment, 205–206
   speciation, 90
   total, 90

Bioavailability (*Continued*)
   unavailable, 90
   in water, 204–205
Biochemical changes, fish, 159
Bioconcentration factors, 104, 109, 203
Biological materials, sample decomposition, 242–243
Blanks, errors related to, 240–241
Blood plasma:
   changes, goldfish, 160
   ion balance, fish, 175–176
Bluegills, behavioral changes, 158

Calcium:
   influence on uptake, 110–111
   levels and uptake, 93
   toxicity and, 191
Canadian drinking water, levels, 8
Carbonate extraction, 66
Carbonate phases, sediment, 60
Cation exchange, sediment, 60
Cd(II), species in seawater, 10
Chelating agents, toxicity, 129
Chemical forms, *see* Speciation
Chemical leaching, 54–55
Chloro complexation, 80
Chronic toxicity testing:
   fish, 154–158
   invertebrates, 123, 126–128
   phytoplankton, 120–122
Cladocera:
   acute toxicity, 124
   sublethal responses, 126
Clam:
   bioaccumulation, 207–208
   bioavailability, 205–206
Clearance, *see* Elimination
Combustion residues, 65
Complexation, effect on toxicity, 128–129
Complexing agents, uptake effects, 111
Concentration:
   anthrogenic sources, 23–24
   associated with behavioral, physiological, and structural responses, 182–184
   atmosphere, 3–5
   atmospheric precipitation, 5
   fish, 106–108
   freshwater, 8–9, 21, 36–39, 233
   Great Lakes, 36–39
   lethal, 164–165
   marine mammals, 213
   marine organisms, 233
   Neckar River, 24–25
   ocean waters, 233
   polluted water, 162
   rain, 4
   relationship with salinity, 30–31
   rivers, 9, 21–22
   rocks, 2, 5–6
   St. Lawrence Basin, 21, 23
   St. Lawrence River, 37–39
   sediment, 15, 24–25, 30, 39, 41, 78, 203
   snow, 26
   soils, 6–7
   techniques, 244–246
      coprecipitation, 245
      evaporation, 244
      ion exchange, 245
      solvent extraction, 245
Confined land disposal, speciation, 75
Contamination, 24
Contamination control:
   analysis, 238–241, 256–257
   analyst, 239
   dealing with blanks, 240–241
   laboratory, 238–239
   laboratory ware, 239
   monitoring, 241
   reagents, 239–240
Copepod, 184–186
   acute toxicity, 124
   temperature and toxicity, 189
Copper, mixtures, toxicity, 153
Crab, 184
   bioaccumulation, 208, 211
   toxicity:
      salinity and, 188
      temperature and, 189
Crayfish, bioaccumulation, 102–103
Crustacean, 122
   acute toxicity, 124, 177–186
      subadult and adult, 177–178, 181, 184
      young stages, 179–180, 185
   bioaccumulation, 208–209
   lowest concentration associated with behavioral, physiological, and structural responses, 182–184
   sublethal responses, 126
   toxicity, 122–123
      salinity and temperature, 190
Cycle, marine environment, 201
Cycling, *see* Transport

Decapoda, acute toxicity, 124
Denitrification, 78
Deposit, atmospheric, 3

Depuration, aquatic insects, 100
Desorption, from particles in estuaries, 14
Detection, limits, 235
Determination:
 electroanalytical methods, 249–251
 flame atomic absorption spectrometry, 237, 246–247
 graphite furnace atomic absorption spectrometry, 237, 247–249
 inductively coupled plasma atomic emission spectrometry, 237, 251–252
 inductively coupled plasma mass spectrometry, 237, 253
 neutron activation analysis, 237, 252
 spark source mass spectrometry, 237, 252–253
Detroit River, 37–38
 particulates, 42
Dialysis bag, for sampling, 57
Diatom, 204
Diptera, acute toxicity, 125
Distribution:
 bacteria, 94
 body, 216
 fish, 103–105, 109, 161
 freshwater environment, 5–9
 insect larvae, 99–100
 macrophytes, 91–93
 marine organisms, 203–204
 oceans, 9–10, 20
 phytoplankton, 95–96
 surficial sediment, 40
 vertical:
  nutrient level in water column, 200
  oceans, 11–12
 zooplankton, 97–98
Distribution coefficient:
 dissolved particulates, 44
 equilibrium, 45
 between particulate and dissolved cadmium, 21

Echinoderms:
 acute toxicity, 187
 subadult and adult, 178
EDTA, 67, 129, 149, 204–205
Eggs, fish, susceptibility, 172
Elbe River, 75–77
Electroanalytical methods, 249–251
Electron microprobe, 55
Electrothermal atomic absorption spectrometry, 247–249

Elimination:
 bacteria, 95
 fish, 105, 108, 110
 insect larvae, 100
 macrophytes, 93
 phytoplankton, 96–97
 zooplankton, 98
Environmental matrices, 232
Enzyme:
 activity, effect of cadmium, 121–122
 suppression, fish, 173, 175
Ephemeroptera, acute toxicity, 125
Equilibrium, 72
 dissolved particulates, 44
 distribution coefficient, 8
 phytoplankton, 95
 sediment-water, 53
Estuarine processes, 14
Estuary, 75
Ethylenediaminetetraacetic acid, 67, 129, 149, 204–205
Evaporation, 244
Excretion, 219–220
Exoskeleton, uptake, 100
Exposure:
 ambient, 108
 intraperitoneal injection, 108, 110
Extracellular traps, 219
Extraction methods:
 metals from chemical components, 60
 problems with procedures, 66–67
 sediment, 54–55, 58–69
 selectivity, 65
 sequential, 63–65
 single-leaching, 59–62
 specificity, 65–69

Fish:
 acute toxicity, 172–174
 bioaccumulation, 209
 biochemical changes, 159
 circulatory problems, 160
 concentration levels, 106–108
 convulsive uncoordinate swimming, 152
 destruction of gill tissue, 151
 distribution, 103–105, 109, 161
 eggs, susceptibility, 172
 elimination, 105, 108, 110
 hyperirritability, 152
 larvae, susceptibility, 172–173
 nonsalmonid, toxicity, 143–144, 146–147
 physiological responses as modifying factors, 150–151

Fish (*Continued*)
  respiratory/coughing rates, 159
  respiratory responses, 159
  sublethal effects:
    enzyme activity, 173, 175
    ion balance, blood plasma, 175–176
    metallothioneins, 176
    threshold, 155
  sublethal-lethal ratios, 155
  temperature and toxicity, 189
  tissue residues, 161
  uptake, 103–104, 109
Fixation, in seawater, 14–15
Flame atomic absorption spectrometry, 237, 246–247
Flux, *see* Transport
Food, bioaccumulation from, 210–212
Fraction:
  carbonatic, 78
  exchangeable, 78
  hydroxylamine-extractable, 79
Fractionation, 64–65
  diagenetic stages, 73
  patterns, 72–73
Freshwater environment:
  cadmium chemistry, 7–8
  concentration, 20, 233
  distribution in, 5–9
  equilibrium with soil, 8
  sample storage, 243–244
Fungi, effects on, 118–119

Galena, 68
Gastropods, toxicity, 123
  acute, 125
Georgian Bay, concentration, 37
Goldfish, plasma changes, 160
Graphite furnace atomic absorption spectrometry, 237, 247–249
Great Lakes, 23
  bottom sediments, 39–42
  concentration, 36–39
    dissolved, 39
  description, 35–36
  mass balance budget, 47–48
  partitioning, 44–46
  surficial sediment, 40–41
  suspended particulate material, 42

Half-life, 219–220
Hardness, toxicity effects, 130, 140, 142–143, 145, 148, 154–155, 164–165
Heamatite, 68

Helena River, concentration, 9
Human activities, *see* Anthropogenic sources
Humic acid, 149, 205
Hydrochloric acid, 62
Hydrogen peroxide, 67
Hydroxylamine, 61–62

Igneous rocks, content, 2, 6
Inductively coupled plasma atomic emission spectrometry, 237, 251–252
Inductively coupled plasma mass spectrometry, 237, 253
Inorganic interferences, electrochemical determination, 250
Insect larvae:
  distribution, 99–100
  elimination, 100
  sensitivity, 123
  uptake, 99–100
Insects, acute toxicity, 125
Instrumental neutron activation analysis, 237, 252
Interstitial waters:
  analysis, 53–54
  extraction from sediment, 56–57
  modeling, 54
Intracellular electron-dense granules, 218
Invertebrates:
  acute toxicity testing, 122–125
  benthic, 98–103. *See also specific invertebrates*
  chronic toxicity testing, 123, 126–128
  metal mixture, 191
  toxicity, 122–128
    interactions with other metals, 133
Ion exchange concentration techniques, 245

Kidney damage, high levels and, 152

Lake Erie:
  concentration, 37–38
    dissolved, 37
    sediment, 39, 41
  mass balance budget, 47–48
  particulates, 42–43
Lake Huron, concentration, 37
Lake Michigan, concentration, 36
  sediment, 39, 41
Lake Ontario:
  concentration, 38
  sediment, 41
  mass balance budget, 47
  particulates, 43

Lake Superior, concentration, 36
Lead, chemical extraction, 65
Light, toxicity effect, 130
Limpets, bioaccumulation, seasonal variation, 215
Lobster:
  bioaccumulation, 208–209, 211
  excretion, 219

Macrophytes, 76
  clearance, 93
  effects on, 122
  uptake and depuration, 91
Magnesium:
  toxicity and, 191
  uptake and, 187–188
Marine mammals, concentration, 213
Marine organisms:
  concentration, 233
  distribution in, 203–204
Marine sediments, 14–16, 202
Mass balance:
  budget, Great Lakes, 47–48
  oceanic, 26–29
    carrier phase data, 27–28
Matrix modifiers, analyte solutions, 248
Measurement anomalies, 20, 38, 43
  sediments, 55
Measurement methods, 53–55
Membrane-limited vesicles, storage mechanisms, 218–219
Mercury, cadmium antagonism, 191
Metal accumulation, 58
Metal-binding proteins, *see* Metallothioneins
Metallothioneins, 96, 108–109, 151
  association with particulate structure, 218–219
  fish, 176
  storage mechanism, 217–218
Metal mixtures, toxicity, 131–133, 153–154, 191–192
Microorganisms:
  effects on, 118
    growth, 118–119
  toxicity, 118–120
    interaction with other metals, 131
Minerals:
  authigenic, 74
  clay, toxicity effect, 129
  crystalline, 68
  secondary, 74
  sequential extraction, 68–69
Mississippi River, concentration, 9, 20

Mobilization:
  from sediments, 15
  world rivers, 21–25
Mollusks:
  acute toxicity, 125, 176–181
    subadult and adult, 176–177
    young stages, 176, 179–181
  bioaccumulation, 101–102, 207–208
    lowest concentration associated with behavioral, physiological, and structural responses, 182–184
Molting, clearance by, 103
Mudflats, tidal, 76–77
Mussel:
  bioaccumulation, 207
    seasonal variation, 216
  excretion, 219

Neckar River, concentration, 24–25
Neutron activation analysis, 237, 252
Niagara River, 41, 43
Nitrate, levels, 76, 78
Nitrification, 76
Nitrilotriacetic acid, 129
Nodules, ferromanganese, 68
Nonlabile cadmium, 149
NTA, 129, 204

Ocean waters, *see* Seawater
Odonata, acute toxicity, 125
Organically bound cadmium, 150
Organic interferences, electrochemical determination, 250–251
Oxidation, 71, 78
  chemical, 67
  microbial, 74
  organic matter, 67, 74
Oxidizable phases, 60, 64
Oxygen, toxicity effects, 130–131, 150
Oxyhydroxides, 68
Oyster:
  bioaccumulation, seasonal variation, 216
  excretion, 220

Particles, 44
  atmospheric, 3
  dissolved phases, 21
  freshwater, 42–44
  measurement, 52–55
Particulate-dissolved metal exchange, 14
Partitioning:
  anoxic mud, 70
  Great Lakes, 44–46

Partitioning (*Continued*)
  measurement methods, 53
  mechanistic approach, 58
  pattern, 63
  phase approach, 58
  sediment, 58
Pelagic clays, content, 15
pH:
  toxicity effects, 129–130, 145
  uptake effects, 111
Phases, mineral, 68
Phosphate, relationship with cadmium, 11, 13
Phytoplankton:
  bioaccumulation, 207
    seasonal variation, 215
  distribution, 95–96
  elimination, 96–97
  toxicity, 120–122, 187–188
    interaction with other metals, 131–133
  uptake, 95–96
Plants, bioaccumulation, 91–92
Plecoptera, acute toxicity, 125
Pollution, 24
  Great Lakes, 36–39
    sediments, 39
  water concentration, 162
Polychaetes, bioaccumulation, 209
  sediment levels and, 209–210
Precipitation, 79
  concentration, 5
  to ocean sediments, 13
  see also specific types of precipitation
Precipitation-dissolution, 72
Protein:
  binding, 96, 108
  cadmium-binding, 103, 108, 110, 217
  high-molecular-weight, 95
  low-molecular-weight, 108, 110. *See also* Metallothioneins

Quebec City, particulates, 43–44

Rain, concentration, 4
Rainbow trout:
  biochemical changes, 159–160
  toxicity:
    acute, 148
    curve, 143
Readsorption, 66
Reagents:
  contamination control, 239–240, 256–257
  extraction, 60–62

Redox potential:
  discontinuity, 76
  sediment-water system, 70
  speciation, 76
Reducible phases, sediment, 60, 64
Reference materials, 238, 253–255
Regeneration, in oceans, 12
Remobilization, 25, 79–80
  bacterial influence, 80–81
  sediment, 75–76
Reproduction, effect on, 127
  metal mixtures, 153
Resin-labile cadmium, 149
Respiration, fish:
  changes, 159
  respiratory/coughing rates, 159
Rhine River, 23–25
Rivers:
  concentration, 9, 21–22
  discharge fluxes, 29
  runoff, 29
  transport, 21, 28
  see also specific rivers
Rocks, content, 2, 6

St. Lawrence Basin, concentration, 21, 23
St. Lawrence River:
  cadmium discharge, 23
  concentration, 37–39
  particulates, 43–44, 46
  partitioning, 45
  suspended particulate material, 42
Salinity, 10, 79–80
  bioaccumulation and, 214–215
  relationship with concentration, 30–31
  temperature and, 189–190
  toxicity and, 188–189
Salmonid, toxicity, 163
  acute, 140–143
  LC_{50}, 142
  sublethal effects, 159
  threshold, 154
Sample:
  decomposition, 242–243
  storage, 243–244
  sediment, 56
Sampling, 242
  sediment, 56–58
Seals, bioaccumulation, 212–213
Seasonal variation, bioaccumulation, 215–216
Seawater, 9–14
  anthropogenically augmented influx, 30–31

budget, 26–29
distribution, 9–10, 20
    vertical, 11–12
fixation in, 14–15
marine organisms, 233
mass balance, 26–29
natural sources, 13
regeneration, 12
relationship of cadmium and phosphate, 11, 13
removal from deeper layers, 12–13
sample storage, 243–244
sediments, 25
Sediment, 95
  anaerobic, 74
  analysis, defined, 52
  anoxic, 70
  associations, under different physiochemical conditions, 70–73
  bioaccumulation, 209–210
  bioavailability, 205–206
  cation exchange capacity, 202
  components, 60
  concentration, 203
  contamination, 92
  cores, 76–77
  extraction methods, *see* Extraction methods
  freshwater, 24
  frictional dissipation effects, 75
  Great Lakes, 39–42
  levels and bioaccumulation, 209–210
  marine, 14–16, 202
  measurement, 52–55
  metal enrichment, 58–59
  oceans, 25
    concentration, 30
  organic-rich, 74
  particulate association, 52
  pelagic, as sink for oceanic cadmium, 27–28
  phases, 60, 64–65
  polluted, 52
    release from, 78
  sample:
    handling, 56–58
    storage, 56
  sampling, 56–58
  selectivity, 59
  shelf, as sink for oceanic cadmium, 27–28
  single-leaching, 59–62
  suspended, 57–58, 76
  see also Interstitial waters

Sedimentary rocks, content, 2, 6
Sediment-water system:
  interface:
    release, 74–81
    sediment, oxidation effects, 75
    redox potential, 70
    sorption characteristics, 202
Sequential extraction procedures, sediment, 63–65
Sewage sludge, 64–65
Shrimp, 184, 208
Single-leaching, 59–62
Snow, concentration, 26
Soil:
  content, 6–7
  extraction, 7
  roadside, 64–65
Solid waste materials, fractions, 65
Solubility, hardness and, 145
Solvent extraction, 245
Spark source mass spectrometry, 237, 252–253
Speciation, 44, 69
  bioavailability, 90
  change, 70
  confined land disposal, 75
  freshwater, 7–8
  marine environment, 201–203
  redox potential, 76
  seawater, 10–11
  sediment, 59, 78
  studies, 57
Storage, 216–219
  extracellular mechanisms, 219
  subcellular mechanisms, 217–219
Street dust, 64–65
Stress, toxicity and, 150
Sulfide-organic fraction, 71
Surfactants, 67
Suspended particulate material, 42–44
Suspended sediments, sampling, 57–58

Temperature:
  influence on spectrometric determination, 248
  salinity and, 189–190
  toxicity effects, 130, 189
  uptake effects, 111
Thionein, 217. *See also* Metallothioneins
Toxicity:
  behavioral reactions, 158–159
  calcium-magnesium content, 191
  chelating agents, 129

Toxicity (*Continued*)
  complexation effect, 128–129
  crustaceans, 122–123
  environmental factors, 128–131, 144–145, 149–151
  gastropods, 123
  hardness effects, 130, 140, 142–143, 145, 148, 154–155, 164–165
  inorganic substances, 145, 148–149
  invertebrates, 122–128
    interactions with other metals, 133
  light effect, 130
  mechanisms, 151–152
  metal mixtures, 153–154
  microorganisms, 118–120
    interaction with other metals, 131
  modifying factors, 188–192
    physiological responses of fish as, 150–151
  multiple, 191–192
  nonsalmonid fish, 143–144, 146–147
  organic substances in water, 149–150
  oxygen effect, 130–131, 150
  pH effect, 129–130, 145
  phytoplankton, 120–122, 187–188
    interaction with other metals, 131–133
  salinity and, 188–189
    and temperature, 189–190
  salmonid fish, 140–143
  secondary mechanisms, 152, 155
  size and sex effects, 190–191
  stress, 150
  sublethal effects:
    enzyme activity, 173, 175
    ion balance, blood plasma, 175–176
    metallothioneins, 176
    thresholds, 154–155, 158
  surface water levels, 161–162
  temperature effect, 130, 189
  tissue residues, 161
  zooplankton, 126–127
Transport:
  atmospheric, 3, 26, 48
  flux, 78–79
    pore water, 78
  freshwater, 23, 26, 44
  Great Lakes, 47
  oceanic, 11–12, 14–15, 26–28
  phytoplankton, 97

  river, 21, 28
  runoff, 75
  sediment-water, 80
Trichoptera, acute toxicity, 125

Uptake:
  antagonistic, 92
  bacteria, 94–95
  calcium effects, 93, 110–111
  desorption, 97
  environmental factor effects, 110–111
  equilibrium, 91, 93
  exoskeleton, 100
  fish, 103–104, 109
  insect larvae, 99–100
  intracellular, 94
  macrophytes, 91–93
  Mg^{2+} and, 187–188
  modes, 95–96
  organic material, 92
  phytoplankton, 95–96
  salinity and, 214–215
  sequestering, 96
  synergistic, 92
  from water, 204–205
  zooplankton, 97–98
  see also Bioaccumulation

Washing technique, 256
Waste material, 65
Water:
  bioaccumulation from, 206–209
  bioavailability in, 204–205
  extraction, sediment, 56–57
  inorganic substances in, toxicity and, 145, 148–149
  levels and toxicity, 161–162
  organic substances, toxicity and, 149–150

Yeasts, effects on, 118–119

Zinc:
  antagonism to cadmium, 132–133, 153
  chemical extraction, 65
Zooplankton:
  distribution, 97–98
  elimination, 98
  toxicity, 126–127
  uptake, 97–98